ENVIRONMENTAL SCIENCE, ENGINEERING AND TECHNOLOGY SERIES

# PROCESS ENGINEERING IN PLANT-BASED PRODUCTS

# Environmental Science, Engineering and Technology Series

**The Amazon Gold Rush and Environmental Mercury Contamination**
*Daniel Marcos Bonotto and Ene Glória da Silveira*
2009. ISBN: 978-1-60741-609-8

**Environmental Effects of Off-Highway Vehicles**
*Douglas S. Ouren, Christopher Haas, Cynthia P. Melcher, Susan C. Stewart, Phadrea D. Ponds, Natalie R. Sexton, Lucy Burris, Tammy Fancher and Zachary H. Bowen*
2009. ISBN: 978-1-60692-936-0

**Nitrous Oxide Emissions Research Progress**
*Adam I. Sheldon and Edward P. Barnhart (Editors)*
2009. ISBN: 978-1-60692-267-5

**Agricultural Runoff, Coastal Engineering and Flooding**
*Christopher A. Hudspeth and Timothy E. Reeve (Editors)*
2009. ISBN: 978-1-60741-097-3

**Fundamentals and Applications of Biosorption Isotherms, Kinetics and Thermodynamics**
*Yu Liu and Jianlong Wang (Editors)*
2009. ISBN: 978-1-60741-169-7

**Conservation of Natural Resources**
*Nikolas J. Kudrow (Editor)*
2009. ISBN: 978-1-60741-178-9

**Forest Canopies: Forest Production, Ecosystem Health and Climate Conditions**
*Jason D. Creighton and Paul J. Roney (Editor)*
2009. ISBN: 978-1-60741-457-5

**Directory of Conservation Funding Sources for Developing Countries: Conservation Biology, Education and Training, Fellowships and Scholarships**
*Alfred O. Owino and Joseph O. Oyugi*
2009 ISBN: 978-1-60741-367-7

**Process Engineering in Plant-Based Products**
*Hongzhang Chen*
2009. ISBN: 978-1-60741-962-4

ENVIRONMENTAL SCIENCE, ENGINEERING AND TECHNOLOGY SERIES

# PROCESS ENGINEERING IN PLANT-BASED PRODUCTS

HONGZHANG CHEN

**Nova Science Publishers, Inc.**
*New York*

For permission to use material from this book please contact us:
Telephone 631-231-7269; Fax 631-231-8175
Web Site: http://www.novapublishers.com



**LIBRARY OF CONGRESS CATALOGING-IN-PUBLICATION DATA**

Chen, Hongzhang.
Process engineering in plant-based products / Hongzhang Chen.
p. cm.
ISBN 978-1-60741-962-4 (hardcover)
1. Biomass chemicals. 2. Biochemical engineering. 3. Chemical processes. 4. Energy crops. I. Title.
TP248.B55C475 2009
661'.8--dc22 2009020641

*Published by Nova Science Publishers, Inc. ✝ New York*

# CONTENTS

# PREFACE

Plant-based biomass resources, including cereals, crop residues, seed of oil crops, sugar crops, forage crops and various woody crops, can be converted to several of bio-products through physical, chemical and biological conversion methods. Although the components of different plant-based raw materials are varied, there are four main compounds: carbohydrate (sugar, starch, cellulose and hemicellulose), lignin (poly-phenol), esters and protein. Because of the diversity of plant-based raw materials, during product development, we should not only consider its functions to fit certain requirement, but also select the suitable raw material and conversion approaches to effectively link products, raw materials and production processes together.

The exploitation and production of plant-based products are restricted by the plant components, growth season, growth area *et al.*. The development of plant-based product must connect with agriculture and forest since that it supplies wth the raw material for plant-based products. On the one hand we must decide whether the target product can be produced from plant-based material and how to produce it, basing on the components of the plant-based materials and the function of the target product. In fact one special product needs only one or two components of the plant. Therefore, sometimes to obtain one special component of the plant biomass we can rebuild the plant by the methods of traditional breeding and genetic engineering. To choose the right season for planting and cultivation sites agriculture or forestry planning is needed. On the other side, any plant is constituted by lots of components. If only one compound is used it would be difficult to generate significant economic benefits. It is necessary to design products and process for utilizing all components of the plant biomass according to the diversity and characteristic of plant biomass components. At last, a chain of ecological production will be formed, all the components of the plant being converted to relative products and no waste discharge. Moreover, the support of plant raw materials is affected by the growth season and site of plant cultivation, which will also restrict the production of plant-based products. The production site and production scale would be selected according to the specific situation. Multi-purpose equipments are necessary to adapt to the different seasons of different plant raw materials supply.

This book aims to analyze the common problems of the conversion process of plant-based products starting with the key problems of the complexity of the plant-based raw materials, the "process integration" of process engineering, "structure and function" of product engineering. In this book, the concept of "plant-based products process engineering" is advanced based on the integration of multidisciplinary knowledge, and the system of multilevel conversion is expatiated from the three areas: raw materials, conversion process and product development. These afford a theoretic and technologic support for development of plant-based products process engineering. This book also introduces the key technology platforms for plant-based products process engineering, the main approaches for using these platforms to achieve a multilevel utilization of plant-based raw materials, and the constitution of a novel mode for the ecological industry of plant-based product.

To realize the value increment for the plant-based raw materials it is necessary to convert them multi-levelly and directionally according to the characteristic of them and the production process using various conversion methods. In this book we have advanced a novel plant-based raw material multilevel conversion system facing to raw materials, process and products. As for facing to raw materials, it means that studies the mechanism and method for separating the components of the raw materials on the base of the diversity and heterogeneity, and realize the effective separation of components. Facing to process is to extract the common key problems of the production process of plant-based products, study deeply the characteristic of utilization of plant-based materials and the technical bottlenecks, establish key technical platforms for the plant-based products, create clean and efficient process, flow and equipment under the direction of the process engineering idea. Facing to products is to sum up the fruits harvested in laboratories and magnify them, solve the bottleneck problems of how to transform the fruits harvested in laboratories to industrialization and obtain the prospective economic benefit. Ultimately, multi-layered and multi-level integrated utilization of plant-based raw materials and eco-industrial of production plant-based products will be realized.

Nowadays, plant-based materials are widely used in pulp industry, textile industry, processing of Chinese herbal medicine, organic chemicals industry *et al.* But some problems still exist in the utilization of these resources, such as aiming at single product and single application technology, which lead to waste of resources and environmental pollution. The common key problems of the production process of plant-based products must be abstracted and the clean production process must be established in order to utilize the plant-based biomass completely, promote the plant-based products industry developing harmoniously and offer theoretic and technologic supports for the pulp industry, textile industry, processing of Chinese herbal medicine, organic chemicals industry *et al.*

Our research is financially supported by National Basic Research Program of China (2004CB719700) and Project of Knowledge Innovation Program of Chinese Academy of Science (KSCX1-YW-11A1). In addition, the works of my masters and doctors are the essential precondition for publishing this book. Dr. Hongqiang Li and Dr. Xiaowei Peng participate in writing Chapter 1 and compiling the whole book. Master Xiaoguo Fu, Dr. Yejun Han, Dr. Weihua Qiu, Master Ye Yang and Dr. Lan Wang participate in writing Chapter 2, 3, 4, 5, and 6, respectively. In this book we quote lots of references of our predecessors and colleagues. I wish to express my sincere thanks to all of them.

Some errors may exist in this book. I sincerely hope to receive criticism and guidance letters from readers.

Hongzhang Chen
National Key Laboratory of Biochemical Engineering, Institute of Process Engineering,
Chinese Academy of Sciences
Zhongguancun
Beijing, PR China
E-mail: hzchen@home.ipe.ac.cn

*Chapter 1*

# INTRODUCTION

## 1.1. A BRIEF INTRODUCTION TO PLANT-BASED PRODUCTS

At the beginning of the last century most non-fuel industrial products, such as dyes, inks, paints, medicines, chemicals, clothing, synthetic fibers, and plastics, etc, were made from plant [1]. By the 1970s, petroleum-based organic chemicals have substituted plant-based products largely, capturing more than 95% of chemicals in the markets. On the other hand, the energy consuming proportion of petroleum accounted for more than 70%. However, recent developments are raising the prospects that many petrochemical products would be replaced by plant-based products because of the increasing depletion of petroleum, the greenhouse effect caused by the large-scale use of fossil fuels, and environmental pollution. Scientists and engineers keep making progresses in research and development of technologies that decrease the processing cost and extend the application scope of plant-based products. At the same time, government and people are gradually aware of the importance of exploiting plant resources and promoting the enterprises and scientific institute to develop new plant resource through the environment and legislation measures. Economy based on the plant-based products re-sprouted [2], whose developing trend depended on the exploration of new market-competitive plant-based products.

Numerous opportunities are emerging due to the industrial requirements of plant-based chemicals and materials. Today's plant-based products include commodity and specialty chemicals, fuels, and materials. From direct physical or chemical processing, more and more products could be produced from plant components, such as cellulose, starch, oil, protein, lignin, and terpene. In China's traditional Chinese herbal medicine processing industry, direct extraction of plant-based products accounts for a considerable part of Chinese herbal medicine industry in extracting active ingredients. Other lignocellulose in Chinese herbal can be effectively integrated into the production system of plant-based fuels and plants Materials. Other products can be indirectly processed from carbohydrates by biotechnologies such as microbial and enzymatic processing. Table1.1 shows worldwide markets for several plant-based products made from microbial and enzymatic conversion of carbohydrates. From this table we can see that the sales of plant-based products are exciting. Analysis of historical and present market growth rates suggests that the worldwide market for plant-based products will grow 15% per year.

**Table1.1. Increase in worldwide sales of biotechnology products[a]**

| Products name | 1983 ($ millions) | 1994 ($ millions) | 2000 ($ millions) | 2007 ($ millions) |
|---|---|---|---|---|
| Fuel and industrial ethanol | 800 | 1500 | 17775 | 25390 |
| High-fructose syrups | 1600 | 3100 | 6994 | 6222 |
| Citric acid | 500 | 900 | 731 | 963 |
| Monosodium glutamate | 600 | 800 | 1125 | 2142 |
| Lysine | 200 | 700 | 1000 | 2000 |
| Enzymes | 400 | 1000 | 1500 | 5500 |
| Specialty chemicals | 1300 | 3000 | 29125 | 42217 |

a : Statistical information from the Internet

A worldwide plant-based industrial products and technologies will be gradually introduced into the industrial markets. Ethanol and oxygenated chemicals derived from fermentable sugars will be the key precursors to other industrial chemicals that traditionally depended on petroleum. In the long term, taking the advantages of genetic engineering large-scale fuel production from lignocellulosic plant materials may have more obvious cost superiority than petroleum fuels [3, 4]. In other cases, biological technologies, such as enzyme catalysis, would be widely utilized in the hazardous petroleum chemical process. Increasingly, niche markets will be sought for wide array of custom-engineered plant polymers which are not available in petrochemical-based products.

### 1.1.1. Plant-Based Energies

Plant-based products include commodity chemicals, transportation fuels and chemical intermediates using for industrial processing. Ethanol is critical product because it can serve as a transportation fuel and also is a precursor to many other industrial chemicals [5]. For example, corn starch fermentation yields ethanol, which then can be dehydrated for production of ethylene that is the substrate for most of the petroleum-based commodity chemicals. In the United States, in 2007, 55 million tons of corn was used for ethanol production, corn ethanol reached 14.8 million tons. In China, in 2006, 4.75 million tons of wheat and corn was used for ethanol production, ethanol yield reached 1.44 million tones. However, it should be noted that, plant-based raw materials for the production of energy mainly focus on the utilization of starch and sugar, which not only increases the comprehensive costs of raw materials, but also put serious pressure on the environment. It is urgent to propose the concept that plant-based materials are compound resources, not just a simple fuel.

#### *1.1.1.1. Fuel Ethanol*

Large imports of crude oil in 1960s and 1970s stimulated interests in fuel ethanol in USA. Ethanol is primarily used as additive of gasohol, usually the ratio of gasoline and ethanol is 9:1. Researchers found that ethanol and its derivative, ethyl tert-butyl ether, can enhance the octane number and burning efficiency of gasoline [6]. Similar interest in ethanol occurred in Brazil, and, with subsidies from the government, Brazil forged ahead with ethanol production (ethanol yield is 13.43 million tons in 2006). Gasoline, blended with oxidative fuel such as ethanol or methyl tert-butyl ether (MTBE), can increase the combustion efficiency of gasoline and decrease carbon monoxide emissions in cold weather.

In recent years, market demand of starch-ethanol was affected by the price of corn, which became higher and higher in the whole world. Therefore, more and more countries opposed the ethanol production using grain as raw material, while promoted the cellulose ethanol. In the long term, large-scale production of fuel ethanol from lignocellulose materials is technically feasible and economically favorable. The key problem of present is that realize the larger scale demonstration of representative raw materials [7]. The producing cost of bio-ethanol must drop significantly if it wants to account for a large market proportion of the transportation fuel. This change will occur only if economical conversion technologies of lignocellulose are achieved. Use of these alternative feedstocks with new conversion processes may reduce production costs sufficiently to allow access to the commodity fuel market, even without subsidies or tax incentives [8]. The case study of lignocellulose-ethanol processing is described in chapter 6.

### *1.1.1.2. Bio-Diesel*

Bio-diesel is vegetable oil, which can produce fatty acid methyl ester through esterification in the existence of catalyst. The used methanol can be easily obtained from biomass, natural gas or coal. Vegetable oils used for producing bio-diesel include soybean oil, rapeseed oil, and so on. If the length of fatty acid chain of the used vegetable oil is similar to that of the ordinary diesel, the combustion characteristics are also equivalent. Usually the bio-diesel is blended with ordinary diesel of 20% [9].

Vegetable-based diesel fuels are appealing partly because these bio-based fuels confer some potential environmental benefits. Compared with diesel produced from petroleum, bio-diesel produced from waste oil, vegetable oil or oil-plant is still lack of competition. Bio-diesel is a fuel that cannot be an economically product in the near future. The main problem is that the source of bio-diesel is not abundant which directly results in the high cost of producing bio-diesel, therefore finding new sources of raw materials becomes an important aspect of bio-diesel development. Institute of Process Engineering, Chinese Academy of Sciences in this regard, using steam exploded straw as substrates to cultivate plant entophytic fungi and produce microbial oil in solid-state fermentation, brings the bio-diesel production to a new step[10, 11]. Further research and development may increase the use of bio-diesel[10]. In USA, bio-diesel would be unlikely to completely replace petroleum-based diesel. Even if all of the vegetable oil currently produced in USA went into bio-diesel production, plant-based diesel production could provide only 6.4% of the nation's annual diesel consumption of 10.73 billion liters. Introduction of bio-diesel as a blend with conventional diesel fuel is a more feasible goal in the United States [12].

### *1.1.1.3 Butanol*

As a new bio-fuel, butanol has the significant environmental efficiency compared with the transportation fuels derived from petroleum refining since that it can reduce the emissions of greenhouse gases to the environment, On the other hand, bio-butanol has many other advantages such as low steam pressure, high tolerance of water when mixing with gasoline, suitable for utilization of existing fuel supply and distribution system. Compared with the existing bio-fuels, there has no necessary to modify the vehicles because bio-butanol has higher mixing proportion with gasoline. Meanwhile, bio-botanol has higher economy, burning efficiency and running mileage than mixture fuel. Of alcohol, acetone is sugar cane, sugar beet, wheat, cassava, maize as raw materials, production processes and alcohol similar to ethanol production facilities can be modified to turn the production of biological alcohol [13]. Early in Group C, a grain alcohol produced by fermentation, the 1970s was due to the development of chemical, fermentation of the phasing out of production technology. In recent years due to the rapid rise in oil prices, coupled with the growing shortage of oil resources, food production fermentation acetone, alcohol technology to display its advantages, especially fermentation production acetone, alcohol is renewable alternative non-renewable resources The oil-based raw materials manufacturing, energy security in line with the state's long-term strategic considerations[14].

#### *1.1.1.4. Hydrogen Energy*

Because hydrogen has the following characteristics by widespread concern: (1) the product of combustion is water, no pollution, and water also can be a source of hydrogen production, not destruction of the natural cycle. If the use of catalytic fuel or fuel cell will not produce NOx, can achieve the true sense of "zero emission" (2) light weight, high calorific value, high efficiency and extensive application (3) can be stored and cost-effective transportation, metal hydride as the storage medium, but also with chemical energy, thermal and mechanical energy into the function of each other [15].

Hydrogen is a new kind of clean energy with hot research. The world's annual production of hydrogen is more than 40 million tons, of which 4% by electrolysis of water and the rest is used chemical method from oil, coal and natural gas and other fossil fuels or gas production plant. This requires that there was already a shortage of consumption of fossil fuels or a lot of power, its economic practicality is not strong. Biomass is an important raw material for hydrogen future development of the technology, but subject to a variety of factors, for example difficult to conduct large-scale, gasification technology for hydrogen producing is the lowest-cost method and becomes the focused technology [16].

### 1.1.2. Plant-Based Chemicals

#### *1.1.2.1. Intermediate Chemicals*

Intermediate chemicals play an integral role in economy. Organic chemicals are synthesized primarily from petroleum for production of numerous non-fuel industrial products such as paints, solvents, clothing, synthetic fibers, and plastics. Without these products USA could not maintain its modern way of life. When petroleum supplies are interrupted, price fluctuation occurs in international petroleum markets. These events can have widespread economic consequences on oil-importing nations. Increasing the diversity of strategic feedstock with bio-based raw materials could help mitigate economic downturns created by oil shortages. Thus, intermediate chemicals are an important market target of the bio-based industry.

*Ethylene*

Ethylene is perhaps the most important petrochemical because of the value of its numerous derivatives such as polyethylene, ethylene dichloride, vinyl chloride, ethylene oxide, styrene, ethanol, vinyl acetate, and acetaldehyde [17]. Before the mature lignocellulose conversion technology is appeared, the ethylene market was considered inaccessible to bio-based production. Today, bio-based ethylene production base on ethanol derived from corn stover still is not cost competitive with petrol-ethylene sources. In the near future, fermentation of lignocellulose-based ethylene could move into the margin of competition against petrochemical sources. Along with the prices soaring of oil, price for oil-based ethylene is 1,200 U.S. dollars/ton, but the price for bio-based ethanol is 670 U.S. dollars/ton. With the increase of petroleum prices or further improvements in the bio-based ethylene route, the cost advantage of petrol-ethylene could erode.

*Acetic Acid*

Acetic acid could be a large-volume chemical target for the bio-based industry. It is used primarily as a raw material in the production of vinyl acetate, acetic anhydride, cellulose acetate, acetate solvents, terephthalic acid, and various dyes and pigments which are used as a solvent in the chemical processing industry. The food, textiles, and pharmaceuticals industries also use acetic acid in their manufacturing processes. In 2007, there were 8.18 million tons of acetic acid produced all over the world. Acetic acid may be combined with dolomite lime to produce calcium magnesium acetate, an important deicing agent for the transportation industry. Bio-based acetic acid may be produced by fermenting corn starch or cheese whey waste or as a byproduct of the sulfite wood pulping process. A better understanding is needed for the relative costs of acetic acid from renewable resources compared with petroleum-based feedstocks.

*Fatty Acids*

Fatty acids, readily available from plant oils, are used to make soaps, lubricants, chemical intermediates, such as esters, ethoxylates, and amides. These three important types of intermediates are used in the manufacture of surfactants, cosmetics, alkyd resins, nylon-6, plasticizers, lubricants and greases, paper, and pharmaceuticals. 400000 tons of fatty acids produced in China, about 250000 tons came from importation. Most fatty acids were derived from vegetable and nature oils; the remaining were produced from petrochemical sources.

#### *1.1.2.2. Specialty Chemicals*

Specialty chemical markets represent a wide range of high-value products. These chemicals generally sell for more than \$4.00/kg. Although the worldwide market for these chemicals is smaller than those for bulk and intermediate chemicals, the specialty chemicals market now exceed \$380 billion dollars and is growing 10 to 20 percent annually. Examples of bio-based specialty chemicals include bio-herbicides and bio-pesticides; bulking and thickening agents for food and pharmaceutical products; flavors and fragrances; nutraceuticals (e.g., antioxidants, non-caloric fat replacements, cholesterol-lowering agents, and salt replacements); chiral chemicals; pharmaceuticals; plant growth promoters; essential amino acids; vitamins; industrial biopolymers such as xanthan gum; and enzymes.

Specialty chemicals can be produced through fermentation and enzymatic processes directly extracted from plants genetic engineering has now made possible microbial fermentations that can convert glucose into many products and can yield an essentially unlimited diversity of new bio-chemicals. Furthermore, industrial researchers are discovering that plants can be altered to produce molecules with functionalities and properties not present in existing compounds. It is anticipated that advances in biotechnologies will have significant impacts on the growth of the specialty chemicals market.

#### *1.1.2.3. Enzymes*

Fermentation of biological materials will continue to be a primary source of most enzymes used today and new enzymes produced in the future [18]. Enzymes serve two major purposes. Some act as biological catalysts in industrial processing of food ingredients, specialty chemicals, and feed additives. Others are components in end products such as laundry detergents, diagnostics, laboratory reagents, or digestive aids [19].

Worldwide enzyme sales summed up $5-6 billion in 2007 and the last 10 years, almost unanimously to maintain the 6.5 percent growth rate per year. European companies dominate world enzyme production; the largest company, Novo Nordisk, currently supplies 40 to 50 percent of world sales. The three largest markets for enzymes are the detergent, starch, and dairy industries. With the recent wood cellulose ethanol development trend of increasingly clear, there will be a huge growth trend of cellulose in the near future [20].

*Soaps and Detergents*

Industrial production of soaps and detergents in USA totaled $14.9 billion in 1993 (Ainsworth, 1994). Almost half of the laundry detergents in the United States and 90 percent of those in Europe and Japan contain enzymes. Along with the ban of many countries that forbidding phosphates-containing detergents in 1982, the application of enzyme was increased in soaps and detergents (Ahmed, 1993). The replacement of traditional chlorine bleach with peroxygen-based bleach additives (such as perborate bleach) also has enabled enzymes to play an important role in the soap and detergent industry due to their compatibility with the newer additives (Ainsworth, 1994).

Enzymes are naturally diverse and function in various cleaning agent roles. Protease, lipase, and cellulase enzymes are used in soaps and detergents to break down and help remove dirt stains. Celluzyme, a Novo Nordisk product, removes microfibrils that emerge from cotton fibers after use and cause an “old and gray” appearance (Falch, 1991). In addition, detergent enzymes reduce energy use because they are effective in much cooler wash waters.

*Food Processing Enzymes*

The largest use of enzymes as catalysts is in the production of high-fructose syrup from starch [21]. Amylases break down starch to glucose; then glucose is isomerized into fructose by glucose isomerase. The resulting mixture of glucose and fructose is used as a sweetener in soft drinks. Enzymes also have several uses in the dairy industry. The enzyme rennin coagulates milk protein and is used to make dairy products such as cheeses. Lactase is used to produce lactose-free milk.

*Enzymes for other Uses*

Various industrial enzymes are used as end products or biocatalysts at a smaller scale. The leather manufacturing industry has traditionally used lime and sodium sulfide mixtures to dissolve hair on animal skins, a process that has serious pollution. Proteases provide an alternative treatment that loosens and removes the hair, allowing it to then be filtered off. Proteases also result in a better-quality end product [22]. The pulp and paper industry also uses enzyme technologies, especially xylanase for bleaching to replace chlorine.The textile industrial cellulase for making "stonewashed" jeans (Wrotonowski, 1997) [23].

### 1.1.3. Plant-Based Materials

#### *1.1.3.1. Bio-Plastics*

Renewable resources such as industrial starches, fatty acids, and vegetable oils can serve as sources for bio-plastics. Biodegradable thermoplastics, such as starchy esters, cellulose acetate blends, polylactide, thermoplastic proteins, and poly(hydroxybutyric acid) (PHB)—show great promise for replacing he plastics derived from petrochemicals that generally are not biodegradable [24]. Graft plastic polymers (plastics based on plant materials and petrochemicals) are less bio-degradable than plant-based bio-plastics. The market share of bio-plastics (including polymer, plastics, and resin) is increasing by the rate of 20-30%.

At present there are three important foreign biodegradable plastics, the first category is based on the natural polymer, the second is based on the natural monomer synthesis of the polymer, and the third is based on the polymer of the fermentation process. The first category has already reported by large number of papers, this paper will only introduce the manufacturing, characteristics, application and development of current situation and prospects of second and third category of biodegradable plastics.

*Bio-Degradable Plastics Polymerized by Nature Monomer*

This type of biodegradable plastic is a natural low-molecular-weight biological molecule from the polymer-based polymer in the system, including thermoplastic and thermosetting plastics. This polymer is an important representative of polylactic acid and triglyceride polymer.

*Polylactic Acid*

Polylactic acid (PLA) is a kind of polyester, which is a natural existence of biological monomer. PLA is thermoplastic, which can be used in ordinary processing of plastics processing for the productions of thin film, sheet and fiber. PLA can be soft or hard. It is transparent itself, but it can also be used to produce opaque materials. Because of its high strength, it can be used to produce thin sheet. PLA is insoluble in water and has a good water resistance and anti-oily. By changing the molecular weight and crystallization, or polymerizing the lactic acid with glycolic acid copolymer or caprolactone, the mechanical properties and other properties of PLA can be improved.

PLA can be used as plastic sheeting and biodegradable packaging materials, besides it can also be used as disposable tableware, beverage containers and sporting goods. PLA is widely used in the biomedical area, such as drug delivery materials, bone growth and repair materials. Lactic acid and glycolic acid copolymer can support the growth of new cells and links, since that it can make porous materials with large areas of continuous surface, so that the cells can multiply in the whole matrix. PLA can be fully degradabled in the human's body in a certain period of time, this time depends on the implantation of the shape and size, usually not more than two years.

LA can be produced by chemical synthesis or fermentation manufacturing, but mainly by the fermentation using glucose as raw material. LA fermentation manufacturing yield can be more than 90%, and its production costs mainly due to its multi-step process of purification [25].

*Polymer of Triglycerides*

Triglyceride is a rich source of biological molecules, which recently caused a great deal of concern of people because its polymers can be used to create new biodegradable plastics.

Firstly, converting triglyceride into the intermediate high activity, and the latter is oxidized to liquid polymer with low-molecular-weight. Then, this polymer is mixed with the catalyst and the promotion of hybrid to help cross-linking reaction. Finally, the mixture is loaded into a mold with the increase in fiber, and solidified by heating in the mold to form the thermosetting plastics.

Glass fiber reinforced resin of Soya bean oil is a durable thermosetting materials, can be used for agricultural machinery, automotive, construction and other fields. This resin is obtained from renewable resources with high strength and non-formaldehyde. Besides glass fiber, other vegetable oil and the strengthening of materials, such as plant fiber, and even straw and hay, also can be used to produce the above thermosetting materials. And soybean-based resin for natural fibers has a strong affinity. If using straw instead of wood fiber to create pressure plastic composites (such as the construction industry in the current widespread use of the MDF), there will discover new using approaches to save wood fiber. In the future, nanotechnology may also be used to replace glass fiber cellulose. It can be expected that because of increased vegetable oil thermosetting composite materials development it possible for cheap and durable system of biodegradable materials.

*Bio-Plastics Polymerized by Fermented High Polymer*

A fermentation process, large-scale production of biological polymers, and the fermentation process the raw materials are generally natural resources, derived from plants. Polyhydroxyalkanoates (PHA) as a recent successful development of biological polymers can be used in the fermentation of polyester, in the industrialization of bio-plastic has played an important role [26]. British company ICI in 1978 started to produce PHB. 1987, ICI produced PHB and copolymer and sold in Europe called as Biopol. In the ensuing years, biodegradable polyester production of ICI increased rapidly. Since 1991, many countries performed a lot of research on the biodegradable polyester, especially in the aera of DNA recombinantion. In the mid-1990s, USA used recombinant DNA in plants (such as soybeans) to produce PHBV that the copolymer of hydroxybutyrate (HB) and hydroxy acid (HV).

Starch-polyester mixture (including starch, PHBV and PHA mixture) is being actively studies aiming at producing biodegradable plastics with excellent physical properties of polyester but lower prices. In the future, it is still unknown that whether PHBV will be produced largely-scale and achieve prospective market, but its excellent performance in the future may make it occupy an important place in the field of biodegradable plastics.

Now, PHA can be industrialized by the two-step fermentation method which conducted in the bioreactor by adding microorganisms and organic carbon matrix (PHB production when glucose or sucrose, when the acid production PHV, acid or oil waste from wood pulp in the fermentation). The first step is cell growth and the second step is the accumulation of polymer. The productivity of PHA from the cells is 80% to 90% of the dry cell weight, which can produce the final products through purification. PHA is the raw material for biodegradable plastics due to its biocompatibility and biodegradability.

At present, the fermenting efficiency of PHA is low which results in the high price of products. Now, people are to turn on genetic engineering to improve the PHA production stain, if such a high enough stain production, the price of PHA may have competition to the synthetic polymers. But this growing rape still has to use a lot of land. According to rough estimates, even if 10% of the existing land is used to plant rape, the produced PHA can only meet about 7% packaging materials consumption of USA. Biodegradable plastics produced by fermentation process have excellent physical properties, but the price is too high. However, the price can be considerably decreased if the fermentation process can be easily expanded.

#### *1.1.3.2. Plant-Based Inks*

Soybean oil-based inks were introduced to USA markets in the 1970s in response to the oil shortages. Most recently increased emphasis on improving worker safety and reducing environmental emissions has spurred interest in alternatives to petroleum-based inks. Soybean oil is a carrier for a pigment in ink formulations. Plant-derived inks require less use of hazardous chemicals during equipment maintenance, produce low evaporative emissions of volatile organic hydrocarbons, and are biodegradable. Soy-based inks have a desirable character that the pigment color in this ink seems more real because the lighter color of soybean oil. Black soy-based inks typically require a larger proportion of oil than pigment in comparison to colored printing inks. Because soybean oil costs more than petroleum, black soy-based inks are at a cost disadvantage. Some research indicates that the dispersibility of soy inks is 15% higher than petroleum-based inks, which on some certain degree can offset the differences in cost.

#### *1.1.3.3. Forest Products*

Forest products currently comprise the largest source of renewable resources for the bio-based materials industry. USA annually produces more than 260 million tones of lumber, paper, and derived wood products with a combined value exceeded $131 billion. Approximately half of the wood harvest was initially sent to solid wood products mills. The mill residue was sent to pulp mills or used for fuel. Overall, approximately 28% of the harvest ended up in solid wood products. Preexisting uses of silviculture crops in the pulp and paper industry may provide predictable markets and sustain production of woody silviculture crops while new uses develop in the manufacture of high-value materials and chemicals.

#### *1.1.3.4. Natural Fibers*

Cotton is one of the most promising industrial crops, which is composed of 90 percent cellulose. The long fibers of cotton make it an ideal material for weaving and spinning into cloth. Cotton confers qualities on fabrics that are difficult to duplicate with synthetic fabrics. Demand of cotton products has resurged in recent years. In 2007 cotton output of China summed up to 7.6 million tons, and more than 25.3 million tons of the worlds. Advances in biotechnology and genetic engineering are now enabling the development of cotton cultivars with improved pest resistance, yield, and quality, thereby potentially reducing production costs and better matching cotton characteristics to specific applications.

Natural fibers other than cotton occupy various niche markets, such as specialty fabrics, papers, cordage, and horticultural mulches, and so on. Natural fibers can heighten the environmental concerns and benefit for finding new markets. Jute, hemp, sisal, abaca, coir fibers, and products derived from these fibers will find extensive applications.

#### *1.1.3.5. Natural Polymer*

In the nature, natural high maleculars are periodically polymerized using carbon dioxide, water and sunshine, which possess the excellent performance and are microbial degradable by participating in the natural ecological cycle. Meanwhile, the wonderful properties of these materials can be passed on from generation to generation by biological genetic technologies. Therefore, how to use biotechnology to synthetic polymer materials attracted the concerns of more and more researchers, they keep to the areas of research, and have made significant progress. DuPont of USA, as the largest synthetic fiber maker of the world, has shifted its focus to the development of biological technology, who introduced three items of dawn plan and declared that the biotechnology will consolidate the world's leading science company's position of DuPont. Through about 20 years of unremitting efforts in this area, DuPont discovered that biotechnology was more environmentally friendly, low cost and secure than the traditional method of polymer materials.

PLA is a polyglycolic acid. *Lactobacillus* produced lactic acid is a carbohydrate, is organisms (including humans) in the common natural compounds. Lactic acid through Central dimer of the chemical polymer or lactic acid can be the direct polymerization of the PLA polymer. Polylactic acid as raw materials to get the products, has a good biocompatibility, biological absorbability and good biodegradable, and in the degradation of thermoplastic polymer material in the PLA has the best heat resistance [27].

PLA fiber is a new type of fully biodegradable synthetic fiber using grain as raw material. The waste products in the sea or soil can be decomposd into carbon dioxide and water by microbe, which will not emit poisonous gas or polluted the environment during burning. At present, there have many academic communities on the PLA fiber, especially the Spinning Company of Japan. Spinning Company produces PLA, called as Lactron, throng the following steps: extracting lactic acid from corn, sugar cane or sugar beet through the fermentation and distillation method; polymerizing PLA from lactic acid; getting PLA fiber by solution spinning of PLA. By comparasion, PLA fiber and polyester fiber has almost equal strength and elongation, rather low Young's modulus, and relatively soft of fabric. So the PLA fiber is a fine raw material for cloth. Lactron can be processed into short fiber, single-wire and wire form, and cotton, wool or rayon and other decomposable blended fibers, also can used to produce silk-like fabric for underwear and shirts. Lactron not only possess durability, good moisture absorption, processing stability and good anti-wrinkle property.

#### *1.1.3.6. Biomedical Materials*

Biomedical textiles both belong to the field of technical textiles, medical equipment belonging to areas, it is the life sciences and materials science, the product of cross-cutting, has become a hot research. A biomedical material is an important material foundation for the development of modern clinical medicine. Although the industrial scale of biomedical materials is rather small, but it belongs to the knowledge-intensive and high output industry. Now, countries around the world attach great importance to its development. Biomedical materials used textiles includes short fiber, single-wire, machine-woven silk and knitted fabrics and composite materials. The representatives of biomedical materials are chitin fiber, bone collagen fibers and alginate fibers.

*Chitin Fiber*

Chitin from shrimp, crab and other animal's foot section of the crustacean is the second lager amount natural high molecular materials only inferior to cellulose, with the annual production of 1 billion tons from marine crustaceans.

Chitin is a natural polysaccharide material, owing to its better and more crystal structures of the hydrogen bond, so the poor performance of its dissolution. The crustaceans deacetylation become chitosan (also called deacetylation chitin) the dissolution of chitin than good. Will be refined or the chitin chitosan dissolved in the appropriate solvent, wet spinning through a system of chitin chitosan fiber or fiber.

As chitin or chitosan has a good biocompatibility and adaptability, also has some functions like anti-inflammatory, hermostasis, analgesia and the promotion of body tissue growth. All these properties make chitin as an ideal material for the protection of the wound. Chitin as a low-fiber component of the animal has the both biological function of collagen in higher animal and cellulose in higher plant, which result in the excellent biological characteristics.

*Osteocollagenous Fibers*

Bone glue is a protein, which responsible for the individual protection and morphologic preservation of skin, bones, tendons, blood vessels, intestines, cornea and teeth. Molecule of bone glue is constituted of more than three peptide chain.

Bone collagen fibers through the fabric to the flexor tendon of bovine collagen made from the suspension. Firstly, clean sheet with the tendon of Jieyuan processed to remove bone collagen fibers yuan flexibility hard to make it easy expansion. After removing the bone collagen protein and redundant enzymes, tendon slices is expanded by dipping in the cyanide leaching acid and methanol - water mixture (pH2~3). Then the mixture is treated by uniformization and filtration, and then solidified in the appropriate solidification bath into thread with tensile strength of about 2.7 cN/dtex.

Bone collagen as a medical material has the following features: adaptive biological fine, without antigen, a good absorbent, such as organisms, are at home and abroad for its development and the protection of the wound.

*Alginic Acid Fibre*

Alginate is polysaccharide extracted from seaweed plant. Alginate fibers can be prepared by wet spinning that squeezed out strong alkaline solution of sodium alginate spinneret and send into acid coagulation bath containing calcium ions, sodium alginate acted the ion exchange with calcium, that is, formed water-insoluble calcium alginate fibers. The shortcoming of this fiber is lower breaking strength.

When calcium alginate fibers used for wound contact layer, the interaction between the wound and it will produce alginate, calcium alginate gel. This gel is hydrophilic, which can not pass through the bacteria and can promote the growth of new organizations.

Alginate fibers are usually prepared by co-spinning with non-alginate polymer, which is often water-soluble or soluble organic solvents and they contained negative charge of $COO^-$. For example Carboxymethyl cellulose (CMC), pectin (including galacturonic acid), N-, O-carboxymethyl deacetylation chitin (NOOC), O-carboxymethyl deacetylation chitin (OCC), ammonia-day winter Acid, polyethylene and polypropylene acid glutamate. These non-

Alginate polymers can be improved by adding alginate fibers such as the absorption properties.

### 1.1.4. Plant Extract

Plant extracts are the materials extracted from parts of plant by solvents. The used plants often are medicinal materials, such as Ginkgo biloba, licorice, hawthorn leaves and ginseng, and so on. Some crops such as beans embryo, embryo of corn, soybean meal, and black soybean hull can also be used as raw materials for plant extracts. The used solvents include water and some organic solvents such as ethanol, methanol. The used purification methods include extraction, chromatography, ultrafiltration and molecular distillation. The used drying methods include spray-dried, vacuum drying, microwave drying, freeze-drying and so on. The plant extract are usually powders [28].

The plant extracts are mainly used as food additives, raw materials of functional food, and raw materials of medicine. Most products are exported to foreign countries, such as the United States, Europe, Japan, South Korea, India and other countries. In recent years, the yield of plant extracts in China kept an annual increase rate of 20%, and with a good foreground. In 2007, the total export amounts of plant extracts in China achieve $ 600 million.

China's large-scale production and export of plant extracts based on the rise of "return to nature" in Europe and USA. In this boom, many kinds of plant extracts replaced the chemically synthetic drugs on the market and became the best-selling health products. The ratio of plant drugs in Germany, Britain, France, Spain and Italy all exceed 45%. USA has the most stringent Medicine Management System in the world, but it has the most plant drugs in the market. According to the latest reports, in 2007 USA became the world's second-largest consumer of plant drug with the total sale amount of $ 7-8 billion.

China has the world's largest production base of natural ingredients and has using traditional Chinese medicine for thousands of years. So, plant extracts became one of major export products of China. At present China has about 80-100 kinds of plant extracts with large-scale production and export. They generally fall into three categories: Chinese medicine extracts such as schisandra, gsnoderma (including ganoderma lucidum), medofenoxate, ginkgo biloba, ephedra, licorice, etc.; Chinese drug extracts such as ginkgo, composite salvia and Buzhong Yiqi Compound; extracts of plant active ingredients such as tea polyphenols, soy isoflavones and ginsenoside.

Today, in USA about 100 plant extracts are selling at market. According to the amount, last year, the five most plant extracts imported from China to USA are licorice extract, ginger oil, garlic products (garlic powder and garlic oil), ginkgo biloba extract and ephedrine.

Below, a brief introduction is given to some important plant extracts.

***Ephedrine***

Ephedrine belongs to alkaloid and sympathetic medicine. Ephedrine can be used to treat bronchial asthma, colds, allergies, nasal congestion, edema and hypotension and other diseases. Research shows that ephedrine has remarkable effects on cardiovascular system, central system, smooth muscle system and other aspects. Ephedrine has a lasting and gentle vasoconstriction effect. Ephedrine can expand coronary vessels and increase coronary blood flow. Pseudoephedrine can excite the peripheral nervous system and the central nervous system directly [29]. China's export of ephedrine once occupied 90% of the global market, but shrunk in recent years, which shrunk from the several thousands of tons at last several years to 1000-2000 tons of 2007.

***Glycyrrhiza Extracts***

Similar to ephedrine, glycyrrhiza is also a traditional plant extracts for export in China. Glycyrrhiza in China is mainly used for medicinal, but in western countries it is mainly used as a sweetener alternative to sugar in the food, cosmetic products, cosmetics and tobacco products. USA FDA had listed glycyrrhiza and glycyrrhiza extract in the list of safe and non-toxic substances for a long time. WHO and food additives regulations committee of FAO also included glycyrrhiza into usable list. Japanese scholars have already found that glycyrrhiza extract is a very good anti-virus product. Japan has developed liver protecting products containing glycyrrhiza extract and most other drugs also containing glycyrrhiza extract. Every year, the import of glycyrrhiza extract in Japan is more than 10000 tones from China, Pakistan and Afghanistan. In recent years, Japan began to reduce the import of raw glycyrrhiza by importing extracting intermediate of glycyrrhiza. USA is the largest export market of China for glycyrrhiza extract, it is estimated that in 2007 China exported about 1000 tons glycyrrhiza extract to USA.

***Silymarin***

Silybin is the main active ingredients of silymarin. The main active ingredient extracted from the seeds of purple silymarin includes silybin and isosilybin. Silymarin can be dissolved in methanol, ethanol, acetone and other organic solvents, but not in water. It has the effects of protecting liver, improving liver function and enhancing liver regeneration. It has a better effect on acute and chronic hepatitis, cirrhosis of the liver toxicity and metabolism in the liver damage. It is the main raw material of silybin medicinal derivatives.

In recent years, silybin extract is very popular in the international market. The United States and Europe are the most important silybin consumer market. In 2007, china's total export of silymarin and silybin was bout 100 tons, in which silybin export was about 20 tons. India is another important silymarin exporter of china. As the demand of silymarin and silybin increasing in the international market, China is the largest planting country of silymarin in the world, the silymarin and silybin have very bright market prospects. The export growth maybe keeps at 15%.

***Gingko Extract***

Gingko is valuable tree specie unique to China. It is the most ancient gymnosperm left after the Quaternary Glacial movements. In Three-State Period, ginkgo has been widely planted in the South of Yangzi River Region. From then on, ginkgo is planted throughout China. From the end of the 17$^{th}$ century, ginkgo tree has also been introduced into some European and American countries. It is validated by modern science that the extract of ginkgo leaves, flavonoids, has efficiency on hyperlipidemia, cardio-cerebrovascular disease. The development of ginkgo products has great significance to human health. Some ginkgo products such as tea, tablet and soft capsule have been developed. The medical value of ginkgo biloba includes expansion of blood vessels, protection of vascular endothelial tissues, regulate blood fat, protection of low-density lipoprotein, improving hemorheology, inhibiting platelet activating factor, inhibiting thrombosis, improving memory, eliminating free radicals, anti-anoxia and preventing artery spasm.

Since the 1980s, China ginkgo biloba extract always kept high-speed growth. China's annual ginkgo dry leaves output has reached as much as 15000 tons and it was first place in the world. The ginkgo extract of China mainly used for the production of oral health food products. When used in drugs production, the ginkgo biloba extract must reach special high quality requirement that can not be reached by many Chinese plants extract factories, which makes they choose the way of exporting low-price raw materials. Tianjin Peak Group and Zhejiang Kang Enbei Group are the largest producers and exporters of ginkgo biloba extracts, which export about half ginkgo biloba extracts all of the China. At present, the ginkgo biloba extracts average price has risen to around 750 $/kg.

### *Asiaticoside*

Asiaticoside is mainly used to promote wound healing, stimulate the growth of granulation, and remove the skin spots. It is the main raw materials of cosmetics industry in Western. The international market demands for great asiaticoside. China exports about 10 million U.S. dollars asiaticoside every year. Shanxi Tiancheng Group has become the largest domestic manufacturer and exporter of asiaticoside due to the large areas of wild asiaticoside.

### *Echinacea Purpurea*

Echinacea purpurea is a medicinal plant origin from USA. It is the traditional Indian herbal medicine. In North America and the EU, Echinacea purpurea is an immune accelerator and enhancer used earlier and wider. It is a hot plant drug in western market. Echinacea purpurea and astragalus are currently the world recognized best anti-virus plant drugs. Echinacea purpurea extracts, used as the promotion and regulation of immunity in Germany as early as 1989, has been listed into the most commonly used 2000 prescription, ranked 131. In 1982, the single and complex preparations included Echinacea purpurea reached 300, and there were many formulations had been used extensively for clinical. In 1995, Echinacea purpurea reached the top of American best-selling health foods list. From 1992, Echinacea purpurea product sales have more than 7.8 billion U.S. dollars. In the current cold treatment, people use a lot of antibiotics. Due to microbial resistance, people have very careful use of antibiotics, Echinacea purpurea is used in clinical treatment of a virus or bacteria caused acute and chronic respiratory infections, supporting various antibiotics the treatment of serious bacterial infection. In addition, Echinacea purpurea are used for health care because it can enhance human immune function. Therefore, the use of Echinacea purpurea will become increasingly widespread and the selling market will continue to expand.

### *Paclitaxel*

Paclitaxel is mainly used for treatment of metastatic ovarian cancer, advanced breast cancer, and non-small cell lung cancer. In both developed and developing countries, these three categories of cancer are quite common. From the growth of domestic and foreign markets, foreign markets are generally large scale and more stable, but the domestic market is smaller but with more growth potential. So, the supplies of paclitaxel are far less than the international market demand [30].

Companies like BMS and Ivax of the United States, P-PR Company of France, Immune Company of Canada, yem Tree Company of Netherlands all are paclitaxel manufacturers.

These companies have stronger scientific research institutions and long-term cooperative relationship with governments. They not only have advantages in domestic market but also expand worldwide markets with the support of governments. At present, the companies of China have not enough strength to competitive with these companies. From the domestic market, the resources are obvious constraint factor, especially the low paclitaxel content and slow growth of yew make the resources scarcer. Therefore, the fermentation method attracted a large amount of interest [31].

### *Shikimic Acid*

Shikimic acid is a monomeric compound extracted from star anise, a Chinese medicine. It is the anti-cancer drug intermediates because of its anti-inflammatory and analgesic effect. Beijing University of Traditional Chinese Medicine Pharmacological Research Center firstly found that Shikimic acid has obvious anti-thrombosis effect, which can inhibit the artery and vein thrombois and cerebral thrombosis. In 2005-2006, avian flu has been the hot issue of media and the difficulties of governments. “Tamiflu”, patent medicine of Roche Company, was designated as the drug of fighting against avian flu, which greatly promoted the investment boom of domestic Shikimic acid. So China’s specialty spice plant, star anise, has become the investment hot of enterprises and business.

According to statistics, star anise annual output is about 50000 tons. The main producing area is Guangxi province. The province produces more than 70% star anise. Star anise is mainly used for food seasoning and extraction star anise oil. According to reports, more than 90% of the world star anise is from China. Wild star anise mainly grows in Yunnan, Guizhou and Sichuan mountains that about 400-600 m above sea level. Limited by the natural conditions, only one harvest can be obtained and the annual output is about 6000 tons. Wild star anise has certain toxicity, which can not be eat directly. Because of its Shikimic acid content as high as 10% or more, the wild star anise is mainly used to extract Shikimic acid.

### *Tea Polyphenols*

At present, China’s output of tea polyphenols occupies the first place in the world, with about 90% of the tea polyphenols for export and domestic sales minimal. According to foreign scholars, there are a variety of pharmacological effects of tea polyphenols such as anti-cancer, anti-cardiovascular disease, anti-hypoglycemia and weight loss, etc. Today, health food containing tea polyphenols has become a best-selling product in the international market. China’s biggest export company of tea polyphenols is Wuxi Taiyo Green Power. It is estimated that in 2007 about several hundred tons tea polyphenols are exported.

### *Soybean Isoflavone*

Soybean isoflavone is known as phytoestrogens in western countries. It can be used to treat hormone-related diseases such as menopause syndrome, osteoporosis and Prostatic Hypertrophy. In the United States and Europe, soy isoflavone has become one of the most popular health products with annual sales of hundreds millions of U.S. dollars. Because of the low production cost, the soy isoflavone of China has certain competitiveness in foreign markets [32].

Soy isoflavone is found in soy seeds. Because of its important role in human health, it has aroused extensive attention in recent years in the scientific community. It is shown by clinical studies that soy isoflavone has a unique effect on breast cancer, prostate cancer, heart disease, cardiovascular disease and climacteric syndrome. Because of the role of soy isoflavone in many health care, USA, Japan, Germany, the UK have appear a upsurge of soy isoflavone. The health food mainly included soy isoflavone has become a popular new type food. Dozens of soy isoflavone containing health food have appeared in foreign markets. Various preparations of soy isoflavone, such as tablets, oral liquid and powder were appeared. High-purified soy isoflavone has been applied to the development of pharmaceutical products. As early as 1996, soy isoflavone had been approved by FDA as a health food. According to statistics, in 2007 the global health food products contained soy isoflavone sales volume reached 200-300 million U.S. dollars. The largest manufacturer of soy isoflavone is ADM Company. At present, the demand of soy isoflavone in international market reached 1500 tons, but the actual output was only 300 tons.

***Arteannuin***

Arteannuin is easily dissolved in benzene, chloroform, ethyl acetate, acetic acid and acetone, and can be dissolved in ethanol, methanol, ether and hot petroleum ether. Because of containing peroxide bridge and peracetic acetal structure, artemisinin is instability in acid and base solution, particular strong base. When heated to the melting point, artemisinin will be decomposed rapidly. Artemisinin is a new antimalarial drug, which has the characteristics of low toxicity, efficient and quick-impact. It has effect on plasmodium falciparum and malaria. Artemisinin is the first new drug self-developed by Chinese scientific researchers. It is also the only two Chinese medicines into world pharmacopoeia. The WHO thinks artemisinin is the only truly effective drug current in worldwide for the treatment of malignant malaria. Since 2001, 51 countries and regions have specified artemisinin-related drugs as antimalarial drug.

## 1.2. Process Engineering Development Strategy of Plant-Based Products

### 1.2.1. From the Perspective of Raw Materials

Lignocellulose includes three major components, cellulose, hemicellulose and lignin, also contains a small amount of pectin, fat, wax and other organic compounds and all kinds of metal elements. The composition, structure and distribution of these components are different from the type, origin and season, and so on. Take straw as an example, the results of Chen showed that from the cell level or molecular level, the physical and chemical properties of rice straw as well as enzyme and fermentation performance are different significantly [33]. As for corn stalk, it also showed the same result. Aim at the uneven of straw, Institute of Process Engineering, Chinese Academy of Sciences invented low-pressure steam explosion carding classification technique. By this technique, steam exploded straw can be divided into long fiber fraction, short fiber cells and epidermal cells. Because the long fiber can reach the indicators of wood fiber, it can be modified to produce new cellulose derivatives, such as acetyl cellulose, regenerated cellulose fiber carbamate. The obtained short fiber cells can be enzymatic hydrolyzed into sugars, then be fermented into energy, chemical products, materials, and so on. Fermented residues can be used to produce bio-oil by pyrolysis or produce biological fertilizers by fermentation. The ash of epidermal cells contains 60 percent $SiO_2$ and can be used to produce nano-$SiO_2$.

## 1.2.2. From the Perspective of Process

In order to solve the problems in biomass energy conversion process, Institute of Process Engineering, Chinese Academy of Sciences built three key technical platforms that non-polluting steam explosion and fractionation platform, water-saving and energy-saving pure seed solid state fermentation platform, solid-phase enzymatic hydrolysis coupled with liquid ethanol fermentation and steam-exploded straw membrane recycling enzymatic hydrolysis coupled with fermentation platform.

### *1.2.2.1. Non-Polluting Steam Explosion and Fractionation Technology Platform*

The non-polluting steam explosion technology means that without adding any chemicals. During the steam explosion process, 80% of hemicellulose can be separated from biomass only though controlling the water content of straw. The cellulose left in the biomass can be enzymatic hydrolyzed more than 90% [34, 35]. Based the laboratory study, this technology has been scaled up to 50 $m^3$ and successfully applied into clean pulping, clean ungluing of cannabis, humic acid producing and xylooligosaccharide producing. In addition, long fiber separation technology has achieved by coupled steam explosion with mechanical carding or wet superfine [36].

### *1.2.2.2. Pure-Culture Solid State Fermentation Technology Platform*

Fermentation technology can be divided into submerged fermentation (SmF) and solid state fermentation (SSF) in general. Compared with the SSF, SmF is easy to realize pure culture and scale-up. However, SmF produces a lot of wastewater which caused seriously environmental problems. SSF has the advantages of energy-saving and water saving. However, it is difficult to overcome the heat and mass transfer resistance. It is easily to cause local cells death and yield is low. In addition, SSF is difficult to scale-up and pure culture. However, gas double dynamic solid state fermentation bioreactor (GDSFB) can solved the major problems of large scale pure SSF. In the GDSFB, without mechanical agitation, but only gas controlling is used. On the one hand, the air pressure is pulsation; on the other hand, gas also flows in the bioreactor to improve the transfer of heat and mass. The gas controlling promotes the growth and metabolism of cell. 100 $m^3$ GDSFB has been designed and manufactured to produce cellulose by SSF. This bioreactor is by far the largest SSF bioreactor in the world. In this bioreactor, using steam-exploded straw as main medium, the average activity of cellulase reached to 120 FPA/g dry medium, which means that it can truly realize the large-scale and low-cost production of cellulose [37]. GDSFB can shorten the fermentation period of one-third and increase yield 50-100%. It also plays an important role in mixed fermentation [38].

#### *1.2.2.3. Straw Solid Phase Enzymatic Hydrolysis/Liquid Fermentation Ethanol Coupling Technology Platform*

Cellulose solid-phase enzymatic hydrolysis coupled with the liquid fermentation technology can effectively improve the efficiency of cellulose enzymatic hydrolysis and ethanol fermentation. It can also solve problems such as the large consumption of water and the inconsistent temperature of enzymatic hydrolysis and fermentation. So it can reduce the cost of cellulose ethanol. In this technology, the airlift high-concentration ethanol fermentation and separation coupling system are integrated with the airlift double circulation tower fermentor, vacuum return, $CO_2$ gas stripping and activated carbon adsorption technologies. It realized triple coupling of enzymatic hydrolysis, ethanol liquid fermentation and adsorption separation. The equipment drops the dosage of cellulose to 15 IU/g straw and reduces the wastewater in residues. Because the saccharification and fermentation in different region of the reactor, it can facilitate the coordination of saccharification and fermentation temperature. It overcomes the shortage of solid state ethanol fermentation. In this equipment, the yield of ethanol is 15%, straw cellulose conversion rate is 80 percent, activated carbon adsorption and desorption ethanol concentration is about 50%. All of these greatly reduce the production cost of cellulose ethanol [37].

#### *1.2.2.4. Steam Exploded Straw Membrane Circulation Enzymatic Hydrolysis Coupling Fermentation Sugar Platform*

The cost of cellulase accounts for 50-60% of the entire cost of biomass conversion, which is a limitation for the industrialization of lignocellulose enzymatic hydrolysis. Using membrane bioreactor system to hydrolyze steam exploded straw and reuse cellulose is a better way. Appropriate molecular membrane can retain cellulase and unhydrolyzed cellulose, while the hydrolyzed products can pass through the membrane. Thus, it can eliminate the end products suppression and improve the hydrolysis rate of cellulose.

As for the traditional membrane bioreactor, the reducing sugar concentration from the enzymatic hydrolysis of cellulose is relatively low, which not conducive to the consequent processes [39]. An ultrafiltration membrane bioreactor consisted of 16 enzymatic vessel was used to increase the reducing sugar concentration. The effects of enzyme loading, the number of enzymatic vessel and the dilution rate on enzymatic hydrolysis of steam-exploded rice straw were studied. The results showed that the best condition is: 20 FPU/g, four enzymatic vessels and 0.075/h (Dilute rate). Compared with traditional batch reaction, the conversion rate of substrate increased significantly to 39.5%. Compared with the only one membrane reactor, the yield of reducing sugar increased 60% and the average concentration of reducing sugar rised from 4.56 g/L to 27.23 g/L.

### 1.2.3. From the Perspective of Products

In general, it takes many steps to convert biomass into plant based products. Biotransformation integration technology integrates different methods into one reactor or one step, which benefit for the simplification of processes and the increase of productivity. At present, the research contents of biotransformation integration technology mainly include: 1) simultaneous saccharification fermentation, 2) construction of new strain with integration advantages, 3) integration processes of reaction and separation, 4) integration of bioreaction and process model.

During the biomass transformation process, more attention was paid into the development of the main fermentation products and less attention was paid into using the by-products produced in fermentation. If these by-products can be processed into high value products, the economy of fermentation will be improved significantly. Take straw transformation as an example, after the steam explosion, the hemicellulose dissolved in water can be used to produce xylooligosaccharides, the residue can be divided into long fiber and short fiber. Long fiber can be used for paper making; short fiber can be used for fermentation substrate. The residue of fermentation can be further transformed into bio-oil and nano-silica. Figure1.1 is the comprehensive utilization diagram of steam exploded straw.

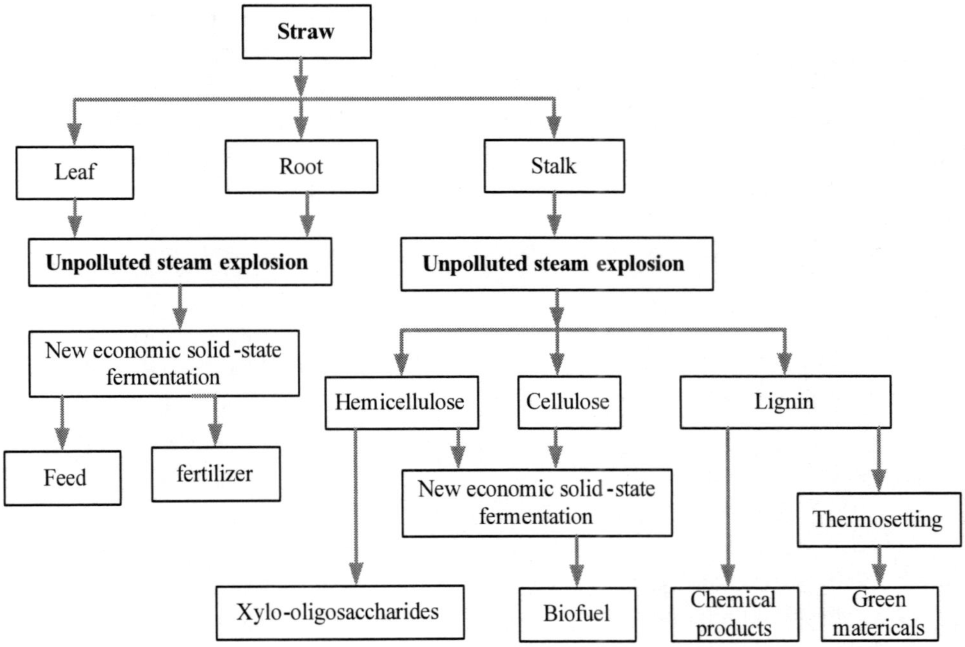


Figure 1.1. Multiple utilization of steam-exploded biomass after fractionation process

## 1.3. The Development Strategy of Plant-Based Products

### 1.3.1. The Complexity of Raw Material and Corresponding Development Strategy

Plant resources are wide variety. It include cereals, crop residues, seed of oil crops, sugar crops, forage crops and various woody crops. Different plants have different components, for example the main component of cereals is starch, of straw is cellulose, hemicellulose and lignin, of the seed of oil crops is fat. Component of same kind of plant is different form the time, growing place and different parts of the plant. However, plant resources can be divided into four basic chemical structures: carbohydrates (sugar, starch, cellulose and hemicellulose), lignin (poly-phenol), esters and protein.

Besides of these basic structural compounds, there are hundreds of species of organic compounds with commercial value came from plants, which include biomedical materials, nutrients, natural products and industrial products. For example, the effective ingredients in Chinese herbal mainly include alkaloids, glycosides category, flavonoids, terpenes, organic acid and polysaccharide. There are also many important industrial products such as the lacquer, gumresin and natural rubber.

Faced with such complex plant materials, what strategies should we develop? We must break the traditional concept of production one single product. Under the guidance of the thoughts of total utilization of biomass, fraction separation and stepwise utilization [40], we utilize all components of biomass and transform them into different products, which can realize the full use of raw materials and maximize the efficiency of land.

First, it is important to fully understand that all components of the plant can be utilized.

Second, it is supposed to make a breakthrough in processing technology of plant resources. The integration of biological process and ecological utilization will further enhance the conversion efficiency of materials and energy, improve product value, and reduce production costs. The improvement of new technology and new processes will promote the biomass energy ratio in the renewable energy. The combination of modern biotechnology, information technology, and engineering technology will enhance the technological content of existing technologies.

Third, we should comform to the resources' advantages and economic development level of different region and combine the modern science with traditional agricultural technology. In accordance with the principle of "overall, coordinate, recycling, renewable", using the system engineering approach to optimize the combination of various technologies, to build plant resources standard system and technical security system, to realize the ecological environment and rural economy virtuous circle and to achieve economic, ecological and social benefits reunification.

Finally, it is necessary to make great extension on the development and utilization theory of solid phase complex materials, which is benefit for promoting the key process of transforming biomass. It is more important to further explore the characteristics of plant utilization and technological bottlenecks and establish a biomass high value systematic theory.

Because of the diversity of raw materials, during product development, we should not only consider their own performance to meet certain requirements, also select the appropriate transformation method and joint products, raw materials and production processes together. Sometimes, in order to produce a specific product, it should attach more and more importance to using plant genetic engineering from the source to modify plant resource and expand the culture area. In order to meet the demand of mass production of fuel ethanol, many research institutes are working on sweet sorghum breeding, which mainly include [41]: reducing height of plant and increasing diameter of stem to prevent lodging and increasing stalk juice; increasing stem production and sugar concentration; breading new varieties and hybrids; enhancing storability, solving storage problems.

## 1.3.2. The Complexity of Process and these Interrelation

Fossil-based products are generally processed by physical and chemical methods, without any relations to bio-reaction. However, plant-based products are processed by physical, chemical, biological methods alone or integrally. Figure 2 describes the transformation of plant based on raw materials methods and products. These methods include the following:

***Separation of Nature Ingredients in Plants***

To separate plant nature ingredients, organic solvent extraction is a commonly used method which is used in the extraction of plant oils and medicinal ingredients. The basic principle of organic solvent extraction is the choice of appropriate organic solvent. The active ingredients can easily be dissolved in this solvent and other ingredients insoluble. So the active ingredients can be dissolved into the solvent from the plant tissue. When the solvent is added to raw materials, because of proliferation and infiltration, the solvent come into the cell through the cell wall. The dissolution of the soluble substances makes a concentration gradient, which drives substances diffusing into the solvent body. This process will carry out continuously until the equilibrium of cells inside and outside. Separating this solvent and adding new solvent into plant materials for several times, the ingredients in plant can be obtained nearly completely.

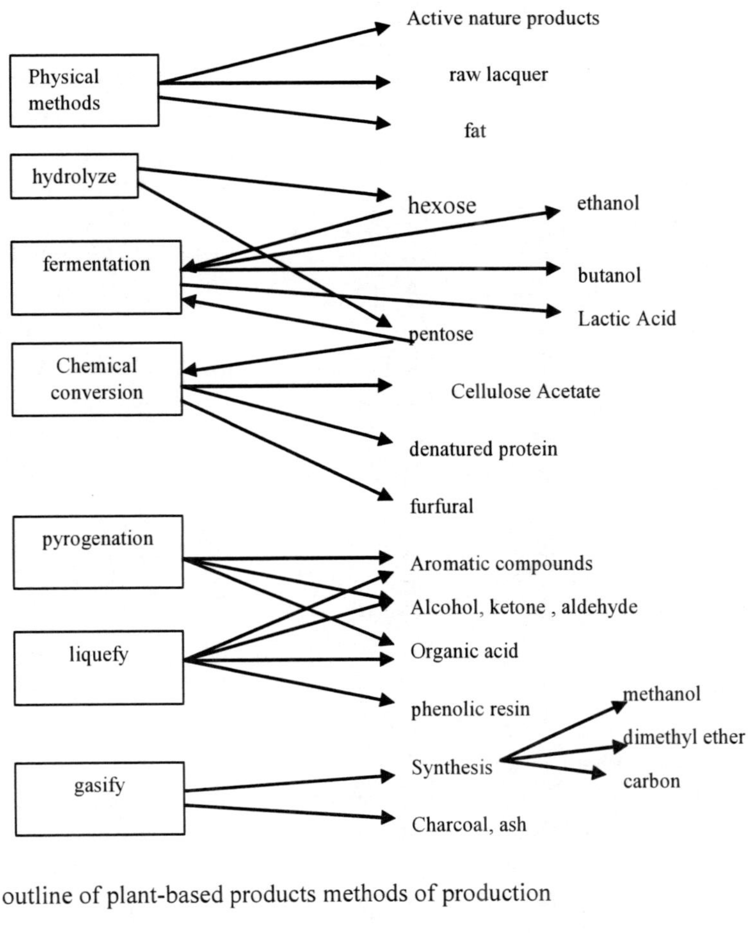


Figure 1.2. The outline of plant-based products methods of production

In recent years, many extraction methods have been developed for natural products. Supercritical fluid extraction is an example, which is a new technology using supercritical fluid to replace conventional organic solvents. Supercritical Fluid Extraction (SFE) is the process of separating one component (the extractant) from another (the matrix) using supercritical fluids as the extracting solvent. Extraction is usually from a solid matrix, but can also be from liquids. SFE can be used as a sample preparation step for analytical purposes, or on a larger scale to either strip unwanted material from a product (e.g. decaffeination), or collect a desired product (e.g. essential oils). Carbon dioxide ($CO_2$) is the most used supercritical fluid, sometimes modified by co-solvents such as ethanol or methanol. Extraction conditions for supercritical $CO_2$ are above the critical temperature of 31°C and critical pressure of 74 bars. Addition of modifiers may slightly alter this. There is also the ultrasonic extraction technology, HSCCC extraction technology and film extraction and separation technology.

***Hydrolysis***

Hydrolysis of polysaccharide by acid or enzyme and get monosaccharide used for fermentation or further processing is the key step of plant based products process. The polysaccharide (carbohydrates) such as starch, cellulose, and hemicellulose is the main component of plant resources. These polymers must be hydrolyzed into monosaccharide so that microorganism can convert them into all kinds of products.

Acid hydrolysis [42] can be divided into concentrated acid hydrolysis and dilute acid hydrolysis [43]. The principle of concentrated acid hydrolysis is that crystalline cellulose completely dissolved in 72% concentration of sulfuric acid, 42% concentration of hydrochloric acid or 77-83% concentration of phosphoric acid at lower temperatures, resulting in homogeneous hydrolysis of cellulose. The most commonly used concentrated acid is sulphuric acid, which has the particular advantages like high recovery rate of sugar that about 90% of sugar from hemicellulose and cellulose can be recovered. The key of concentrated acid hydrolysis is acid recovery from the sugar solution. Ethanol fermentation from the concentrated acid hydrolyzed lignocellulose had been put into commercial production in former Soviet Union and Japan. But it is only feasible when energy crisis occurrences. In normal circumstances, concentrated acid hydrolysis need a lot of acid, resulting in high production cost and lacking of market competitiveness.

The principle of dilute acid hydrolysis is that acid lead the cellulose micro-structure change, such as polymerization degree, fiber density and crystallization [44]. When dilute acid hydrolysis is used, the yield of sugar is lower (about 50-60%) and some fermentation inhibitors will be released. As for dilute acid hydrolysis, Germany, Japan and Russia have been running ethanol plant using infiltration hydrolysis in the past 50 years and have accumulated a wealth of experience. Some companies have recently started to find more economic acid hydrolysis process to enhance the commercial viability [45].

Both concentrated and dilute acid hydrolysis asks strict demand to reactors. The material of reactor must be acid resistant and pressure resistant. During hydrolysis process, additional energy needs to be provided to keep the high-temperature and high pressure conditions and ensure the smooth reaction. The main study object of acid hydrolysis is wood with the most important goal of maximum transformation rate of cellulose.

Compared with acid hydrolysis, the cellulase hydrolysis is more promising direction. The conditions of cellulase hydrolysis are mild and without chemical pollution. At present, many countries and regions are carrying out this kind of research. Take the cellulose ethanol as an example, replacing acid hydrolysis with cellulase hydrolysis has become a developing direction. The initial fermentation process of cellulose ethanol is step-by-step hydrolysis and fermentation [46], which includes two steps that first enzymatically hydrolyzed cellulose into glucose, and then produce ethanol by fermentation. The advantage of this process is that the cellulose hydrolysis and fermentation all are kept at own optimum conditions, 45-50 °C for enzymatic hydrolysis and 30 °C for ethanol fermentation. The disadvantage of this process is the glucose inhibiting β-glycosidase activity. To solve this problem, the biomass concentration must be reduced and cellulase usage must be increased greatly. This reduces the efficiency of the cellulase hydrolysis. Low concentration of biomass leads to low ethanol concentration, which increases the recovery costs of fermentation. To solve the reducing sugar feedback inhibition, simultaneous saccharification and fermentation (SSF) process was proposed. This process coupled the cellulose enzymatic hydrolysis with ethanol fermentation. In this process, the hydrolytic glucose can be utilized by yeast immediately, which decrease the concentration of cellobiose and glucose [47] and further lift the feedback inhibition of cellulase. This method can increase efficiency of enzymatic hydrolysis, reduce the amount of cellulase, and also reduces the equipments and pollution. The main problem of SSF is the inconsistencies of hydrolysis and fermentation temperatures. In order to coordination, the temperature is set at 37-38 °C, if thermotolerant yeast is used, the temperature can be set at 40 °C.

### *Fermentation*

Glucose is the most frequently used fermentation medium. Besides this, pentose, glycerin and other hydrocarbons can be used as fermentation medium. Plant based raw materials can be transformed into all the compounds that can be synthesized by microbes. So plant based products are more abundant than fossil based products. Fermentation can be divided into solid state fermentation and submerged fermentation in general. The submerged fermentation is much popular than SSF in industry. After many years of development, submerged fermentation is more mature. But there are still many difficulties waiting to overcome, such as high energy cost and large amount of wastewater.

Compared to submerged fermentation, solid state fermentation has some advantages [38]: SSF technology decreases the demand to medium; SSF technology decreases the demanding of aseptic resulting in low cost; low downstream processing cost for the high concentration of products; due to SSF reduced water consumption, so no wastewater produced and reduce pollution. More and more attentions have been paid on SSF, particularly in the sweet sorghum and straw utilization.

### *Chemical Conversion and Modification*

On the one hand, the plant materials extracted from the biomass can be directly used to produce required products by chemical synthesis, modification or transformation. For example, cellulose can be used to produce acetate cellulose and nitrate cellulose by reacting with different reagents. The fat can be made of bio-diesel, soap, lubricants by chemical methods. The natural products isolated from plants can be used to form special chemicals such as drugs. On the other hand, plant-based fermentation products such as ethanol and lactic acid can be modified into various materials or fine chemicals.

Starch is a polymer with good biodegradable performance, which can be used to make starch-filled general biodegradable plastic resin plastic, two-degradable plastics and all starch thermoplastics to replace fossil-based plastics. Based the characteristics of glucose hydroxyl, cellulose can be etherified and used to produce cellulose ethers. Cellulose ethers have the excellent performance such as thickening, suspension, filming, emulsification, colloid-protectivity, water-maintenance, agglutination, anti-enzyme, and metabolic-inertia, etc. It is widely used in paints, polymerization, construction materials, oil production, textiles, food, medicine and electronic components and plays an important role in socio-economic development.

Wood pulp or cotton pulp react with acetic anhydride, the hydroxyl of cellulose was substituted by ester, then get acetate cellulose. Acetate cellulose is the first industry products. It is the second largest product of cellulose fiber and only second to viscose fiber. Trichloroacetate cellulose is widely used as film base, X-ray film base [48], plastics [49] and textile fibers. Diacetate Cellulose is used as film materials [50], filter cigarettes [51] and textile materials.

### *Pyrolysis*

Pyrolysis is a thermal degradation process under completely anaerobic or oxygen limited circumstance. Biomass pyrolysis is a complex chemical process which includes molecular bond fracture, heterogeneous and small molecules polymerization reaction. Hydrocarbons in biomass can be converted into energy by pyrolysis which include liquid (bio-oil), solid (coke) and gas (combustible gas). Researchers generally focus on the production and usage of bio-oil. Pyrolysis has been applied to the industrial production of fuels, solvents and other products using biomass as raw material [52].

1990s is the rapid development period of pyrolysis. With the gradual maturation of pyrolysis test equipment and the development of demonstration and commercialization pyrolysis devices, various scale and type of pyrolysis system were established in the world to produce bio-oil. In the fourth non-fossil fuel organization conference held in 1996 at Ukraine, pyrolysis was cognizanced as a prospect technology which has accelerated the pyrolysis research process.

The core of pyrolysis technology is pyrolysis reactor. Reactor type, mode of heat transfer, stay time determines the final products distribution. In recent years, biomass fast pyrolysis research has made tremendous progress. Many pyrolysis technology and reactors have been developed. The fast pyrolysis reactors can be broadly classified into the following types: mechanical contact pyrolysis reactor, radiative heat transfer pyrolysis reactor, vacuum pyrolysis, mixed-pyrolysis reactor, microwave pyrolysis and plasma pyrolysis reactor [53, 54].

### *Gasification*

Gasification technology [55] is one of the important biomass transformation technologies also is the main essential technology of thermal conversion of liquid fuel, power generation and co-production. Biomass can produce synthesis gas through gasification, which can further produce liquid fuel through Fischer-Tropsch synthesis or liquid oxygen fuel such as methanol, dimethyl ether. Using biomass gas to produce liquid fuel is a kind of indirect liquefaction technique. First, the biomass was gasified to gas, the gas be purified and adjusted to a synthesis gas ($CO+H_2$), then pressurized and catalytic synthesized to get syngas fuel, which is alternative to fossil fuels. The products include synthetic gasoline, kerosene, diesel and oxygen-containing compound liquid fuel [56]. Biomass can also used to get fuel gas through gasification which combusts to generate electricity and heat through internal combustion engine, gas turbine and existing power generation boilers[57]. In addition, the biomass gasification can also be used to produce hydrogen fuel by technology adjustment and reactor optimization.

At present, many countries have made a lot of research work on biomass gasification technology and developed different biomass gasification technologies. Such as Foster wheeler gasification technology of Finland, Babcock Borsig gasification technology of Austria, Battelle gasification technology of the United States, GmbH Freiberg gasification technology of Germany[58].

***Liquefaction***

Liquefaction is an important research aspect of biomass energy [59, 60]. Scientists have done a lot of work and hope to find new liquid fuels and chemical raw materials. Plant liquefaction is the process of converting plant into liquid products directly or indirectly. Indirect liquefaction refers to the process that first gasified the biomass and then synthesized the gas into methanol or dimethyl ether and other liquid products. This process can obtain high-quality products. But the overall thermal efficiency is very low. Direct liquefaction is the process that using water or other organic solvents for the media biomass is catalyzed to a small amount of gas, a large number of liquid products and s small amount of solid residue at a relatively low temperature. According to the used temperature and pressure, direct liquefaction can be divided into high temperature and pressure direct liquefaction and low pressure direct liquefaction. The former is the traditional liquefaction, whose products are generally used as fuel oil but must be modified before using just like the pyrolysis oil [61]. The oil contains a large number of oxygen-containing compounds and acid, the oxygen content is about 15-20 percent. Although less oxygen content in this oil than the pyrolysis oil, direct using it as the raw materials of fuel oil is difficult. The refining process of liquefaction oil is similar to pyrolysis oil. In recent year, more and more research interests were paid to low pressure and normal pressure moderate liquefaction. The products of moderate liquefaction are generally used as raw materials of polymer such as adhesives, phenolic resin, polyurethane foam, or as a fuel additive. However, there are still some problems: the difference of composition between biomass will impact on the quality and stability of final products; certain performances of products have a slight disparity comparing to the traditional products; the mechanism and specific compounds of moderate liquefaction are still not very clear. As a result of further study, the application of moderate liquefaction products will be more extensive [62].

### 1.3.3. The Complexity of Products and these Interrelation

The diversity of plant materials and methods will inevitably lead to the diversity of plant-based products. Compared with the petroleum components, plant contains much more components which provide more opportunities for the development of new products. The main components of petroleum are linear hydrocarbons and the chemical structures are the polymer of –$(CH_2)$-. Generally speaking, petrochemical products are made by adding a few organic functional groups such as hydroxyl, aldehydes, acids, esters, or other elements such as N, Cl and so on. The chemical synthesis of most petroleum products, such as ethylene, propylene, butylenes, etc are based on adding functional groups. For example, benzene exists in crude petroleum, but most benzene compounds are received through catalyzing n-hexane reconstruction. The major components of plant material are carbohydrates including hydroxyl and carbonyl oxygen-containing groups. So it is more suitable for the chemical transformation. In addition, carbohydrates can be used as the medium of microbe to ferment more bio-products. At the same time, the abundant combination of compounds can produce more prescription products and increase the product diversity. The plant-based product flows can be found in Figure 1.3 [63].

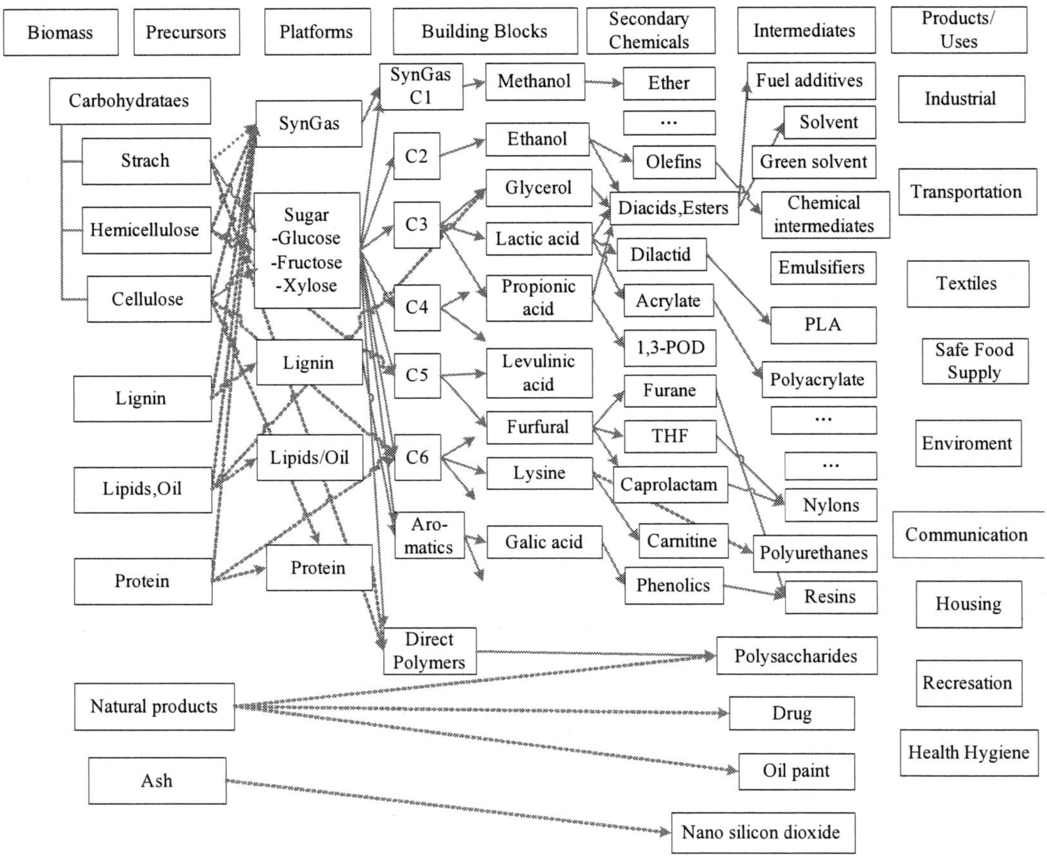


Figure 1.3. The flow graph of Plant-based products [63]

## 1.3.4 The Demand of Green Chemistry

In the late 1980s, in order to consistent with the global strategy on environment and the adjustment of industrial configuration, the concept of green products was proposed. It is generally believed that green products must meet the requirements of the following aspects [64]: first, in its whole life cycle, satisfactory to the demanding of environment protection, harmless to the human body and less or without effect on the environment. Second, the product structure as simple as possible without reducing the products functions; using as little as possible materials without affecting the life; using as little as possible energy in the course of manufacture and use without affecting its effectiveness. Third, after using life, its components can be refurbished, recovery, reuse or be handled safely. The environment-friendly extent of green products reflected in the whole product life cycle, "cradle-to-renewable". Apart from the design, manufacture and use of the products are concerned, whether the raw materials are green, technology is clean, the by-product is less, the used products can be recovery and reused are concerned too.

Plant-based products are in full compliance with the requirements of green products. The green plant itself is a natural chemical synthesis plant. Photosynthesis fixes the carbon dioxide in the atmosphere and generates all kinds of natural organic matters for human. All plant components can be processed into the corresponding products. Manufacturing process of plant-based products can be nearly non-polluting emission. After the life, plant-based products return to nature for the use of plant growth, thus form a virtuous cycle.

Plant materials and plant-based products are closely related each other. The compositions of plant materials, the growing season, producing area restrict the development of plant-based products. Plant materials are provided by the agriculture and forestry. The design and development of plant-based products need interaction with the agriculture and forestry. On the one hand, at the designing period of product, it is necessary according to the composition of raw materials and products features to decide whether the product can be produced by this plant and how to produce it. Usually, specific products just use one or several components of raw materials. We can reform the plant from source according to the actual requirements. Traditional breeding or genetic engineering can be used to make the plant to synthesize more required components. Through agriculture or forestry planning, we can find more suitable cultivation sites and cultivation season. On the other hand, any plant has more than one component. It is difficult to obtain significant economic benefits by utilization one component. Corresponding components and products comprehensive utilization process must be designed to form eco-industrial chains, so that all components of plant can be transformed into valuable products and without waste emission. Plant as raw materials is easily affected by production area and season and restricts the production of plant-based products. It is important to choose the proper site and scale of production. In addition, the generality of equipment must be taking into account to adapt to different materials supply of different seasons.

## 1.4. The Essential Content of Plant-Based Products Process Engineering

Plant-based raw materials have great many of types, compositions and structures which result in variety of plant-based products. With the depleting of oil, plant-based products economic emerged. However, the producing technologies of plant-based products are generally single technology and tend to use single component of raw materials. Based on experiences from many years of research and development on plant-based products, the author focused the common problems of plant-based products and combined the idea of process engineering and advanced the concept of "Plant-based products Process engineering".

The main research contents of plant-base products process engineering include the following four aspects: First, fractionation of plant-based raw materials. The components of plant-based raw material are very complex. In order to use these materials to produce different plant-based products, the concept of fractionation must be introduced. In chapter 3, from the concept, considering the developing history and limitations of plant-based pretreatment, the author introduced the concept and necessity of plant-based raw materials fractionation. At the same time, the author combined the resource science into the plant fractionation. In this chapter, combining with the practical work, the author introduced plant-based raw material fractionation process engineering principle and techniques from various angles with some examples. The chapter reflects the "full utilization of biomass" idea and solutions completely. On the basis of fractionation and multi-level conversion, in chapter 4, plant-based chemicals types and fundamental ways are introduced firstly. Take furfural and other 6 important plant-based products as examples, the process engineering idea and method for designing new plant-based products production system is introduced. So we can produce low-cost plant-based chemicals without new waste and pollutions. Chapter 5 focuses on plant-based materials process engineering and applications. In which includes new ideas of author on combining plant-based materials manufacturing and fractionation. It provides new inspiration to plant-based materials manufacturing engineering. Chapter 6 focuses on plant-based energy process engineering and applications. Although the focus is energy, it was pointed that plant-based materials can be used as energy but much more than energy. Based on this idea, "Utilization of whole", "multi-level utilization" and "process integration" is reflected in every plant-based energy production technologies.

### 1.4.1. The Applications and Fractionation in Plant Raw Materials

Plant raw materials are multi-component materials consisting of cellulose, hemicellulose, lignin and the functional chemical composition, with mutually entangled complex structure. Taking the utilization of one group as purpose will resut in the waste and pollution of other components. So, it is necessary to fully understand the different components of plant-based raw materials of the differences and propose the commonly essential aspects of fractionation, then further guide the development of industries, such as fiber, Chinese herbal medicine processing, and so on. Centering on this theme, first, based on the analysis and conclusion of existing problems about the scope of plant-based resources and developing process, we proposed the general key issues of plant-based raw materials fractionation, which described the existing resources of wood cellulose component separation from the physical, chemical, biological and comprehensive perspective. Finally, it summarized the principle and processing of separating active ingredients from the plant-based raw materials by direct extraction.

### 1.4.2. The Applications and Process Engineering in Plant-Based Chemicals

Plant-based raw materials can be transformed into many chemicals, but the industry mostly using acid hydrolysis, which causes serious environmental pollution and the acid residues are difficult to re-use. Therefore, this book first analyzes the existing problems in the industry transformation process, and then proposes the developing thoughts on the processing and engineering of plant-based chemicals under the guidance of processing and engineering principles. Basing on the three typical plant-based platform compounds: furfural, acid and xylitol as the representative, this book elaborated on the production process, integration methods and the future direction of development of plant raw materials. Finally, through the production of high-temperature liquefied plant-based chemical reaction mechanism of the production process and a brief introduction.

### 1.4.3. The Applications and Principle of Process Engineering in Plant-Based Materials

Plant-based raw materials can be transformed into many types of production materials, but the industry is mostly a single product, other components were burning or abandoned. Under the premise of analyzing the existing problems of industries in the process of transforming and under the guidance of processing and engineering principle, this book sums up the developing ideas and methods of plant-based materials. On this basis, it mainly introduces the technology and developing prospects of plant-derived products, such as carbon fiber material, rosin, the paint, cellulose materials, sheet metal, sandy soil improvement materials, wood cellulose and starch material liquefied modified material production principle.

### 1.4.4. The Applications and Principle of Process Engineering in Plant-Based Energy

Plant-based raw materials can be transformed into high-grade fuel, but the industry has the problem of single conversion technologies, other components were burned or abandoned. In this book, the type of plant-based energy was divided, and then described in detail the source of biomass energy efficiency rating system, conversion of various forms of conversion efficiency were compared. Finally, the key producing technologies, energy efficiency analysis and prospects analysis of several main biomass energy fuel, such as ethanol, bio-diesel, butanol and hydrogen. So that readers can have a comprehensive knowledge and understanding of plant-based bio-energy.

## References

[1] Arntzen, C. E.; Dale, B. E., *Biobased Industrial Products, Priorities for Research and Commercialization*. In National Academy Press, Washington, DC: 1999.

[2] Scheller, J.; Conrad, U., Plant-based material, protein and biodegradable plastic. *Current Opinion in Plant Biology* 2005, 8, (2), 188-196.

[3] Chen, H.; Wang, L., Research Progress on Key Process and Intergrated Eco-industrial Chains of Biobased Products——Proposal of Biobased Product Process Engineering. *The Chinese Journal of Process Engineering* 2008, 8, (4), 86-91.

[4] Lin, Y.; Tanaka, S., Ethanol fermentation from biomass resources: current state and prospects. *Applied Microbiology and Biotechnology* 2006, 69, 627-642.

[5] Goldemberg, J., Ethanol for a Sustainable Energy Future. *Science* 2007, 315, (5813), 808.

[6] Lynd, L. R., Overview and evaluation of fuel ethanol from cellulosic biomass: Technology, Economics, the Environment, and Policy. *Annual Reviews in Energy and the Environment* 1996, 21, (1), 403-465.

[7] Chen, H., *Biomass science and engineering*. Chemical industry press: Beijing, 2008.

[8] Chen, H.; Xu, J.; Yao, J.; Li, Z., Studies on multi-level conversion of liquid fuel with straw. *Chinese Journal of Bioprocess Engineering* 2005, 3, (1), 9-13.

[9] Zhu, J.; Zhang, G., Current Situation and Development Trend of the Research and Production of Biodiesel. *Chemical Industry Times* 2004, 8, (1), 23-27.

[10] Peng, X.; Chen, H., Single cell oil production in solid-state fermentation by *Microsphaeropsis sp.* from steam-exploded wheat straw mixed with wheat bran. *Bioresource Technology* 2008, 99, (9), 3885-3889.

[11] Peng, X.; Chen, H., Microbial oil accumulation and cellulase secretion of the endophytic fungi from oleaginous plants. *Annals of Microbiology* 2007, 57, (2), 239-242.

[12] Demirbas, A., The Importance of Bioethanol and Biodiesel from Biomass. *Energy Sources, Part B: Economics, Planning, and Policy* 2008, 3, (2), 177-185.

[13] Qureshi, N.; Blaschek, H. P., Butanol Production from Agricultural Biomass. *Food Biotechnology* 2005, 525-549.

[14] Ezeji, T. C.; Qureshi, N.; Blaschek, H. P., Bioproduction of butanol from biomass: from genes to bioreactors. *Current Opinion in Biotechnology* 2007, 18, (3), 220-227.

[15] Yuan, C.; Yan, Y.; Cao, J., Study of hydrogen production from biomass. *Coal Conversion* 2002, 25, (1), 18-22.

[16] Li, D.; Chen, H.; Li, Z., Research and Development of Hydrogen Production by Biological Technology. *Biotechnology Information* 2003, (4), 1-5.

[17] Gong, L.; Han, C.; Tan, T., Research on preparation of bio-ethylene. *Modern Chemical Industry* 2006, 26, (4), 44-45.

[18] Pandey, A.; Selvakumar, P.; Soccol, C. R.; Nigam, P., Solid state fermentation for the production of industrial enzymes. *Current Science* 1999, 77, (1), 149-162.

[19] Hu, X., The enzyme industry and its application *Industrial Microbiology* 2001, 31, (3), 41-46.

[20] Ji, C.; Du, F.; Shi, J.; Xu, Z.; Han, S.; Dong, Q., Research Advance in the Use of Cellulase to Produce Fibrous Ethanol. *Liquor-Making Science & Technology* 2007, (7), 118-121.

[21] Xiao, M.; Chen, L., Application and Prospect of Development of Enzymatic Techology in the Food Industry. *Food Science* 2006, 27, (12), 916-919.

[22] Ma, X.; Yu, C.; Wang, Y.; Zhang, C.; Zhang, R., Application of Enzyme in Cleanliness Technology of Leather Process. *Leather Chemicals* 2004, 21, (5), 30-33.

[23] Wang, P.; Fan, X., Application of biotechnology in the pretreatment processing of textiles. *Shanghai Textile Science & Technology* 2005, 33, (4), 4-6.

[24] Braunegg, G.; Koller, M.; Varila, P.; Kutschera, C.; Bona, R.; Hermann, C.; Horvat, P.; Neto, J.; Pereira, L., Production of Plastics from Waste Derived from Agrofood

Industry. In *Renewable Resources and Renewable Energy: A Global Challenge*, Miertus, S.; Graziani, M.; Fornasiero, P., Eds. CRC Press: Amsterdam, 2007.

[25] Lunt, J., Large-scale production, properties and commercial applications of polylactic acid polymers. *Polymer Degradation and Stability* 1998, 59, (1), 145-152.

[26] Lee, S. Y., Plastic bacteria? Progress and prospects for polyhydroxyalkanoate production in bacteria. *Trends in Biotechnology* 1996, 14, (11), 431-438.

[27] Geng, Q.; Hu, X., The developement of Polylacticacid fiber and its products. *Beijing Textile Journal* 2005, 26, (1), 31-33.

[28] Zeng, J., Development with questions and suggestions for medicinal plant extract industry in China. *Chinese Traditional and Herbal Drugs* 2006, 37, (1), 2-12.

[29] Yuan, Y.; Chen, H., Application of Steam Explosion in Ephedrine Extraction. *Journal of China Pharmaceutical University* 2005, 36, (5), 414-416.

[30] Shen, Z.; Wu, L., Advances in Taxol Research. *PROGRESS IN CHEMISTRY* 1997, (1), 15-19.

[31] Wang, S.; Ma, X.; Ping, W.; Zhou, D., Research Advances on Taxol Producion by Microbe Fermentation. *Microbiology* 2007, 34, (3), 561-565.

[32] Liu, K.; Zhang, E., The Exploitation of Soybean Isoflavones and Its Market Prospect. *China Pharmacy* 2006, 17, (2), 92-94.

[33] Jin, S.; Chen, H., Structural properties and enzymatic hydrolysis of rice straw. *Process Biochemistry* 2006, 41, (6), 1261-1264.

[34] Chen, H.; Li, Z., No-Contaminative Steam Explosion and its Application. *Journal of Cellulose Science and Technology* 2002, 10, (3), 47-52.

[35] Chen, H.; Liu, L., Unpolluted fractionation of wheat straw by steam explosion and ethanol extraction. *Bioresource Technology* 2007, 98, (3), 666-676.

[36] Chen, H.; Liu, L., *Steam explosion technology and application*. Chemical Industry Press: Beijing, 2007.

[37] Chen, H.; Qiu, W., The Crucial Problems and Recent Advance on Producing Fuel Alcohol by Fermentation of Straw. *Progress in Chemistry* 2007, 19, (7-8), 1116-1121.

[38] Chen, H.; Xu, J., *Modern times Solid State Fermentation Principle and Application*. Chemical Industry Press: Beijing, 2004; p 16-17.

[39] Yang, S.; Ding, W.; Chen, H., Enzymatic hydrolysis of rice straw in a tubular reactor coupled with UF membrane. *Process Biochemistry* 2006, 41, (3), 721-725.

[40] Chen, H.; Li, Z., Lignocellulosic microorganisms and biomass total utilization *Biotechnology Bulletin* 2002, 2, (2), 25-29.

[41] Zou, J.; Song, R.; Lu, Q.; Zhu, C.; Zhu, K., A New Green Renewable Energy Crop—— Sweet Sorghum and Its Breeding Tactics. *Rain Fed Crops* 2003, 23, (3), 134-135.

[42] Zhang, Y.; Yang, J.; Lu, X.; Ma, P., Research Processes in Acid Hydrolysis of Lignocellulosic Biomass. *World Sci-tech R & D* 2007, 29, (1), 48-54.

[43] Lee, Y. Y.; Iyer, P.; Torget, R. W., Dilute-acid hydrolysis of lignocellulosic biomass. *Advances in Biochemical Engineering/Biotechnology* 1999, 65, 93-115.

[44] Nguyen, Q. A.; Tucker, M. P.; Keller, F. A.; Beaty, D. A.; Connors, K. M.; Eddy, F. P., Dilute acid hydrolysis of softwoods. *Applied Biochemistry and Biotechnology* 1999, 77, (1), 133-142.

[45] Carrasco, J. E.; Saiz, M. C.; Navarro, A.; Soriano, P.; Saez, F.; Martinez, J. M., Effects of dilute acid and steam explosion pretreatments on the cellulose structure and kinetics

of cellulosic fraction hydrolysis by dilute acids in lignocellulosic materials. *Appl Biochem Biotechnol* 1994, 45, (46), 23–34.
[46] Wilke, C. R.; Yang, R. D.; Stockar, U. V., Preliminary cost analyses for enzymatic hydrolysis of newsprint. *Biotechnol Bioeng Symp* 1976, 6, 135-75.
[47] Philippidis, G. P.; Smith, T. K., Limiting factors in the simultaneous saccharification and fermentation process for conversion of cellulosic biomass to fuel ethanol. *Appl. Biochem. Biotechnol* 1995, 51, (52), 117–124.
[48] Woodings, C., *Regenerated Cellulose Fibres*. Woodhead Publishing Ltd: 2001.
[49] Mei, J.; Chen, J.; Ou, Y., The Present and Future Trend of Cellulose Acetate. *Journal of Cellulose Science and Technology* 1999, (4), 56-62.
[50] Idris, A.; Yet, L. K., The effect of different molecular weight PEG additives on cellulose acetate asymmetric dialysis membrane performance. *Journal of Membrane Science* 2006, 280, (1-2), 920-927.
[51] Ma, X.; Shen, L.; Yang, Z.; Cao, J., Study on Synthesis of Cellulose Acetate used in Cigarette with Domestic Wood Pulp. *Synthetic Fiber in China* 2004, 33, (5), 10-12.
[52] Yaman, S., Pyrolysis of biomass to produce fuels and chemical feedstocks. *Energy Conversion and Management* 2004, 45, (5), 651-671.
[53] Krieger-Brockett, B., Microwave Pyrolysis of Biomass. *Research on Chemical Intermediates* 1994, 20, (1), 39-49.
[54] Bridgwater, A. V.; Peacocke, G. V. C., Fast pyrolysis processes for biomass. *Renewable and Sustainable Energy Reviews* 2000, 4, (1), 1-73.
[55] Reed, T. B., *Biomass gasification: principles and technology*. Noyes Data Corp., Park Ridge, NJ: 1981.
[56] Wang, T.; Chang, J.; Zhu, J.; Fu, Y., Dimethyl Ether Synthesis from Syngas Derived from Biomass Gasification and Reforming. *Journal of Fuel Chemistry and Technology* 2004, 32, (3), 297-300.
[57] Bridgwater, A. V., The technical and economic feasibility of biomass gasification for power generation. *Fuel* 1995, 74, (5), 631-653.
[58] Maniatis, K., Progress in biomass gasification: an overview. *Progress in Thermochemical Biomass Conversion* 2001.
[59] Zheng, J.; Zhu, X.; Guo, Q.; Zhu, Q., An Overview of Technology for Liquid Fuel From Biomass. *Engineering Science* 2005, 7, (4), 5-10.
[60] Chornet, E.; Overend, R. P., Biomass liquefaction: an overview. *Fundamentals of Thermochemical Biomass Conversion* 1985, 967-1002.
[61] Zhang, Q.; Chang, J.; Wang, T.; Xu, Y., Progress on Research of Properties and Upgrading of Bio-Oil. *Petrochemical Technology* 2006, 35, (5), 493-498.
[62] Zhang, S.; Yan, Y.; Ren, Z.; Li, T., Analysis of Liquid Product Obtained by the Fast Pyrolysis of Biomass. *Journal of East China University of Science and Technology* 2001, 27, (6), 666-668.
[63] Werpy, T. A.; Holladay, J. E.; White, J. F. *Top Value Added Chemicals From Biomass: I. Results of Screening for Potential Candidates from Sugars and Synthesis Gas*; PNNL-14808, Pacific Northwest National Laboratory (PNNL), Richland, WA (US): 2004.
[64] Zhang, S.; Zhang, X., *Green Process System Integration* China Petrochemical Press: Beijing, 2006.

*Chapter 2*

# PROCESS ENGINEERING PRINCIPLES OF PLANT-BASED PRODUCTS

## 2.1. HISTORY AND CONNOTATION OF PROCESS ENGINEERING

### 2.1.1. Establishment and Development of Chemical Engineering

Chemical engineering science is the scientific basis of the chemical processing technology dealing with the common discipline of the mass and energy conservation and transfer under different scales. At the same time, chemical engineering science is also a powerful tool and adjuster for converting laboratory research achievements on mass and momentum conservation and transfer into practical production.

Chemical engineering originated from industrial chemicals and chemical technology [1]. Some typical techniques of chemical products appeared in the last part of the 19th century, such as the calcinations of limestone, the extraction of coal tar, the preparation of soda ash and caustic soda, the Lead Chamber Process for sulfuric acid production and the preparation of nitroglycerin, etc, had become relatively standardized processes [2]. However, these techniques were just early records for chemical techniques without the consciousness of finding their common principles. Later, the increased requirements on output and quality of the basic chemical raw materials (such as: sulphuric acid, soda ash, caustic soda, etc.) made the chemical industries shift from the original small, intermittent technology to large, continuous new technology. To accomplish these changes, the engineering viewpoint was used to solve the developing problems of modern industrial. For example, caustic soda production based on Solvay process and synthesis of sulfuric acid based on the contact process were successful projects through a series of breakthroughs. Since then, people gradually realized that the emphases of chemical industry development had begun to shift on the aspect of engineering. Times were calling for the chemical engineering [3].

In the late part of 19th century, the development of higher mathematics, classical physics, general chemistry, thermodynamics and chemical dynamics provided more direct theories to establish a new discipline. Davis and other pioneers expounded the difference between applied chemistry, chemical engineering and chemical technology, and made great efforts to design the factory production process as a series of operational projects, such as mixed, evaporation and filtration, etc, which promoted the appearance of "Operational Unit" concept.

In 1950, the book, Unit Operation, written by G. G. Brown etc, illustrated the development course of chemical engineering in nearly 40 years from 20th century [3].

Based on the extensive and deep research of various unit operations, people gradually realized that the unit operations of physical operation manifest itself in the following processes: heat transfer, mass transfer and momentum transfer, which itemed as "Three Transport". Transfer Phenomenon, compiled by R. B. Bird, W. E. Sterwart and E. N. Lightfoot in 1960, provided a thought to study and analyze various transfer process in term of mathematical method, which made the chemical engineering theory more mature.

The progress of theoretical and the rapid development of chemical industry, especially the development of the petrochemical and natural gas chemical engineering, raised a more thorough and systematic research upsurge on the industrial chemical reaction processes in the 1950s. People found that the intrinsic link not only between heat transfer, mass transfer and momentum transfer, but also in the chemical reaction process. During the industrial chemical reaction, with macro-characteristics, the "Three Transport" and "One Reaction" was essentially interwoven, and then a new branch discipline "Chemical Reaction Engineering" was born. Main representative works of this period include: "Reaction Kinetics of Chemical Engineer" written by S.M.Walas in 1959, "Chemical Reaction Engineering" written by O. Levenspiel in 1962, and "Chemical Engineering Dynamics" written by J.M.Smith in 1970, etc. Based on transfer process, chemical kinetics, thermodynamics and process principle, "Chemical Engineering Dynamics" integrated the "Three Transport" and "One Reaction" together by precise mathematical means, promoting the chemical engineering into a new age.

The chemical industries, such as petrochemicals, coal chemicals, basic organic synthesis, inorganic chemicals, fine chemicals and polymer chemicals, bio-chemical, were flourished in the 1950's and 60's which was the developing stage of chemical engineering [4]. In this stage chemical engineering act as discipline and chemical technology act as technology mutually promoted and improved together with the development of chemical industry.

Chemical engineering kept developing in the last part of the 20th century. Some new interdisciplinaries were appeared by closely integrating chemical engineering with the science of biology, material, resource, and environment and microelectronic. Chemical engineering discipline entered into extension stage which embodied in the following aspects [5-12]:

(1) Chemical Engineering permeated with some new technical disciplines to form cross and front disciplines.

For instance, biochemical engineering by integrated with biochemistry and biology (including microbiology), material chemical engineering by integrated with material physics and chemistry, microelectronic chemical engineering by integrated with electronics and microelectronics, energy and resource of chemical engineering by integrated with energy chemistry, petroleum chemistry and coal chemistry, environmental chemical engineering by integrated with environmental chemistry and ecology.

(2) Chemical engineering was closely linked with the natural sciences, such as mathematics, physics, and chemistry.

(3) The research contents of chemical engineering disciplines were deepening and extension. For example, the research object was developed from the simple system to the complex system, from small molecule to polymer and macromolecule, from macroscopy to microcosm, from stationary state to non-stationary state, from ideal solution to non-ideal

solution and from non-life system to life system. The study route was developed from simple process to complex, from singleness to coupling, etc.

### 2.1.2. The Proposal of Process Engineering

With the development and application in other industrial processes, chemical engineering develops gradually into process engineering [13].

Process engineering [14] is the process technology that changes the status of raw materials, micro-structure and/or chemical composition and its matching application through a series of physical and chemical separations and chemical reactions (including catalytic, electrochemical and biochemical reactions). As early as 1955, H. E. Schweier proposed the concept of process engineering in his book “process engineering economics”. In Zurich, the Swiss Federal Institute of Technology established the Department of Mechanical and Process Engineering and the Institute of Process Engineering. French National Centre for Scientific Research Institute of Chemical Engineering Science clearly stated that the research purpose of process engineering was to study the conversion process of material and energy, develop and achieve the necessary concepts, theories and methods for the design, optimization, control and industrialization of process. In some sense, the process engineering can be understood as "Universal Chemical Engineering", but made an upgrade in scientific sense.

### 2.1.3. Connotation and Function of Process Engineering

Process engineering is a general term for processing technology with wide range including the traditional production of acid, alkali, metals, fuel and drug, disposal of nuclear fuel, manufacturing of information materials and components, protection of environmental, etc[13]. In big science era, sometimes there was not the strict dividing line between process science and process engineering just like materials science and materials engineering.

The research connotation of process engineering, including the research of process principle, the investigation and organization of data information, and the system integration of process, is shown in Fig.2.1. The research of process engineering is based on the research of physical, chemical and biological transformation process. The task of process engineering is to solve the bottlenecks during the conversion from laboratory results to industrial transformation, create clean and highly efficient technology, processes and equipment by resolving the common problems of process in different areas. Process engineering achieved the quantification, design, magnification and optimization of operations during the material conversion process. However, the process science emphasizes particularly on theoretical research and innovation [15].

The essential difference of process engineering research and chemical experimental research lies in that the chemical reaction system of the latter is operated in the laboratory scale, which is often accomplished under ideal conditions. While the former is accomplished on the scale of industrializations, the systematic conditions corresponding to experimental research, such as physical, chemical, biological changes and the process, are quite different, and require the enlarged laboratory.

The traditional "Enlargement Step by Step" which depended on the experience of the traditional might have the possibility to explore the industrial discipline of the entire system, but it has the disadvantages just as consuming a large amount of time, manpower, material and financial resources. Thus one basic research content of process engineering is to study low-power high-efficiency amplification method, which demonstrates the epistemological status of the process engineering.

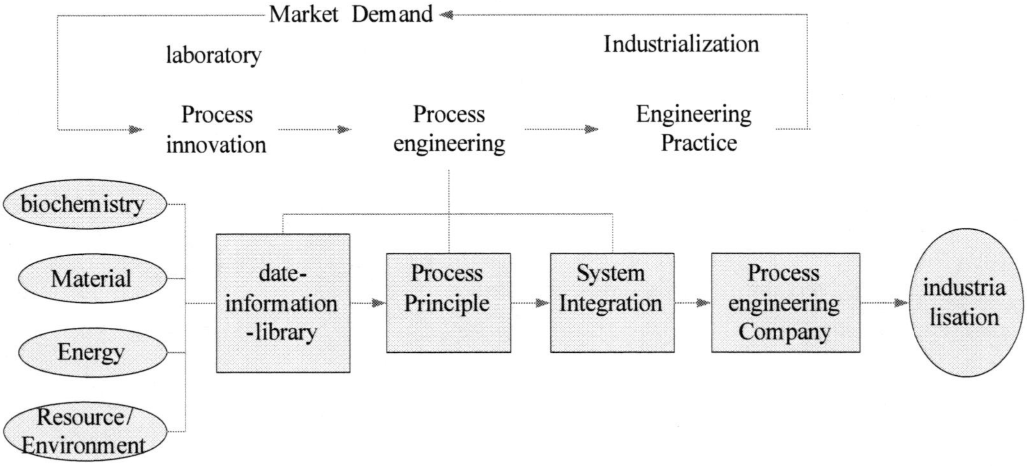


Figure2.1. Process engineering and its connotation [15]

Process engineering is also different from industrial design. Industrial design is based on the existing mature technology, while the process engineering is the innovative researches on new technology, new processes, and new equipment. The principles of process innovation, system integration and innovation are all new industrial technologies and provide a reliable basis for design. Accomplishment of a process engineering project often needs cooperation by different departments such as the scientific research institutions, design departments and economic management departments.

To implement the research project, it must form a macro structure to deal with the value chain, which includes an innovative team innovating the technique according to the market demand, the common problems aiming at the transition from the laboratory results to the industrialization, the research superiority on data and information, process principles and system integration, and the ability implementing projects in cooperation with enterprises. In an integral value chain, if any link is not success, the bottlenecks can not be resolved.

Process system engineering is a sub-discipline of process engineering recently widely referred [16]. The research object of process system engineering is process system, which primarily deal with the material-energy flow system on the conversion process. Process systems engineering provides a decisive scientific basic for the unit operation, optimizes the industrial process and the planning, design, operation, control of process, and also provides the possibility for the development of new process.

### 2.1.4. Knowledge Base of Process Engineering

Along with the division of process industry and product industry, process engineering is urgently needs to form its own technology theoretical foundation which studies on principles how to improve productivity and equipment, reduce investment operating costs, develop new process, and eliminate pollution as possible according to the basic principles of sustainable development.

***(1) "Four Transfers" and "One Reaction"***

Process science originates from the chemical engineering principle of "Three Transfers". The chemical industry is a major branch of the process industrial. The chemical engineering principle is aroused from the classification and conclusion of common operation in various chemical techniques, which includes three basic transfer processes: heat transfer, mass transfer and momentum transfer.

Heat transfer is the physical basic for all heat exchange process including heating, cooling, heat exchange, and so on. Mass transfer is the physical basic for all separation process, such as distillation, absorption, extraction of gas-liquid interface. Momentum transfer is the physical basic for all flowing process, including the transmission of fluid in the pipeline, the two-phase flowing between the liquid and vapor in distillation tray.

During the two-stage summarization of transfer phenomenon, a lot of concept and methods on physics and mathematics were introduced, making the quantitative analysis possible and significantly accelerating the design, forecast and control of chemical industry.

The transfer process was gradually combined with the technology of chemical reaction, after being classified and concluded a new branch discipline, Chemical Reaction Engineering, was appeared to study the chemical industry process with chemical reaction, making chemical engineering become a more comprehensive engineering science.

In the early 1970s, with the rising of modern biotechnology industry, a new area of biochemical reaction engineering was proposed [17]. In 1980s, people has realized the essential difference between chemical reaction and biology reaction that the former is a entropy increased process, while the latter is a entropy reduced process.

With the knowledge accumulation of biology and deeper understanding on the biological innate nature, especially the learning and thinking of self-organization theory, nonlinear theory, modern information and system theory, we proposed that besides the traditional "Three Transfers" there also exists information transfer. We made an attempt at the theoretical basic for biochemical engineering science, and widely exploration and demonstration on the laboratory experiment to industrial applications [17].

***(2) Nonlinear Science***

The introduction of the information transfer increased the complexity of system, so it is necessary for the development of process science to absorb the knowledge of complex systems. The complexity of system determined that nonlinear science is also the scientific foundation of process engineering [14]. As time goes on, thermodynamics as a base of discipline is developed into irreversible thermodynamics process, which indicates that the development of process science inevitability towards to nonlinear science.

The obvious difference between the process science and product science is the amplification effect. The chemical and physical reactions mechanism changed when the chemical reaction system scaled up to industrialization. The characteristics parameters of system with different scale do not increase or remain the same monotonically with the variety of geometry. Therefore, the process science should study the different mechanisms within the different scales system, which explained the inevitability of the process science based on the nonlinear science.

***(3) Process Integration***

Process integration refers to an advanced production technology, which simultaneously performed in one process or equipment by integrated two or more techniques or steps together organically. Process integration aimed at streamlining the integration process, enhancing production efficiency and reducing investment and production cost.

In the last century, the process industry had made tremendous progress and development in the whole world. For example, unit technology reached a very high level, producing scale and efficiency gradually improved, the connotation and discipline basis constantly developed. However, the contradiction between development and pollution and the sustainable development are still the key issue to be resolved in the coming future.

Industrial processes involved the following links [18], the first is the material transformation technology, the second is to achieve the optimization of the process and equipment and scale up of engineering, and the third is to achieve the system integration aiming at establishing a multi-process technique with clean, high efficient and low consumption. The problems existed in foregoing three links are just the common problems in the process of industrial development. To solve the common problems, the process integration is the key issue needed break through.

## 2.1.5. Development Trend of Process Engineering

The principle and knowledge of chemical engineering have been applied into many processing fields such as metallurgy, materials, energy, environment, bioengineering and so on [19]. Therefore the chemical engineering could be renamed as process engineering. Along with the fast development of computer technology and measurement instruments, the research range of process engineering is extending to cover from observation and prediction of the micro-scale structure and interface to the system integration of the macro-scale equipment and plant [20], which refers to the structure, interface, and multi-scale issues. In addition to the shortage of energy and resources, the pollution of the environment has become the challenging problems facing to the mankind. The green operation production and optimal integration of process engineering have become focus for researchers. The principles and methods on the structure, interface, multi-scale, green operation, and optimal integration of process engineering should be the new basic contents for process engineering.

### *2.1.5.1. Multi-Scale Research of Process Engineering*

The research objects of process engineering cover various levels from the molecule, micro-nano particle, clustered particle, unit, factory, to ecological industrial park, which bring forward new challenge for process engineering. The research focus gradually converts to the space-time multi-scale structure of process transformation. Space-time Multi-scale structure [21] refers to the non-uniform distribution of material, pressure, temperature, and flow rate during the process transformation of materials, which demonstrates with the dynamic changes in time and the difference on each tested point of space. The resulting effect is that this structural change of reaction has effects on the reaction and transmission of process, further on the structure and performance of products, which plays a major role on the transformation process [22].

To ensure the material transformation conditions within the technologic requirements of system, the control of chemical process can only be completed in the system and equipment scale. These control measures influence the micro-scale chemical reaction conditions by reacting on the phenomena appeared in various scales. Therefore, the perception of various scales of space-time structure and its variation rules becomes the focus of process engineering science in 21st century [22].

Most chemical processes are non-linear and non-equilibrium, belonging to the so-called complex systems for which multi-scale structure is the common nature. The involved problems are different in different scales on the process of the material transformation. In micro- and nano-molecular aggregates, as well as a smaller scale, it relates to the physical and chemical product engineering issues aiming at process development. The process engineering problems, such as equipment and scale-up, in millimeters to meter-scales mainly refers to the green integration relating to the process systems engineering and ecological systems.

#### *2.1.5.2. Green and Information of Process Engineering*

Green chemistry is the choice under the global environmental and resources crisis. Business needs environment-friendly green technology to achieve the source control of industrial production and keep the sustainable development. The scale-up and optimization research of process, the system integration and the multidisciplinary cross are the extremely important bottleneck in the development of green process engineering [23].

Green process engineering, covering the process of green chemistry and chemical engineering, is becoming an integrated science and engineering that researches the green process of mass transformation process by comprehensively applying environment, resource, material, energy, chemical, biological engineering and information engineering, etc.

To establish the ideology, methodology, and implementation strategy for resources-environmental protection system, and clean production strategy and eco-industrial system for source pollution control and identical theory of resource and environmental.

Atom economy of the chemical reaction process is the core of the green process engineering. The ideal green chemical reaction is that the raw material of atomic is changed for the product, without by-products or waste, and zero emissions.

By using the comprehensive evaluation system of environment and economy, to establish the integrated optimization system on the materials-energy-information process of process industry and to develop the process integration biological technology, the clean energy and renewable resources alternative technologies.

Simulate the symbiosis, mutualism, cycle transferring network of energy and elements of nature species and establish an eco-industrial system with the optimum multi-level recycling utilization of material.

The object of green chemical is to reduce the hazardous substances which come from the chemical product and industry process needed to maintain the living standards of scientific and technological, therefore the chemical activities associated with the foregoing are all the research areas of green chemistry.

The trends of the green chemistry and green chemical engineering include the following areas: green product design, green raw materials platform, new reaction technology, the green preparation of catalyst and new catalytic technology, the green solvent, the new reactor coupled with the process ntensification technology, the new separation technology, green chemical coupling process systems, and the integration of the computational chemistry and green chemistry [24].

The research contents of the information process engineering include the computer aided design, process optimization and automatic control. It is important to simulate all the details from the virtual simulation technology to industrialization and system integration, to achieve quantificational design, to substitute direct scale-up for gradual scale-up.

#### *2.1.5.3. System Integration of Process Engineering*

*1) The Foundation of Process Integration---The Integration of Technical Choice with Knowledge*

Technological integrative innovation process faces the assessment and choice of existing technology. People are always tending to solve new problems according to experience, which sometimes restrict the technical innovation. Therefore, it must get rid of the bound of stereotyped thinking and take full advantage of the opportunities brought about by technological developments. It is the foundation of process integration that integrates the technology choice with new knowledge of different areas.

Technological integration should be a continuously promoting process. It is necessary for enterprises to maintain competitive advantage that keep exploring new areas of knowledge and introduce into the research and development of products. In the technological integration process, sometimes it is difficult to control all areas of knowledge, while the thoughts and methods of technological integration, namely systematic knowledge, often have strong externality. Through the technological integrative innovation, we will built product platforms based on the system. In the subsequent production process, the technical capacity keep soaring through the integration of relative knowledge, particularly new knowledge, and the enhancement of the technical content and differences degree.

With the development of computer and computing technology, the systems integration based on information technology has become an important works and methods of the process engineering. The future of process science will focus on the development of virtual factories.

*2) Products and Process Integration*

Future-oriented technology integration process, focusing on integrating the products and process will be increasingly widely recognized. Integration of product and process, embodied in many specific technologies and theories, is a principle, a way of thinking, a kind of guiding ideology coming from practice and returning to the practice, a basic concept and methodology organizing the technological integrative activities. Integration of product and process technology is rooted in, but higher than numerous technologies and theories.

Integration of product and process is not limited in the narrow concept of the technology, but cover a wider field. Multi-level integration of product and process includes the integration of product and process technology, the integration of product planning and process planning, the integration of product design and process design, the integration of product system and process systems, the integration of product management and process management, as well as the integration of products organization and process organization, and so on. The concept of product and process integration also can be extended to the integrated relationship, such as manufacturing, sale and supply. Product and process integration has been widely used in the industrial manufacturing sector, and is an important development trend for the biomass conversion in the future.

The process integration can achieve the sustainable use of raw materials by embodied itself in the following aspects: improving the utilization efficiency of raw materials, increasing the energy utilization efficiency, reducing emissions and waste minimization, etc.

### Process Integration Considering the Influence of Environment

Furfural is an important chemical raw material. However the discharge of great deal of furfural wastewater not only wastes resources but also causes serious environmental pollution. Data indicate that furfural wastewater contains 1.0~2.5% acetic acid, 0.2~0.5% furfural and many other kinds of organic compounds. Various chemicals can be got from the furfural wastewater just by suitable treatment.

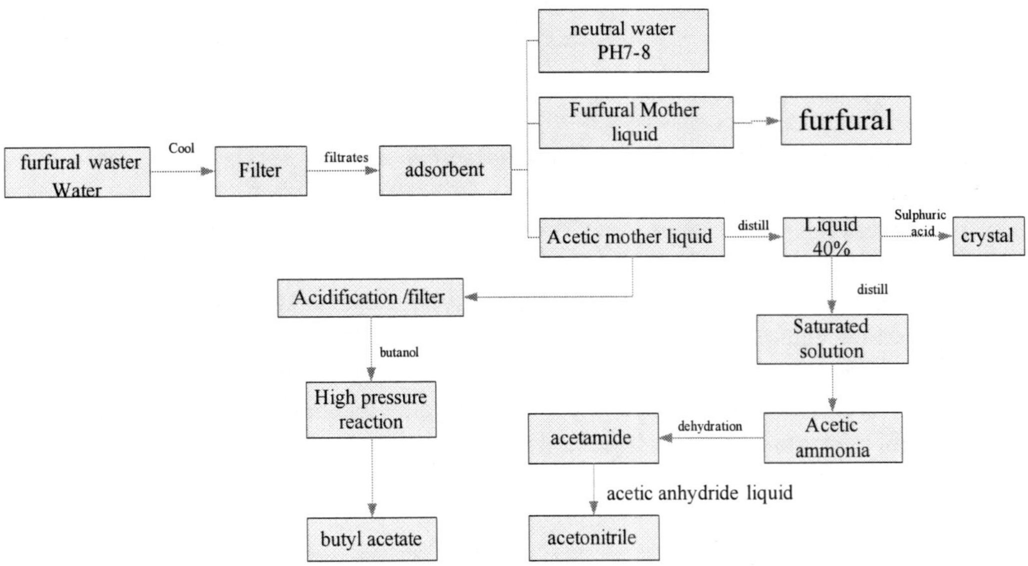


Figure 2.2. The suitable treatment of furfural waster water

### Multi-Products Process Integration

Corn, the third largest crop in China, is an important material for grain, economy and feedstuff. In the 1990s, the world corn used for feedstuff industry, food and industry were 75.5, 12.5 and 12.0%, respectively. Industrial consumption (starch and starch deep processing) increased by more than 13 % of each year. As shown in Figure 2.3, corn can be used to produce various chemicals through different processes and can be value-added utilized through multi-products process integration.

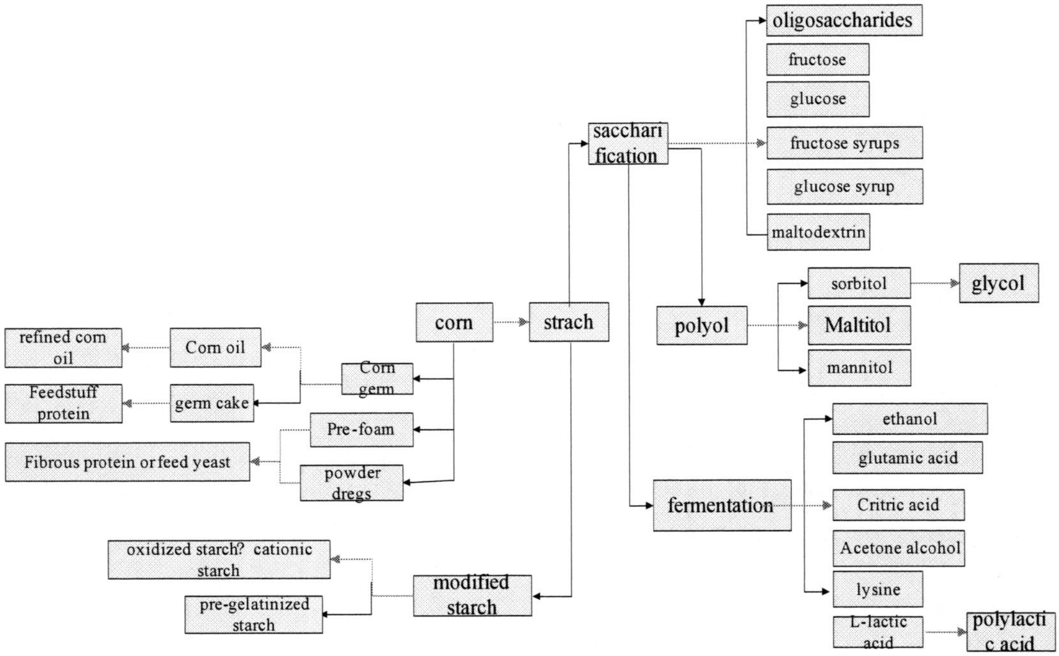


Figure 2.3. Multi-products process integration of corn

*3) The Integration and Optimization of Unit Operation*

Unit operation is the general summarization of basic operation in the chemical industry and other process industries. Since British scholar G. E. Davis firstly proposed the concept of "Operational Unit", it has a history more than 100 years. With the theoretical and practical development of unit operation, researchers have not only deepth studied the improvement of various operation units, the enhancement of efficiency of the process, but also paid much attention to the common rule analysis of various operation units. At present, many outstanding effects were investigated in the new process by combining some unit operations, which brought new vitality for the development of unit operation [25].

## Reactive Distillation

In most cases of chemical engineering, the reaction and separation are two independent units and performed in completely separated equipments. If the reaction and separation processes would be performed in equipment, it will not only reduce energy consumption and equipment investment, but also improve the process efficiency. Therefore, such combination is not just a simple mechanical combination, but a process coupling.

Distillation process is a traditional unit operation, if the reaction process and the distillation process are combined into one process, it would create a new concept -- reactive distillation. The concept of reactive distillation was firstly mentioned by Bacchaus in 1921, which was focused on esterification in the early research stage, such as the synthesis of ethyl acetate by acetic acid and ethanol in the ordinary distillation tower. Afterward, reactive distillation was applied to the transesterification, saponification, hydrolysis, isomerization reaction, etc. In the late 1970s, reactive distillation extended from the homogeneous reaction system into non-homogeneous system, namely catalytic distillation using solid-state catalyst as filler. In recent years, reactive distillation gradually becomes a very active branch.

During the process of reactive distillation, once the reaction started the product would be distilled and separated from the reaction section, which might increase the response and conversion to the complex reaction. Due to the opportune separation of products, the reactant concentration in reaction section was always higher than the product concentration, thus increasing the reaction rate and production capacity. Reactive distillation is energy saving and cost saving process since that the reaction process and distillation process was performed within the same section. Especially for some materials separated difficultly, it will get more pure product using the reaction distillation.

Non-Aqueous Enzymatic Catalysis

Broadly-defined biological catalysts include enzymatic catalysis, biological transformation, biological oxidation and precursor fermentation. The non-aqueous enzymatic catalysis technology mainly studies the catalytic nature of enzyme in non-aqueous solvent which might improve the catalytic property. The biological catalysis process was traditionally divided into upstream process aiming at the discovery and modification of bio-catalyst and downstream process aiming at the separation and purification of products. However, with the high-integration of engineering technology, especially the development of non-aqueous enzymatic catalysis process, the above mentioned classification system become illegible. For instance, the situ product removal technology is the organic integration between downstream separation and bio-catalytic reactions, the two-phase distribution bioreactor is the highly integrated technology of non-aqueous enzymatic catalysis technology with situ removal technology. Nowadays, two-phase distribution bioreactor technologies, such as the microbial transformation, β-lactam antibiotics and synthetic enzymatic hydrolysis, etc, are becoming the hot research of the whole world [26].

*4) Establishment of the Circulatory Economic Industrial Chain*

The recycling economy, in essence, is an ecological economy. It requires the ecological law rather than mechanistic law to guide the economic activities of human society. The traditional economy is a unidirectional liner economy with the developing model of "resource -product- pollution discharge", the characteristics of which is high-exploitation, low-use and high emissions. Different from the traditional economy, the recycling economy, harmonious development of environment and economy, is a feedback economy with the model of "resource-product- renewable resource", the characteristics of which is low-exploitation, high-use and low emissions. High efficient integration of technology and reasonable recycling economy industrial chain are the new requirements for the technology integration model in the future.

## 2.2. Process Engineering Development

### 2.2.1. The Basic Flow of Process Engineering Development

This section introduces how to start the process engineering development and procedures.

Firstly, the referred product should has economic value and can meet the need of the national economy. There also need a possible method for the referred product which has made a comparison test with other methods. Through the preliminary assessment, has proven that the quality of the products can satisfy the national economy, otherwise, should change the technology, to enable it meet the requirements of nature.

The involved reactions should be examined in laboratory. Usually, the laboratory process is intermittent and the equipments are mostly made of glass. The desired reaction and other possible side reactions are well-known. Various possible separation steps can be designed if the natures of the materials are clear. It is demanded that the analytical methods for intermediate products, by-products and purity of products should be predetermined. It is definitively important to the enterprises that whether the laboratory results can be directly applied in the industrialization.

During the process engineering development of a product, the main focus is on converting the producing production of conceives into realities. Therefore, the information gets in laboratory stages is far from enough, the reason can be concluded as follows.

1) Thermodynamic data and kinetic data are not sufficient to explain the transition from intermittent laboratory device to continuous industrial equipment, including many intermediate separation steps.
2) The effect of by-products on the reaction process and final quality of product is still not clear enough.
3) In laboratory, the end-products are often got through the most favorable purification operation, which are totally different from industrial production. Therefore, the characteristics of the final products are not yet conclusive.
4) The experiment is generally completed within the laboratory scope, for example, the pressure of the glass equipment is limited. So even closer to optimize is impossible.
5) It did not consider the corrosion and product contamination problem in the laboratory because the materials, equipment and resources are different from these used in industrial. At the same time, the producing period of laboratory experiment usually shorter than industrial production.
6) The first calculation result based on the above conditions is very unreliable.

Although there is a vision of the process, it is not clear enough. Whether such producing methods are able to achieve in technically? How to determine the sizes of all parts of the installation? The necessary processes of calculating the number of engineering and technical information are still absent. Development of process engineering is to gradually improve and fill those blanks through the relevant steps. To resolve these issues, the task of process engineering development includes the following major contents or steps:

1) Basic development research
2) Conceptual design and initial evaluation
3) Intermittent test
4) Basic design and final evaluation

Those steps are several key elements of the process engineering development, not all of the work.

## 2.2.2. Basic Development Research

The basic development research is the foundation of process development, which is different from the basic theoretical research. The basic research of process development is to explore new methods, new technologies, new reactions and new compounds, and clarify their basic scientific principles.

### *2.2.2.1. The Content of Basic Development Research*

*Propose and Verify the Selected Producing Methods and Technical Conditions*

1) Selection of the Initial Production Methods

A plant-based product can be produced from different raw material through different production process, even from the same raw material through different production process. Meanwhile, the same producing process can also adopt different production process to produce same product. The quality, producing costs and product investment is varied from different producing process. The primary task of process development of plant-based product is to explore effective producing methods, which should be enabled the less investment of the equipments, good economic benefit, advanced and reliable technology, which mainly include the following contents and links: ① Choice of raw materials; ② Selection and preparation of bio-catalyst (bacteria, the enzyme); ③ Selection of reaction methods and the form and material of reactor; ④ Decided project for the separation and refining of raw materials and products; ⑤ Decided analysis methods.

In short, to select the advanced reliable technologies and reasonable production methods, it should try to use quantitative or semi-quantitative manner to estimate the device investment used in the different producing methods, product costs and repayment of investment. When estimating investment and costs, we should try to use process simulation software to improve the accuracy and speed of estimates.

2) Selection of Initial Optimum Conditions

It is necessary to optimize the technical condition of each unit operation after deciding the initial production methods, especially the key unit operation such as reaction process. In particular, we should clarify the control step and find the key factor affecting each unit operation for the purpose of providing the reference for the project amplification. To provide reliable data for industrial production, it is important to perform the detailed research on the optimum reaction condition, the producing method and life span of strain, enzyme and catalyst. The optimizing research of technology conditions should be carried out for obtaining a wide range of optimization and reliable results with the scientific experimental design and data processing methods.

*Determination of Physical, Thermodynamic and Dynamic Data*

Achieving large amount of complete and accurate data in plant-based product development process is also very important works, which mainly include physical data, thermodynamic data and dynamic data; In general they can be obtained through literatures, correlation estimation, and experimental determination.

*Estimate the Effect of Transfer Process and Carry out Large-Scale Cold Model Experiments if Necessary*

Transfer process will affect the temperature distribution in reactor, which is the main reason for the amplification effect in the scale-up process. The cold model experiment is a popular basic research method to determine the effect of transfer process. The flow and mixture of the inner materials movement within the equipment, residence time distribution, branch-flow, interface and various internal components of the equipment have influence on the transfer process. We may measure them through the cold model experiments in the basic development research.

#### ***2.2.2.2. The Difference Between the Basic Development Research and the Traditional Laboratory Research***

*(1) Research Tasks*

The research tasks of the traditional laboratory research are to explore the technology conditions, while the tasks of fundamental development research are to explore the characteristics and principles of the reaction and to forecast the amplification effects based on the engineering knowledge.

*(2) Research Method*

The traditional laboratory research is disjointed with the engineering development research, which often results in contradictive even contrary results. The fundamental development research combined tightly with the engineering exploitation, taking into account the effect of projects and improving the fundamental research based on the results of engineering development. Thus the fundamental development research is a process with short cycle and high success rate.

### 2.2.3. Conceptual Design and Preliminary Evaluation

#### ***2.2.3.1. Conceptual Design***

Conceptual design is a hypothetical design when the fundamental research conducts to a certain stage, in according to the future industrial-scale production facilities from the point of engineering [27]. Conceptual design can discover the existing problems of fundamental research and provide guidance to further process research such as pilot trial and simulation test. It is obvious that conceptual design is not for the purpose on construction, but on detailed evaluation of the process engineering development projects. The significance of conceptual design is to highlight the key technique by proposing and discovering problems during the factory construction.

*(1) The Content and Methods of Conceptual Design*

Based on the concept and date provided by fundamental research and relative deduction and calculating results of technical and economic data, conceptual design means that to design in accordance with the stated methods and steps.

Conceptual design mainly includes the following contents:

[1] Design scale, the operation time, production methods, design basis;
[2] The specifications and requirements for public works, such as raw materials, finished products, catalyst, by-products, assistive chemical reagents;
[3] Installation specifications;
[4] Technical flow chart and simple illustration;
[5] Material and energy balance;
[6] Calculating results of equipment, equipment inventory, equipment design illustration and quality requirements of materials;
[7] Producing control and instrumentation;
[8] Consumption quota, fixed operators;
[9] Labor protection, health and safety, waste handling, and the possible effects on the environment with the installation of production;
[10]Investment and cost estimation, payback period and economic guideline of major technology;
[11]Suggestion and advice for the possible problems appearing in the technical process, amplification technology and industrialization, for the process design, and for the fundamental and processing research and development;
[12]Diagram of major equipment and equipment layout.

*(2) Data of Conceptual Design from the Following Four Areas*

[1] Literatures
[2] Measured data as a reference for plants
[3] Results of small experiment and simulation test
[4] The basic scientific knowledge such as chemical engineering and design experience

*(3) Function of Conceptual Design*

[1] Through the conceptual design, we can test whether the results of basic research can satisfy the requirements, expose the problems in the basic research process, such as the main chemical unit, the structure and material of equipment, control program, environmental protection and so on, propose approach and program solve these problems, and then further determine the additional contents demanded for fundamental research.
[2] Through the conceptual design, we can judge whether the simulation and pilot test is necessary, determine its content, key points and scale, and simulate the problems needed to be solved and the available data.
[3] With a preliminary evaluation, we can estimate the main technical and economic indicators and the possible risk after the implementation of technical carried out, and then determine whether the project have the value of industrialization.

In short, the conceptual design is a summary and identification for the fundamental research. Conceptual design can achieve the combination between the design and early research, make the existing problems on the basic research as soon as possible and solve these problems in timely.

Conceptual design is conducive to guiding the pilot test with technical and economic view, to provide accurate information for the process development, and to form the correct design ideas. Since the conceptual design can ensure the integrality of basic research and avoid faults in the trial, the development of conceptual design can reduce the development cycle and improve the development quality.

#### *2.2.3.2. Preliminary Evaluation*

After the conceptual design, it is necessary to carry out a preliminary technical and economic evaluation on whether the technology of project is advanced, economic and reasonable, then draw a conclusion that how to conduct the research and development in the next step [17, 28].

*(1) The Contents of Preliminary Assessment*

Payback period of project is one of the most popular evaluating criterion in preliminary evaluation. According to the "economic performance and the technical evaluation" promulgated by the Ministry of Chemical Industry Planning Council of China, the payback period of the chemical industry projects is 2-4 years, and for the fine chemical products the lowlimit should be taken.

According to the regulation from a large company, the profitability of the chemical industry projects should meet the following indexes:

1) Initial evaluation: the payback rate $\geq$ 50% -100%;
2) Middle evaluation: the payback rate $\geq$ 30% -40%;
3) Final evaluation: the payback rate $\geq$ 20%.

*(2)The Investment Estimation of Preliminary Evaluation*

Investment estimation is one of the main elements of preliminary evaluation. Investment of an engineering project is the sum of construction fund (known as fixed capital) and operation funds (liquidity). Asset investment also known as fixed capital includes investment in technology equipment, support productive investments and public investment, facilities investment and waste handling investments, among which the investment of technical equipment is the base when calculating the fixed capital of a project.

Due to the lack of detailed equipment and public works data, in the initial phase of preliminary evaluation the investment estimation are generally conducted according to the following two methods.

1) The index method for production scale

This method means that the production capacity and the equipment investment have an index relationship.

$$S_2 = S_1\left(\frac{Q_2}{Q_1}\right)^n \quad 2.1$$

Where $S_1$, $S_2$ are known investment for the plant or planned project, $Q_1$, $Q_2$ are known production capacity for the plant or planned project, n is scale index which usually is 0.6-0.8.

If the production scale is expanded by expanding the size of equipment n values 0.6-0.7, and 0.8-1.0 by increasing the number of production equipment.

2) Ratio methods

Ratio method, basing on the equipment value of equipment after installed according to the technical demand, estimates the cost of the project such as pipelines, controlled instruments, construction and ancillary facilities according to the type of project and proportions of various expenses cost estimate, calculates the entity part value of the project, then calculates the construction costs in accordance with the proportion, unforeseen expenses, and lastly collects to get the total investment.

① The value of equipment transported to plant

The value of equipment transported to plant =equipment ex-factory price × (103% - 108%)

② The value of equipment installed

The value of equipment installed= The value of equipment transported to plant × (110% - 143%)

③ Construction costs of workshop

Construction costs of workshop= the values of equipments installed × (10% -100%), the coefficient is valued as follows: 10% -20% for outdoor equipment; 20% -60% for the equipments arranged both indoor and outdoor; 60% -100% for indoor equipments.

④. Ancillary facilities costs

Ancillary facilities costs = the value of equipment installed × (0% -100%), the coefficient is valued as 0% if all ancillary facilities are the present devices of company, 0-5% if expanded a little; 5% -25% if expanded largely, and 25% -100% if all are new.

⑤ Automatic Instrument costs

Automatic Instrument costs = the value of equipment installed × (3% -20%), the coefficient is valued as 3% -5% if without automatic instrument, 5 %- 12% if only a portion of instruments are automatized, 12% -20% if most instruments are automatized.

⑥ Technical Pipeline costs

Technical Pipeline costs = the value of equipment installed× (7% -20%) which depends on the phase of material. For solid materials, the coefficient is 7- 10 %, and 10- 12% for the materials with solid and fluid and 12% -20% for the fluid materials.

⑦ The entity investment of factories

The entity investment of factories = ② + ③ + ④ + ⑤ + ⑥

⑧ Construction costs

Construction costs = The entity investment cost of factories × (20% -60%) which depends on the complexity of the construction. For a simple construction, the factor is 20% - 35%; while 35% -60% for a complexity construction.

⑨ Unforeseen costs

Unforeseen costs = the entity investment of factories × 20%

⑩ Fixed capital

Fixed capital = ⑦ + ⑧ + ⑨

⑪ Circulating capital

Circulating capital = fixed capital × (10% -20%)

⑫ Total investment

Total investment= fixed capital +circulating capital

3 When doing the preliminary evaluation, Table 2.1 can used to estimate the cost of production products.

**Table 2.1. Estimation the cost of production products**

| Number | Item | Method |
|---|---|---|
| 1 | Material cost | Unit price×consume ration×annual yield |
| 2 | Assistant material cost | Unit price×consume ration×annual yield |
| 3 | Fuel cost | Unit price×consume ration×annual yield |
| 4 | laborage | Annual average laborage×number×1.11 |
| 5 | Depreciation of capital asserts | capital asserts× (8%-12%) |
| 6 | repair cost | capital asserts× (3%-6%) |
| 7 | Plant outlay | (1+2+3+4+5+6) ×2% |
| 8 | Plant cost | 1+2+3+4+5+6+7 |
| 9 | Corporation manager | 8× (6%-9%) |
| 10 | Factory cost | 8+9 |

## 2.2.4. Intermediate Test

### *2.2.4.1. The Necessity of Intermediate Test*

The scientific and technological achievements of laboratory are theoretical and practical value through the transformation of the objective world. Basically, it is still a form of hidden knowledge productivity which becomes true productivity through trial in the secondary period of development. In other words, it is also necessary to solve the contradictions between the hidden productivity of knowledge forms and the realistic productive of industrialization. In addition, the realization of the industrialization of scientific and technological achievements of the process also has great risk from technology, market and fund. To solve these problems, not considered well at the basic development stage, and pave the way for the industrialization, we must carry out the process amplification, product manufacturing and product sales through the intermediate test. Therefore, the intermediate test is a very important to the process development of plant-based products.

The necessities of intermediate test are shown in the following four aspects.

[1] Only through intermediate test can we obtain the key data from the circulation of materials, the accumulation of impurities, the corrosion of materials and the equipment running.

[2] Only at intermediate test equipment can we obtain the data of transfer process.

[3] Only at intermediate test equipment can we obtain a certain batch of industrial products to conduct experiments and provide for consumers.

[4] The equivalence of the mathematical model and the continuity and stability of equipment must be tested in the intermediate test.

### *2.2.4.2. The Tasks of Intermediate Test*

The tasks of intermediate test are described as follows.

[1] Test and improve the mathematical model of process, check the feasibility of technology in the conceptual design;
[2] Examine and identify the reliability and security of long-term operation system, determine the control methods of producing process;
[3] Provide more comprehensive and accurate data for the fundamental design;
[4] Inspect the effects of trace impurities on the process and verify the pretreatment program of raw materials;
[5] Inspect the form of equipment and applicability of equipment materials of key equipments;
[6] Verify the feasibility of the separation, purification and handling program;
[7] Provide sufficient products and side products for experiments and users;
[8] Identify the program to treat with and comprehensively utilize the waste gas, waste water and waste residue.
[9] Determine the actual consumption scale and the technical and economic indexes;
[10]Provide operation experience for a new process, investigate the feasibility of control-conditioning system.

#### *2.2.4.3 The Basic Establishment Principles of Pilot Plant*

*(1) Scale-Up of Pilot Plant*

The traditional chemical process development is enlarged on experience step by step. To ensure the reliability of industrial installations, the trial scale is expected as large as possible. Modern Carfer Strategy recommended the "Narrow the technologies" that the conceptual design on the narrow test. Along with the development of computer information science and technology, the mutual influence of the most chemical process operation conditions is studied in the smaller model experiments. Therefore, there is a trend of “micro-pilot plant” in the pilot plant.

The micro-pilot plant has some advantages, such as small scale, low investment, quick and easy modification, low operating costs, operational safety, and so on. However, the scale of pilot test can not be too small, otherwise it is difficult to sample and not facilitate to adjust process parameters. Moreover, if the products produced in pilot test are used for application experiments or the trial use of consumers, the pilot test must have a certain scale.

Therefore, the scale of pilot test is determined by the characteristics of the verified mathematic model and the necessary requirements of key equipments, such as reactor. Concretely speaking, the following questions should be specifically considered.

1) The characteristics of reaction process

The type and the smallest size of the test reactor are depended on the special characteristics of reaction process.

2) The experimental equipment of pilot test should meet various requirements, such as the application experiments of products, installation of instruments and apparatus, regulation control, sampling and the evaluation of production safety and environmental protection, etc.

3) Essential conditions for the key equipments which means that the pilot plant equipment size should be able to meet the processing installation requirements for the internal components (such as blenders, heat transfer equipment, spray device for bubbling, large lining, etc.).

4) The market demand

It is necessary to perform the application and market development of a new product. The scale of the pilot test equipment depends on the market demand.

*(2)The Integrity of Pilot Plant*

A complete chemical industrial process includes important four sections: the pretreatment of raw materials, chemical reaction process, separation and purification of products, post-treatment process, and recovery and recycling process.

The traditional pilot plant is a miniature production plant It needs large investment and a long time for experiment. According to developing theories of modern chemical process, pilot plant may not entirely simulate the entire process of industrial production, but carry out the pilot test on only part of the process even partial key equipment according to the mastery degree of technical process and the requirements for the pilot test itself. Whatever form of the pilot test can comply with the following general principles.

1) Pilot test on the whole process must be carried out under the following conditions

[1] Need a comprehensive study of the whole process based on the small trial;
[2] Need to provide a certain amount of samples for application test;
[3] The recycling of materials has unpredictable impact on the production, and has a great influence on the productivity.

2) Chemical reaction process and Fermentation process should be carried out in pilot test.

The reactor is the core of the chemical process development. Therefore, the general chemical reaction process and the fermentation process are needed carry out in the pilot test. As the mathematical model of some chemical process such as distillation, absorption, extraction and heat exchangers, were relatively mature, they has no necessary to be experimented. So, the practical separation equipment and pump cycle can be replaced by the data information from on-line analytical instruments and real-time computer, which contribute to save the following pilot test of separation and purification.

3) No matter what new products mixture, the used separation method and equipment are needed carry out in pilot test.

If the pretreatment of raw materials, the separation, purification and post-processing of product are belong to a new mixture, or the physical properties data are incomplete or unreliable, or new separation methods and equipment are used, or the unit operation such as drying, crystallization, precipitation, filtration is used, all of these should carry out pilot test together with the reaction process.

4) It is necessary to carry out pilot test when investigating the effect of the recycling process on the production process. For example the single pass conversion rate is low that a considerable number of unreacted raw materials need to recover and reuse, the material recycling existing in the processes, such as crystallization, evaporation, absorption, de-absorption, and azeotropic distillation etc, lead to the continuous accumulation of impurities, which will cause catalyst poisoning and made the microorganisms not grow. Also, the accumulation of impurities would reduce the absorption effect and regeneration of solvent, cause the generation of bubble and scaling, impact the growth of crystal, pollute the products, and even cause serious safe accident resulting from explosive composition. Therefore, we should study detailed the effect of the recycling process on the product process, and determine the methods for removing impurity through experiments.

*(3) Running Period and Detection Control of Pilot Test*

1) The running period of pilot test

The running period of pilot test depends on the characteristics of chemical industrial process, attained the reliable data of amplification design through running.

2) The detection control of pilot test

In order to facilitate the detection and control of the process for detailed study, the amount of control points, the detection accuracy and the detection scope of pilot test equipment all should exceed the industrial equipment. At the same time, it is necessary for the pilot test that the equipment should be easily to dismount, assemble, to observe and sample, to achieve the data testing and improving the mathematical model and analyzing the process sensitivity.

#### *2.2.4.4. The Relationship between the Basic Development Research and the Intermediate Test*

From the beginning of basic development research, it is important to consider the process amplification, especially emphasize on determining the characteristics of transfer process and dynamic data. In terms of technical condition test, it is necessary to identify the key reaction process as the basis for the process enlarged. It should be mainly studied the criterion for the process amplification, which would simplify the process.

### 2.2.5. Fundamental Design and Final Evaluation

#### *2.2.5.1. Fundamental Design*

The fundamental design is the final outcome for the development of chemical industrial process and also the foundation for engineering design.

Based on the detail and definite datum of key technologies in basic design process, the design department can make a complete engineering design according to the specific conditions in construction region.

Engineering design, a specific construction project, is the rational combination of many products, which has specific site and utility supply system program. Fundamental design without specific site and public system program is a single product. In addition to the technology condition, the fundamental design includes a large number of chemical engineering data, in particular the reactive process data and calculating results attained according to these data. In fundamental design, the systems engineering theory integrated with engineering experience were applied in the fundamental design, which would conduce to reducing the fixed consumption (especially energy), product costs and project investment, so that the whole process can achieve the best integration effect.

*Design Principles*

(1) Technical and economic indexes

Payback period is less than 2-4 years.

Investment profit is greater than 15%.

(2) The reliability of the production

The reliability of the production refers to the identified design of cushion.

(3) The fluctuation range of production capacity

The productivity of equipment has fluctuation, the lowest operating load normally allows for 50 percent of the normal load. The lowest operating load is used as the lower limit of tower fluid dynamics, and used to determine the turbine compressor surge measures.

(4) The application principle of material

The use of expensive corrosion-resistant materials or cheap intolerance corrosion material is based on the corrosion-resistant hierarchy of entire process.

(5) The level of autoimmunization

(6) Storage time of tank.

(7) Standby principles of equipment

Standby principles of equipment refer that the equipment with the characteristics like rotation, easy to wear and easy to corrosion and easy to coking should have the standby equipment.

(8) Whether the development program is in consideration.

*Foundation of Design*

(1) The annual operation time

To determine the annual operation time, the following factors should be mainly considered:

The corrosion of intermedium;

Whether the process has coking and polymerization reaction;

Reserve coefficient of equipment;

The manage level of operations;

The maturity level of technology.

Generally, the annual operating time is 7000 to 8200 h.

(2) The specifications of raw material

The prices of raw materials are related to the purity, thus, the choosing should be reasonable.

To consider the impact of impurities on the reaction that generates by-product, decreases yield, and increases separation difficulty.

To consider the impact of impurities on the catalyst that deactivates and poisons the catalyst.

To consider the impact of inert impurities that decreases the concentration of active ingredients, does harm to dominant reaction and increases the air emissions.

It is necessary to consider the impact of raw materials the purity on products separation.

It is necessary to expect the value of separated products and the payment for the enhancement of separation purity.

(3) Specification of product

The specification of product should comply with the national standards or requirements of users.

(4) Yield of product and consumption of raw material

The consumption of raw materials should be the compulsive indicators. If the targets were not attained the consumption requirements or the process parameters should be modified to decrease the consumption until the targets were achieved.

(5) Conditions at the edge of handover area

The conditions are the transition temperature, pressure and phase of the raw materials transferred with pipe, intermediate goods and public sector projects at the edge of handover area.

(6) The specification of public engineering

The public engineering include power, water, steam, instrument, compressed air, nitrogen, catalyst, desiccant, additives and other chemicals, waste water, waste gas, waste residues, and so on.

*The Technical Process and the Illustration*

Design and illustrate technology process according to the control point mapped in the preliminary design stage of the production process.

*The Equipment Specification*

The choice of the equipment form and the technical calculation of equipment;

Selection of equipment material

Safety factor

Design temperature

It may not be the normal operating temperature, should consider the fluctuations in production, the maximum and minimum temperature under abnormal circumstances. The heat transfer equipment has to consider scaling from both sides of the wall temperature changes.

*Design Pressure*

The greatest pressure may be aroused under various operating conditions. The design pressure is the 1.1 time of the operating pressure, is the operating pressure added 0.18 MPa.

Vacuum equipment can withstand the maximum pressure inside and outside.

Tank pressure liquid hydrocarbon is designed according to the saturation vapor pressure at 50 ° C.

*Pipeline Flow Chart (PID) with Control Points*

The role of the pipeline flow chart includes:

[1] Determine the detailed pipeline process.
[2] Determine the diameter of pipeline;
[3] Determine the controlling programs;
[4] The security and performance analysis according to the pipeline processes and the control program

#### 2.2.5.2. Final Evaluation

The technical and economic evaluation of chemical industrial process development can be carried out according to the requirements of the feasible study [28], which focuses on forecasting economic benefits (including investment recovery and profit potential extent), and providing the technical and commercial aspects basis for the investment decision-making.

[1] The study of commercial feasibility focused on the study of forecast and market demand.

[2] The study of technical feasibility focuses on the analysis of the advancement, reality, reliability of project.
[3] The study of economic feasibility focuses on the analysis of investment, profit, and risk.
[4] The study of financial feasibility focuses on demonstrating the idea of affording and the lending.
[5] The study of management feasibility focuses on assignment and source of staff.
[6] The study of implement feasibility focuses on the study of construction progress, environmental protection, and legal issues.

## 2.3. Process Engineering of Plant-Based Products

As previously mentioned the task of process engineering is to solve the bottleneck problems from the laboratory results to industrial transformation, to create clean and highly efficient technology, processes and equipments. To achieve the ultimate aim, it is necessary to understand the common problems of process engineering of plant-based products. To sum up, the common problems mainly include raw materials (such as the source, composition, etc.), maximum use of the materials, minimum energy consumption and ecological industrialization production in the product development process. Thus, the process of plant-based products should include plant-based materials engineering, plant-based products transfer engineering, industrialization of plant-based products, as well as technical and economic analysis of plant-based products.

Process engineering of plant-based products is the scale and commercial need for the plant-based products, is the theory and technology for transformation of plant-based raw materials, and also is the integration of plant-based raw materials transformation and engineering in the process design and process development.

### 2.3.1 Plant-Based Raw Materials Engineering

Plant-based raw materials are the basis of plant-based products. Just as the development of other processing industries, with the development of processing industry and mature of related technologies, raw material costs play dominant role in the product costs. It is important to use successfully of the low costs renewable plant-based raw materials in the large-scale commercialization and intensive factories.

Dr. Roessler said the plant-based raw materials should have the following characteristics [29]: economy, price stability and easy access, stable components, low-costs, high value-adding of by-product, diversity of products, conducive to environment, endurable storage and inexpensive.

Plant-based sources are widespread, rich and renewable. But they also have the problems like dispersion, multiform difference of composition and structure between materials. So in order to use and convert plant-based raw materials efficiently, we must investigate the plant-based raw materials from the source, study their structural characteristics and the relation between the process engineering aiming at transforming the plant-based materials into products, set up the plant-based raw materials engineering, and achieve the industrial use of biomass raw materials from the view of structural characteristics.

#### *2.3.1.1. The Structure and Active Ingredients of Plant-Based Raw Materials*

Nature is a forever repeated production chain producing plant-based materials through photosynthesis annually $1.7 \times 10^{11}$t such as cereals, crop residues, seed oil crops, sugar crops, grass, various crops and woody crops. The structure and composition of biomass materials are shown in Table 2.2. The major components include carbohydrates (from the cellulose and hemicellulose), cereal starch, sugar crops, lipid (wax and oil), various types of protein, aromatic compounds (mainly lignin), ash (non-carbon minerals such as calcium, phosphorus, potassium, etc.) and active ingredients applied to the food and drug.

**Table2.2. The category and their major ingredient of plant-based raw materials**

| Category | Representative material | Major ingredient |
|---|---|---|
| lignocellulose | Wood, waster biomass in agriculture and forestry | cellulose、hemicellulose、lignose (4:3:3), a spot of extracted substance and ash [30, 31] |
| sugar | Sweet sorghum、sugarcane、sugar beet | Sweet sorghum [32]: fiber 11.5-12.5%, sugarcane12.5-14.5%, moisture content 70-75%, non-sugar 2-4%, non-sugar includes sugar 1-1.5%, pectin and organic acid 0.3-0.4%, nitrogen compounds 0.4%, lipid and wax 0.2%, ash 0.4-0.6%.<br>sugar beet [33] : sugarcane 17.5%, moisture content 75%, cellulose and hemicellulose 3.3%, pectin 2.4%, protein and nitrogen compounds 1.2%, ash 0.6%. |
| starch | Corn, sorghum, wheat, barley, oat, rice, sweet potato, cassava | Corn [34]: carbohydrates 73%, moisture content 6-15%, protein 8.5%, lipid 4.3%, ash1.7%、crude fiber 1.3%, and a small amount of carotene, riboflavin, calcium, phosphorus, iron, etc.<br>Sorghum [35]: carbohydrates 78%, moisture content 12%, protein 8.2%, lipid 2.2%, ash 0.4%、crude fiber 0.3%, and a small amount of carotene, riboflavin, calcium, phosphorus, iron, etc.<br>Fresh cassava [36] : starch content of 25% to 30%, while the dry cassava starch content can be as high as about 70 percent, in addition to about 4 percent of the sugar. |
| lipid | Rapeseed, sunflower seed, algae with high content lipid [37] | Rapeseed(dry): lipid 40%, cellulose 16%, protein22%,hemicellulose and lignose 22% |
| active ingredients in natural herb | Taxus, ginkgo biloba | pigment, plant steroids, saponin, polysaccharides, polyphenol, protease inhibitors, peptides, plant hormones, phytic acid, unsaturated acid [38] |

#### *2.3.1.2. The Distribution of Plant-Based Raw Materials*

According to statistics about the common crop residues shown in the Table 3, Asia can make use of the largest of crop residues, accounting for about 60 percent of the world, followed by North America and Europe.

According to the statistics of UN Food and Agriculture Organization (FAO) in 2000 years, the world forest area is $38.6 \times 10^8$hm$^2$, being equal to 0.6 hm$^2$ per capita. Although due to artificial forestation the forest area of temperate regions continuously increases, from around the world, the forest area reduced is $9 \times 10^6$hm$^2$ because of the decrease of tropical zone annually $13.5 \times 10^6$hm$^2$. The biomass of trunks of wood category is 350 billion tons (dry weight), of which one-third in South America. Accounting with trunk wood in the world, the average amount of biomass every per hm2 is 92 t (dry weight). The total planting area of the world is 45 × $10^5$hm$^2$ every year and the world total area of artificial forestation is 1.87 × $10^8$hm$^2$, rapid growth eucalyptus trees accounting for 10% (1.786 × 107 hm2) and Acacia trees accounting for 5% (8.32 × $10^6$ hm$^2$).

**Table2.3. The available crop residues in world [39]**

| materials/ hundreds of millions kilogram | Africa | Asia | Europe | North America | Central America | Oceania | South America | Subtotal |
|---|---|---|---|---|---|---|---|---|
| Corn straw | 0.0 | 33.9 | 28.6 | 134 | 0.0 | 0.2 | 7.2 | 204 |
| Barley straw | 0.0 | 2.0 | 44.2 | 9.9 | 0.2 | 1.9 | 0.3 | 58.5 |
| Oat straw | 0.0 | 0.3 | 6.8 | 2.8 | 0.0 | 0.5 | 0.2 | 10.6 |
| Rice straw | 20.9 | 668 | 3.9 | 10.9 | 2.8 | 1.7 | 23.5 | 731 |
| Wheat straw | 5.3 | 145 | 133 | 50.1 | 2.8 | 8.6 | 9.8 | 354 |
| Broomcorn stalk | 0.0 | 0.0 | 0.4 | 6.9 | 1.2 | 0.3 | 1.5 | 10.3 |
| Bagasse | 11.7 | 74.9 | 0.0 | 4.6 | 19.2 | 6.5 | 63.8 | 181 |
| Subtotal | 38.0 | 924 | 217 | 219 | 26.1 | 19.7 | 106 | 1549 |

According to the statistics in 2000, the world total production of starch, sugar produced from plants is about 4.2 billion tons which the total mass of cereal and taro occupies 27 million tons. The worldwide oil production of plant resources, including seeds and fruit, is about 250 million tons annually; the total amount adding the trunk or branches from the oiling coconut trees (dry weight of about 100 million tons) is 350 million tons. In addition, oil produced from the seeds and fruit is about 100 million tons.

#### *2.3.1.3. Collection of Plant-Based Raw Materials*

Plant-based raw materials in particular crop stalks have the characters as light weight, large-size, wide distribution area, seasonal harvest, complex component, low heat value, low energy density, difficult to transport and storage, and so on. So the collection storage of raw materials is always a problem, the cost of collection and transport is a high proportion of the total cost. Due to these features of plant-based raw materials, it ought to study seriously the collection, storage and transportation mode of material to ensure adequate supply of raw materials and then better serve humanity. We take crop stalks as an example to descript the collection methods and the collection costs of plant-based raw materials.

At present, the modes of the collection, storage and transportation can be divided into the following categories:

(1) By converting into high heat value liquid fuel such as alcohol, petrol and diesel oil

As proposed by the Chinese University of Science and Technology, the best technical course for material used as clean energy should be described as follows: firstly the raw materials are pyrolyzed dispersedly into the primary liquid fuel- bio-oil, easily transporting and storing, at an appropriate scale (raw materials collected in the 10-20 km radius) in the producing area of raw materials; then, after being collected, the products are reprocessed further such as refining, producing hydrogen or syngas richened in hydrogen, synthesizing methanol or complex alcohol, synthesizing gasoline and diesel. This technical course will fundamentally solve the bottlenecks of the biomass resources dispersed and seasonal restrictions on large-scale application.

(2) By forming solid biomass fuels

The straw is compressed and formed in rural with the biomass solid briquetting equipment, which avoid long-distance transport and make it relatively easy to collect and store. The density of solid biomass fuels is usually 1.0 t / m3 just equivalent to coal which enable the transport of straw as coal and greatly reduced the transport difficulty. After been compressed and formed, the flame point is relatively increased and the fireproof performance is also relatively increased. The volume of compressed and formed straw is significantly reduced and the density increased markedly, which makes the biomass fuels easily enter into the furnace when biomass is used in the field of power generation. However, the compression and formation of biomass consumes great manpower, power and resource, increasing the cost of biomass fuel. Take an energy company in Heilongjiang province of China, the cost of solid biomass fuel is described as follows: labor costs 20 yuan / t, power costs 40 yuan / t, Straw smash cost 20 yuan / t, materials consumed 5 yuan / t. Therefore, the cost of straw increased by 85 yuan per ton in the compressing and forming technique.

(3) By baling

The density of baled straw is 0.18 t / $m^3$, while the density of straw briquetting is 1.0 t / $m^3$. Compared with the briquetting straw, the baled straw will cost more. The transport cost of briquetting straw is 0.5 yuan/t•km, accounting to the distance of 50 km, so the transport cost is 25 yuan/t, while the transport cost of baled straw is 125 yuan/t, and the cost of straw without baled will be even more. In addition, baled straw generally should set up a purchase station, which will also increase the cost, for example the wages of staff, housing costs for staff, costs of straw weighing equipment, cost of baling machinery and equipment, stacking machinery and equipment costs, straw venue rental costs, straw rain-proofing facilities, equipped with fire engines and firemen, and so on.

**Table 2.4. Questionnaire of straw gathering at some zones in china（yuanRMB/Kg）[40]**

| zone | Rice straw | Wheat straw | Cotton stalk | Cole stalk | Jute stalk | Corn straw | Gather radius（Km） |
|---|---|---|---|---|---|---|---|
| Zhejiang | 0.15 | 0.15 | 0.2 | 0.14 | 0.2 | - | 2 |
| Shandong | 0.2 | 0.2 | - | - | - | 0.12 | 0.7-3.4 |
| Sichuan | 0.15-0.2 | 0.15-0.2 | - | - | - | - | plain< 8Km; mountainous area< 3 Km ; hill<5Km |

**Table 2.5. Cost contrast of straw collection (yuan/t) [40]**

| | He Bei | Shan Dong | Shang Hai | Hu Nan |
|---|---|---|---|---|
| Purchasing Price | 100 | 150-200 | 150 | 102 |

Plant-based resources are widespread, rich and renewable, but there have significant characteristic and structural difference between raw materials from different origin. Therefore, different materials have different collecting ways and costs. Crop stalks are important biomass materials. In this section, the crop stalks are took as the representative to discuss calculation of the conveyor economic radius of the biomass.

The collection radius and collection costs of straw are closely related to the distribution density of crop stalks resources and labor prices. According to the relevant facts and data from the availability and economic analysis of China's crop stalks, the collection radius and cost of crop stalks are shown in Table 6. Table 7 compares the collection cost of stalks from different region, which shows that the regional differences are quite significant.

With the definite integral micro-element analysis method, Liu gang, Hao Haide and Dong Yuping established the mathematic model for collecting costs of stalks under the relevant parameters assumptions and the basic premise of the environment that the distribution of biomass straw should meet the characteristics of widespread, single, uniform, periodic. This model defined the collecting distance as collection radius and neglected the effect of other risk factors (such as climate change) on the straw collection. The straw collection costs mainly include the following parts. (1) Acquisition cost, which refers that the cost pay for straw. (2) Transportation costs, which refers that the cost for transport the straw to the enterprise. (3) Other costs referring to the additional, except the acquisition cost and transportation costs, including handling costs, labor costs and storage costs. Therefore, the collection cost of straw can be calculated as the formula that the collection cost = the acquisition cost + transportation costs +other costs [41], namely:

$$G = X + Y + Z \quad (2.2)$$

Where, G is the collection cost;

X is the acquisition cost, and the acquisition cost = collection amount × acquisition price, that is $X = ACo$ (2.3)

Y is the transportation costs, and the transportation costs = transportation distance× traffic volume= collection volumes ×transport distance;

Z is other costs mainly including handling costs, labor costs, storage costs, etc.

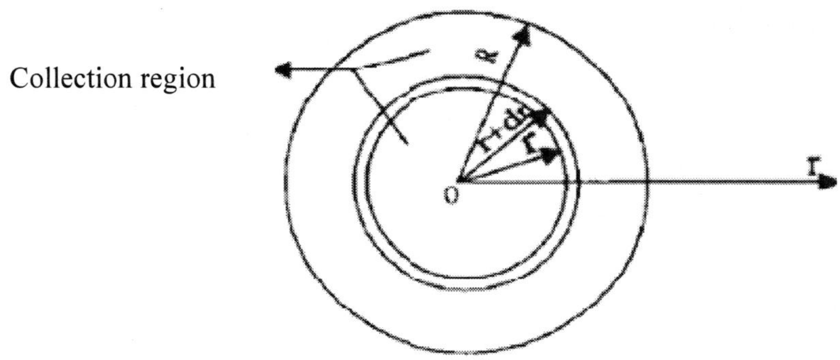


Figure 2.4. The calculation for transportation costs [41]

The different collection regions of straw, correspondingly the different transport distances, results in the transportation costs are also different. Under certain collection amount, considering the lowest transportation costs, the final form of collecting region of straw should be circular, as shown in Figure 2.4. Under certain traffic volumes, transportation costs linearly increased with the transport distances.

The relation of collection radius (R) and the collection volumes (A) as the following [40]:

$$A = \pi R^2 \times K_1 \times a_0 \times K_2 \times K_3 \tag{2.4}$$

Where, $\pi R^2$ is the collection area;

$K_1$is the proportion of the arable land area accounted for the land area;

$a_0$ is straw yield per unit area;

$K_2$ is collection coefficient of straw;

$K_3$ is available coefficient of straw;

Assuming the comprehensive coefficient $K = K_1 \bullet K_2 \bullet K_3$,

$$\text{Then } A = K\, a_0\, \pi R^2 \tag{2.5}$$

Thus, the collection radius (R) can be calculated in accordance with the collection volumes and straw yield per unit area.

$$R = (A / K\, a_0\, \pi)^{1/2} \tag{2.6}$$

In order to solve the storage and transport problems of raw materials, on one hand, we should rely on the planning and arrangements of local governments, based on the point of supporting the development and utilization of biomass energy, to study and establish management policies and mechanisms of straw collection, storage, transport, ensuring the quantities and quality of straw, to establish a new logistics system and new infrastructure, making the very scattered straw and other biomass energy artificially collect for conversion and transport to end-users. On another hand, we should rely on the market, which plays a role in economic leverage, to increase income of farmers and enhance the enthusiasm on straw collection, storage and transport of farmers.

### 2.3.2. Transformation Engineering of Plant-Based Products

The products obtained from plant-based raw materials have been widely used in daily life. Besides the natural carbohydrate polymer, plant-derived oil or simple extracted plant active ingredients, most of the plant-based products are obtained from the conversion of plant-based raw materials. For example carbohydrates raw materials are converted into some chemical products or fermented into some products. Transformation engineering of plant-based products is based on the efficient conversion of raw materials [17]. The methods used in the transforming process can be divided into efficient biological transformation and thermal chemical transformation.

***(1) Efficient Biological Transformation***

Plant-based raw materials can be used by microorganisms except that the long transformation period and low utilization rate of biological transformation is not suitable for industrial applications. The purpose of high efficient biological transformation is to looking for efficient and economic biological transformation way with the current biotechnology. The main study contents of high efficient biological transformation include: ① Separating, screening, training and improving of efficient strains, ② Construction of genetic engineering strains, ③ Research and regulation of microbial metabolism engineering, ④ Development of high value-added products, ⑤ Separation and purification of products, ⑥ Design and enlargement of bioreactors.

***(2) Thermochemical Transformation***

A possible approach used for processing high fabric raw materials is the thermochemical transformation method. Firstly, these raw materials were processed into the same intermediate product that further transformed into the plant-based products. Such alternative production processes are well known as gasification and pyrolysis.

Gasification is a process in which the solid carbon content of fuels is transformed into gas mixture in high-temperature (750-850 ℃). The gas mixture can burn, sometimes referred as synthetic gas or synthesis gas, which consist of CO, $H_2$, $CO_2$, methane, N2, and a small number of senior hydrocarbons. Synthesis gas can not only be used to produce heat and power, also be used as raw materials to product liquid fuel and compounds.

Pyrolysis refers to the process that plant-based raw materials thermal decomposed to produce coke, oil and gas under the conditions of the oxidation medium excluded. Since the chemical composition and components are different from raw materials, the reactive process of pyrolysis, the composition and output of pyrolysis product are also different. The composition and proportion of pyrolysis product are directly determined by pyrolysis process parameters [42]. If the purpose of pyrolysis is to attain liquid product, the pyrolysis condition of the plant-based raw materials should be as low temperature, high heat rate and short gas retention time. If the purpose of pyrolysis is to achieve a high yield of fuel gas, pyrolysis conditions would be high-temperature, low-heat conduction rate and long gas residence time. If the purpose of pyrolysis is to attain a high yield of coke, low temperature, as well as low-heat conduction rate is needed [42]. According to different retention time and temperature of pyrolysis process, the pyrolysis process can be divided into three categories: conventional pyrolysis, fast pyrolysis and flash pyrolysis. The largest liquid yield of fast pyrolysis can be up to 80%. The current pyrolysis technology research focuses on fast pyrolysis.

## 2.3.3. The Industrialization Process of Plant-Based Products

The concept of industrialization developed from the concept of industry. To understand the concept of industry, first of all need to understand what the industry is. The so-called "industry", originally referred to the various production sectors of national economy, and sometimes also referred specifically to industry. Later, as the division and emergence of the tertiary industry, by extension, "industry" was used to refer all kinds of the business and organization providing material products, circulation approaches, and labor services and so on. The concept of "industry" is an integrated conception that lies between the cells of micro-economic and the unit of macro-economic, which means that industry is a collection of business and organizations with same attributes, also is the summation of the parts divided from the national economy by a certain standard. Industrialization refers to making the same attributes of a business or organization set the size recognized by the community, completing the process of quantitative variation to qualitative leap, and truly becoming an important part of the national economy to a certain standard [17].

The industrialization process of plant-based products refers to the entire process of laboratory research results converting into the first set of industrial installations. The research of industrialization process has significant means for achieving the economic efficiency of plant-based raw materials transforming technology [16].

In accordance with the technical course of process engineering development mentioned above, the industrialization process involves not only process engineering, conversion technologies, mechanical equipment, automatic control, material, technical and economic, and other fields, but also a laboratory test, pilot test and building factories, and so on. It is a highly integrated technology projects. The successful implementation of the industrialization process of plant-based products from plant-based raw materials will be a highly embodiment of the green, informing technology and process integration of process engineering. The current industrialized process of plant-based products mainly focus on the fuel ethanol, biodiesel and joint burning, also need more research work on raw materials, new products, new technology and equipment development in the future.

## 2.3.4. The Technical and Economic Analysis of Plant-Based Products

The technical and economic analysis of plant-based products can be evaluated according to the requirements of the feasibility study, which should be theoretically correct, technically reliable, operationally secure, and economically reasonable.

Continuously improving the efficiency of resource utilization, recycling economy is an economic means relying on the development of ecological circulation of resources, which is in the system of human, natural resources and science and technology, also in the whole process of the resources input, enterprise production, products consumption and discard. An important issue in the transformation process of plant-based raw material is the secondary pollution to the environment and the comprehensive utilization of the residue problem after the transformation process. At present, the idea of recycling economy has emerged in the use of plant-based raw materials such as establishment of the eco-industrial park of high value of straw utilization which was proposed by Prof. HongZhang Chen, from Institute of Process Engineering, Chinese Academy of Sciences [17, 43, 44]. The eco-industrial park is a type of economic cycle at the regional level, which makes the material and energy recycling within the region between the various industries and enterprises receive, and makes regional waste minimized.

As for the mainly biomass energy and chemical products, the upstream processes of ethanol and bio-diesel were coupled with the downstream producing processes of PDO, biological ethylene through process integration and industrialization link technology. As for the by-products and waste, it can develop the industry like feed, gas power generation, $CO_2$, degradable plastics through the comprehensive utilization, improve the chain of recycling economy, enhance the utilization efficiency, and increase the added value of by-products. As for the raw materials, it will achieve the conversion from mainly on food to mainly on multi-resources such as aged food, potato, cellulose, oil crops, waste oil and others. As for the public works, such as steam, water, it may be reduce production costs by synergistic effect generated by optimizing the allocation of relying on the key biomass energy and chemical industry output, it will enable many downstream chemical products enterprises to establish industrial cluster with agglomerative cooperation to enhance the value of the industrial chain. In accordance with the development refining-chemical integration mode of petrochemical industry, it may form a complex industrial ecosystem full of diversity forms.

With the simultaneous development of hard technologies and industrial practice, by means of using advanced computer software platform, with the evolution system, we will establish a virtual industrialization park of biomass refinery to meet the needs of managing refining-chemical integration mode, guide the allocation of resources, identify the potential - effective growth points or bottleneck constraints, adjust the business strategy and the production plan timely, form a rapid, flexible and accurate response system.

It is a long and complex process to establish a perfect biomass energy and chemical industry system that subject to policy, technology, personnel, financial constraints, and other factors and many difficulties are needed to be resolved. The biomass energy and its chemical industry application will certainly become a hot domestic development in the next five to ten years because of their fundamentally consistent with the requirements of sustainable development and vast market space of biomass products.

The combination of practical experience of biomass energy and chemical industries combines with the ecological industry and theoretical research of recycling economy, and the development and in-depth study on the biomass energy and economy of chemical recycling will lay a foundation for the sustainable development of biomass industry in the future.

## REFERENCES

[1] K., W., Process and product engineering-achievements, present and future challenges. *Chemical Engineering Science* 1999, 54, (11), 1601-1620.

[2] Liu, Q., History and Logic of Chemical Engineering. *Journal of Nanjing University of Technology (Social Science Edition)* 2002, 1, (1), 92-96.

[3] Chan, M., Past, present and future of Chemical Engineering. *Jiangsu Chemical industry* 1994, 22, (2), 4-5.

[4] Yang, S., Current Situation of Biochemical engineering development. *Chemical Engineering* 1995, 23, (1), 42-48.

[5] Li, C., The research and the prospect of chemical engineering. *Chemical Industry and Engineering Progress* 2000, 19, (3), 4-7.

[6] Jin, Y.; Wang, C., Chemical Engineering in the 21st century. *Chemical Industry and Engineering Progress* 2000, 19, (1), 5-10.

[7] Huang, Z.; Li, H., Development trend of chemical engineering in the 21st century. *Chemical Industry and Engineering Progress* 2001, 20, (4), 1-4.

[8] Fang, D., Chemical engineering for 21st century. *Chemical Production and Technology* 2000, 7, (1), 5-11.

[9] E., F.; V., F.; C., R., Trends in chemical engineering education: Process, product and sustainable chemical engineering challenges. *Education for Chemical Engineers* 2008, 3, (1), 22-27.

[10] J, G.-S.; L, P.-B.; J., C. M., New trends for design towards sustainability in chemical engineering: Green engineering. *Chemical Engineering Journal* 2007, 133, (1-3), 7-30.

[11] E., G. J., Chemical engineering education in the next century. *Chemical Engineering & Technology* 2001, 24, (6), 561-570.

[12] M., P. J., Chemical engineering and the postmodern world. *Chemical Engineering Science* 2001, 56, (12), 3627-3639.

[13] Kwauk, M., Process Engineering. *The Chinese journal of process engineering* 2001, 1, (1), 2-7.

[14] Niu, Q.; Qian, W., Process science and process engineering. *Studies in Science of Science* 2002, 20, (2), 143-147.

[15] Chen, J., Process industry and process engineering. *The Chinese Journal of Process Engineering* 2001, 1, (1), 8-9.

[16] Cheng, S.; Yang, Q., Developing history, status and prospect of process systems engineering. *Journal of Tianjin University* 2007, 40, (3), 321-328.

[17] Chen, H.; Wang, L., Research progress on key process and intergrated eco-industrial chains of biobased products –proposal of biobased products process engineering. *The Chinese Journal of Process Engineering* 2008, 8, (4), 676-681.

[18] Xiao, X.; Zhang, S.; Shi, D.; Li, J., Update and green of process industries. *Bulletin of National Natural Science Foundation of China* 2004, 18, (6), 333-339.

[19] Qi, E., A survey on the development of modern industrial engineering. *Industrial Engineering Journal* 1998, 1, (1), 3-6.

[20] Ouyang, F., Feature and frontier of Biochemical process engineering. *The Chinese Journal of Process Engineering* 2001, 1, (3), 230-238.

[21] Sun, H., Focus of chemical engineering-understanding space-time multi- scale structure. *Chemical industry and Engineering Progress* 2003, 22, (3), 224-227.

[22] C., C. J., Four main objectives for the future of chemical and process engineering mainly concerned by the science and technologies of new materials production. *Chemical Engineering Journal* 2005, 107, (1-3), 3-17.

[23] Zhang, Y., The green process engineering science. *The Chinese Journal of Process Engineering* 2001, 1, (1), 10-15.

[24] Ji, H.; She, Y., Development and research on basic issues of green chemistry and chemical technology. *Chemical Industry and Engineering Progress* 2007, 26, (5), 605-614.

[25] Qin, J., Couple – a new way to study unit operation. *Chemical Industry Times* 2004, 18, (3), 57-58.

[26] Wang, Z., Multidiscipline and process integration of nonaqueous biocatalysis. *Journal of Chemical Industry and Engineering (China)* 2005, 56, (4), 563-569.

[27] M., D. J., *Conceptualized design of chemical process*. Chemical Industry Press: Beijing, 1994.

[28] Li, X.; Lu, J.; Zhang, Q., *Technical Economics*. Wuhan University Press: Wuhan, 2003.

[29] Birgit, K.; Patrick, R. G.; Michael, K., Biorefineries: Industrial Processes and Products. In WILEY-VCH GmbH&Co KgaA: 2006.

[30] Yang, S., *Cellulose Chemistry*. Light Industry Press of China: Beijing, 2001.

[31] M., B.; B., M.; M., V. M., Biosynthesis and genetic engineering of lignin. *Critical Reviews in Plant Sciences* 1998, 17, (2), 125-197.
[32] Lu, Q., Research progress of Sweet sorghum. *World Agriculture* 1998, (5), 21-23.
[33] Wu, Z.; Wang, Y., Analysis of nutritional components in tends leaves and stems of wild plant. *Journal of Plant Resources and Environment* 2005, 14, (1), 60-61.
[34] Gao, Y., Analysis the nutritional constituents and development of the black sweet corn. *Food Science* 2000, 12, 59-61.
[35] Wang, H., Analysis and evaluation of nutritional constituents of mainly miscellaneous grains crop in Gansu Province. *Gansu Agricultural Science and Technology* 2001, (5), 20-21.
[36] Sun, H.; Ma, X., Study on comparison of properties of tapioca and complex modified starches. *Science and Technology of Food Industry* 2008, 29, (6), 82-87.
[37] Miao, X.; Wu, Q., Exploitation of biomass renewable energy sources of Microalgae. *Renewable Energy* 2003, (3), 13-16.
[38] S., O.; C., K. K., Ferulic acid: pharmaceutical functions, preparation and applications in foods. *Journal of the Science of Food and Agriculture* 2004, 84, (11), 1261-1269.
[39] S., K.; E., D. B., Global potential bioethanol production from wasted crops and crop residues. *Biomass and Bioenergy* 2004, 26, (4), 361-375.
[40] Http://www.crein.org.cn/sub-pages/biomass/biomass.htm
[41] Li, J.; Ren, D.; Zhuang, X., The system evaluation methods and examples of renewable energy resources. *Journal of Natural Resources* 2001, 16, (4), 373-380.
[42] Pam, C.; B., V. J.; AM, L. C., Utilisation of biomass for the supply of energy carriers. *Applied Microbiology and Biotechnology* 1999, 52, (6), 741-755.
[43] Chen, H.; Li, Z., Ligncellosical microorganisms and biomass total utilization. *Biotechnology Bulletin* 2001, 24, (5), 17-20.
[44] Chen, H.; Li, Z., Paradiam and new concept for biochemical engineering- development and its theory base of ecological biochemical engineering. *Journal of Chinese biotechnology* 2002, 22, (3), 747-777.

*Chapter 3*

# PROCESS AND PRINCIPLE OF COMPONENTS SEPARATION FROM PLANT-BASED MATERIALS

## 3.1. INTRODUCTION

Plant-based biomass is the organic material formed by solar energy and $CO_2$ through photosynthesis, including agricultural products, crop stalks, wood wastes, energy crops, medicinal plants, aquatic plants and so on. The solar energy is stored in the form of carbohydrates through photosynthesis during plant growth, which can be converted into fuels through appropriate means.

China is one of the countries with most abundant plant resources from tropical, subtropical, temperate and frigid zones, and has about 30,000 species (second to Brazil and Colombia) of higher plants, and more than half of them are China's endemic species. China is the world's most abundant resources of medicinal plants. A recent survey shows that there are 12,807 kinds of traditional Chinese medicine resources in China, of which 21,146 species are medicinal plants (located in 383 families, 2,309 genuses). At present, the natural medicines in the world trade volume has reached 15 billion U.S. dollars, the turnover of medicinal plants and their products, including health products, natural cosmetics and perfumes, and other natural products summed up to 30 billion U.S. dollars.

Crop stalk is the most abundant renewable resource on the planet, according to statistics annual output of the straw in China is $7\times10^8$ tons, the largest amount in the world. Food crop stalks account for 90.5% of total straw, mainly including rice straw, wheat straw and corn stalks, accounting for, 27.5%, 15.2% and 36.7%, respectively [1-3].

However, degradation of the major components of biomass is very difficult, which limits their development and utilization. The annual harvest of straw, about $2.8\times10^7$ tons, is for making paper, $2.13\times10^8$ tons is for feed or feed ingredients, $1.09\times10^8$ tons is for making fertilizer to field and losses in the course of collecting, the remaining approximately $3.761\times10^8$ tons are burned or directly discarded. With the development of industrial modernization, these numbers will be increased year by year, resulting in serious waste of resources and environmental pollution.

Development and utilization of straw biomass are of great importance for easing the severity of energy crisis and environment protection. With the rapid economic development of the world, all the countries are facing problems of resources, energy crisis and environmental pollution. With the existing energy and chemical industry over-dependence on oil, coal and gas-based fossil fuels, the environmental pollutions caused by fossil fuels have increased, coupled with diminishing energy resource reserve, therefore human are facing an increasingly serious energy and resources crisis. Fortunately, the unique chemical composition and renewable characterization of biomass plants provide a possible way to solve the crisis energy and resources [4].

That the crop stalk had not been effectively used is mainly due to its main characterization. Firstly, the structure of straw is heterogeneous, that is to say various parts of the chemical composition and fibre are in very different forms, which result in that different parts have different values. Secondly, the chemical composition of straw is varied in different parts, some contain high hemicellulose and ash content, and the ash content of some straw can be as high as 10%. Thirdly, fiber characteristic is different. The contents of small fiber and parenchyma cell in straw are high as many as 40-50%, the content of fibroblast is as low as 40-70%. For wood biomass the content of parenchyma cell is low, but the content of fibroblast is as high as 60-80% and 90-95% for broad-leaved wood and coniferous wood respectively. Straw is different from wood and both composition and structure have "heterogeneity", therefore it is necessary to investigate the characteristics of straw deeply in view of the component separation [5].

The purposes of separating components is separation of raw materials at the largest recovery, maintaining the integrity of compounds as far as possible, and obtaining the optimal utilization with maximum possibility and the greatest value ultimately. Pretreatment technology is developed on the basis of all types of wood pretreatment. Although much progress has been made, it is still difficult to achieve total use of various components [6].

The efficiency of direct combustion of straw is very low, and the general heat loss is 30-90%, resulting in waste of resources and environmental pollution. As feed of animal, the digestibility and nutritional value of rough straw without pretreatment are low. The main reasons for lower nutritional value of straw are in the following areas: (1) . The presence of lignin. Lignin can not be decomposed by most microorganisms, nor be used for livestock. Lignin, cellulose and hemicellulose intertwined with each other to form cell wall structure, so that the cells bonded together to form a natural barrier of lignin which affects not only the microbial degradation of cellulose and hemicellulose, but also the use of other nutrients on feed. (2) . Low protein content. The protein content of straw is 2-8%. To overcome the problem of inadequate protein content, it is necessary to modulate in feed processing. (3) . The ash content of straw is high, and mainly in the form of silicate, which has no nutritional value for livestock and impacts on digestion and absorption of other nutrients for livestock. (4) . The palatability is poor. The straw texture is hard, and the palatability is poor, so limited for livestock feed. Straw lacks of the necessary vitamin A, vitamin D, vitamin E, and so on that necessary for animal growth. The straw without pretreatment for feed livestock, could only used to full fill stomachs of animals, but can not meet the nutritional needs of livestock. The straw used as feed must be processed to eliminate or reduce the impact of nutrition constraints [7].

The interweaved structure of cellulose, hemicellulose and lignin determines that degradation of any one type of composition will inevitably restricted by other compositions. For example, lignin hinders the degradation cellulose and hemicellulose in straw, which results in the uncompleted decomposition of cellulose in raw materials. The main components and chemical properties of straw are very stable. They can not be dissolved in water or general organic solvents. At room temperature, straw can not be hydrolyzed by dilute acid and alkali. The sugar yield is very low in straw direct cellulose hydrolysis, only 20% of theoretic ratio, and most of sugars are pentose and hexose produced by easily hydrolyzed hemicellulose. The hydrolysis ration of most cellulose of raw materials is below 10% with excessive cellulase. Without pretreatment, the three major components of straw, cellulose, lignin and hemicellulose, are cross-linked closely together and the chemical structure and characteristic of them are different in nature, therefore it is difficult to direct use of them at high value. It is necessary to perform appropriate pretreatment, damaging or changing the structure of the straw to achieve high value utilization [8].

The roots, stems, leaves, flowers, bark, seeds and other different parts of Chinese herbal medicines have inconsistent medicinal value, and the main active ingredients and their content in different parts should be further analyzed. The exports of such products in China are only 1 billion U.S. dollars, accounting for 3.3% of the world. As the highlights of Chinese medicine for cancer and Alzheimer treatment, the Chinese medicine market will have even broader prospects. How to give better play to China's export advantage of the Chinese herbal medicine has become an urgent issue [9].

Earth is rich in plant resources, in addition to providing food and medicine for survival of the human, which also has important applications in industry, environmental protection and bio-energy. With the shortages of resources and energy and the increase of environmental pollution, human pay more and more attention to the development and application of plant resources. However, such a great trove of resources has still not been fully and effectively developed. This chapter summarized the structural characteristics and ingredients of the plant resources, and the existing problems and shortcomings in the development of plant resources. Taking adequate and efficient use of plant resources as the goal, new ideas for the development of resources was put forwarded, providing theoretical and technical basis for the development of Chinese medicine, functional foods, bio-energy and chemicals.

### 3.1.1 Plant-Based Resources

China is one of the countries with most abundant plant resources of the world, nearly 30,000 kinds of higher plants. To develop such a great treasure resources, it is necessary to systematically study the plant resources, and provide a theoretical basis for the development of plants, functional foods and chemical raw material. Plant resources are belonging to five main categories according to the end-use: food plant resources, medicinal plants resources, industrial plant resources, environmental protection plant resources and germplasm resources.

#### *1) Medicinal Plants*

Medicinal plant can be used as medicine in whole or part. The medicinal plant is in wide variety, and their medicinal parts are different with each other, for example whole plant of prunella vulgaris, parts of ginseng and extracted quinine are used as medicine.

More than 2,000 years ago, Zhang Qian was sent to a diplomatic mission to the Western Regions which entrenched the Silk Road and introduced plants from abroad for food and medicine to domestic cultivation such as pomegranate, sesame, walnut, garlic and alfalfa. In the Sui Dynasty, the books about cultivation of traditional Chinese medicine were presented, such as "Drug Cultivation" in "Sui book". In the masterpiece "Compendium of Materia Medica" of Ming Dynasty's medicine experts Li Shizhen, also recorded and narrated for about 180 kinds of cultivation methods for medicinal plant. At present, there are about 250 species of Chinese medicine plants with large cultivation area occupying more than $3.33\times10^5$ hectares, the variety and growing area of which are reached to an unprecedented scale.

Medicinal plants can be divided into two categories:

Chinese Herbal Medicine: At present, there are more than 5,000 species, among them about 400 species are commonly used in China, and some have been cultivated and manufactured for medicines or chemicals as raw materials.

Botanical Pesticides: Which include about 500 pesticides plants, such as pyrethrum. Contain pyrethrum and plant alkali, glycosides and plant hormones.

### *2) Edible Plant Resource*

Edible plant resource, including direct and indirect (fodder, food) consumption of plants, mainly divided into the following seven categories.

Starch, sugar plants: such as acorn, puerarin starch, and other starch wild plants in China.

Protein plant: including chlorella, winged bean, etc.

Oil plants: According to the crude survey, China's wild oil plants with the oil content of more than 15% are of approximately 1,000 species. The woody plants with oil content more than 20% are about 300 species, and more than 100 species could be eat, such as hazelnut, crown, and all kinds of wild fruit tea, walnuts, and so on.

Vitamin plants: mainly in a variety of wild plants, such as the kiwi, hawthorn, begonia, the fruit contain general 200-800 mg vitamin/100 g.

Beverage plants: In addition to the general drinks tea, cocoa and coffee, there are a number of regional beverage plants, such as mulberry leaf tea, chrysanthemum tea, honeysuckle, etc.

Edible flavor and pigment plants: madder, saffron, turmeric and other traditional Chinese food for the edible pigment. Pepper, hot tea, star anise and cinnamon are special cooking flavors of China.

Plant-based feed and bait: including most grasses, legumes and branches of the pod, all kinds of wild banana, etc.

### *3) Resources of Industrial Plant*

Industrial plant resources can be divided into seven categories, which include wood, fiber, tanning branch, aromatic oil, plastic resin, industrial oils and vegetable dyes, and other resources.

Timber resources: Wood is the most common plant resources, the fast-growing, valuable timber is the focus of future development.

Fiber resources: There are more than 190 species of important fiber plants in China, such as monocots-leaf plants, elm, mulberry, reeds, straw and kapok.

Tan-containing resources: tan-containing plants are rich in tannin, which can be used as tanning plastic baking and pharmaceuticals, larch, spruce, catechu, and other raw materials are important tannin plants.

Aromatic oil resources: aromatic oil plant is main raw material for the extraction of perfume and flavor, more than 60 branches of seed plants in China contain aromatic oil.

Plant-based plastic resources: including plant rich in rubber, resins and plastic, such as guar bean, acacia and astragalus.

Industrial oil resources: the industrial oil plants accounting for more than 50% of woody plants with oil content more than 20%, such as the tung tree, Rhus vernicifera, Sapium, wind-nan, and other plants.

Industrial plant-based dyes: such as Morin, redwood indigo leaves, turmeric, etc.

#### *4) Plant Resources for Environmental Protection and Improvement*

Plant resources for environmental protection and improvement can be divided into five categories:

Windbreak and sand fixation plants: such as casuarinas, rice grass, eucalyptus, leucaena, poplar, etc.

Plant for conserving soil and water and modifying barren hills and wasteland: such as leucaena, acacia, hazel chik oil and woody plants.

Plants for nitrogen fixation fertility: such as alder, suaeda, sesbania, milk vetch, azolla, etc.

Green landscaping, environmental protection plants: including all types of turf, street trees, ornamental flowers, potted plants, etc.

Monitoring and anti-pollution plants: such as suaeda that can monitor the content of mercury in the environment.

#### *5) Plant Germplasm*

In accordance with the genetics, each species has different genetic characteristics, which should be treated as a separate germplasm. The known plants in China now are nearly $3\times10^4$ kinds, which are also precious germplasm resources. All kinds of useful plants are vested in different classification units to families, genera, species, often have a large number of relative species, new quality and future generations can be produced through hybridization.

### 3.1.2 The Problems in the Development of Plant Resources

At present, the plant resources have an important value in food, medicine, energy and environmental protection, but the development and utilization of plant resources are still in its infancy, there are still many issues requiring urgent solution, which are described as follows:

1) At present, developing process of resource only stresses on single utilization of cellulose or single technology applications, lacking of pretreatment system technology integration and technical support of the research, which causes environmental pollution, waste of resources and economic problems of higher costs. For example, in the producing process of cellulose ethanol, it only stresses on the use of cellulose while other ingredients like hemicellulose, lignin, and protein are not utilized. Single utilization of cellulose for producing ethanol, not only causes the substantial increase of cost, but also causes the waste of resources and environmental pollution. The economic bottleneck has become a problem to be solved in cellulose ethanol industry.

2) The high cost and serious pollution of ingredients extracting process: the active ingredients in plants needs be extracted from the complex heterogeneous or homogeneous system, and then to remove impurities and refine through separation and purification. Some traditional separation technologies such as filtering, settlement, centrifuges, distillation, extraction, chromatography, crystal, absorption, molecular distillation, ultrafiltration, electrodialysis and reverse osmosis have made a great contribution to the extraction of plant active ingredients.

3) Low-level utilization of raw materials and low yield of products: in the extracting process of plant active ingredients, the lower utilization of raw materials resulted in lower product. To obtain a high yield of products, some extraction methods are used for refinery of raw materials, thus cause increasing waste and high cost.

The use of plant resources involves in multi-disciplinary, multi-sector, multi-level scientific and technical issues, which are difficult to solve the existing problems by any single aspect of research and application. Therefore, it is possible to achieve an economic use of resources by considering the characteristics of plant resources and the development of values on the whole, performing the multi-disciplinary integration, treating plant resources as organic whole, and integrating the research and development of each ingredients organically. The breakthrough of the existing problems in the use of plant resources will promote the use of active ingredients to the direction of safe and efficient, easy to control, low-cost and environmentally friendly.

## 3.2. Ideas and Methods for Fractionation of Raw Materials Component

### 3.2.1 The Necessary for Component Fractionation of Plant Raw Material

Plant resource has complex, uneven distribution, multi-level structure. Taken straw as an example, on the organ level rice straw includes leaves, leaf sheath, festival, rice stubble and roots; on organizational level, rice straw includes vascular organizations, parenchyma, skin and fibrous; on cell level, straw is divided into thin-walled cells, skin cells and duct cells.

1) The heterogeneity of straw. The chemical composition and fibre pattern are different. The fiber characteristics of some parts are superior to some broad-leaved wood fiber, indicating that these sites have high-value potential. The harvest straw is generally has different organs, the whole stalk contains various organs and tissues. 2) The difference in chemical composition. 3) The difference in fiber characteristics, small cell and hybrid components of straw is evidence.

The structural characteristic and composition of various straw biomasses are different from organs, tissues and cells, so the transformation methods are different. Taking corn stalks as an example, the skin of the stalk consists of two parts, the outer layer of the cortex, mainly epidermal cells lining the main fiber layer of skin containing cellulose, which is the raw materials of grass plant fiber and also the main source of paper. At the same time the skin contains the ash, therefore, removal of the corn stalk the outer skin is bound to greatly increase their fiber content, hence providing the basis for the application of cellulose. The core of corn stalks are vascular bundle which surrounded by a large number of thin-walled cells. The core department has relatively high content of the cellulose, hemicellulose and lignin, at the same time the existence of large amount thin-walled cell make the core structure much loose with highly absorbent capacity. So, this part is suitable for the fungus as some large-scale fermentation carrier, such as preparation of caramel. The leaf stalks of corn are consisted of epidermis and mesophyll and a large number of silicon cells in the skin, hence having high ash content. In mesophyll there have little vascular bundle being surrounded by mesophyll cells and therefore leaves the cellulose content is relatively low. At the same time the leaves can curl and open which is not only relative to the movement of cells, but also relative to its low lignified level. Therefore leaves compared with other parts have better palatability and much suitable for livestock feed. Node in corn stalks expands into a circle, the leaves connecting it, has high lignification level, so high of lignin content.

The interwoven structure of cellulose, hemicellulose and lignin determines that the degradation of any composition will inevitably be restricted by other components. For example, lignin hinders the enzymatic degradation of carbohydrates by cellulase and hemicellulase, so that some microbe can not degrade cellulose in raw materials. The chemical properties of straw are very stable, do not dissolve in water, nor dissolve in the general organic solvents. At room temperature, it can not be hydrolyzed in acid and alkali.

In addition, the high crystalline degree of cellulose is also an important factor for the difficult utilization of straw. Cellulose is a chain polymers adopted by β-1, 4- glucoside bond. Cellulose molecules are often gathered together, some of which are relatively tidy, regularity, showing a clear X-ray map, this part called crystallization. The relaxation region called amorphous area. The crystal structure of cellulose often hinders the degradation of cellulose.

The aim of pretreatment is that to destruct the entanglement of lignin and decrease the crystalline degree of cellulose. It is an important tool to achieve high-value application of lignocellulose. Pretreatment could make cellulose easily converted to fermentable sugar. The standard of effective pretreatment requires reducing size of biomass, retaining hemicellulose, decreasing the formation of fermentation inhibit product, reducing energy consumption and production cost, and so on. Pretreatment is the highest cost step in the process of cellulose ethanol, therefore there is still great space to improve technology to lower costs of pretreatment. The current pretreatment only emphasizes on single cellulose component and single technology, lacking of integrated pretreatment technology, thereby causing environmental pollution, waste of resources and economic problems of higher costs.

That study the heterogeneity of stalk in-depth and separation components according to the structure and characteristics is expected to find a breakthrough in straw transformation in high-value. Separation of components is developed on the basis of pre-treatment technology, and it is a concept that aimed at cellulose pretreatment technology only emphasized on single component, other components abandoned. Its purpose is the full use of biomass, the deepening and upgrading technology for pretreatment. Component separation is not limited to cellulose, but on the basis of the comprehensive utilization of biomass, to maximize high-value of each component in biomass [10].

To sum up, the structure and components make the direct and high value-added utilization much difficult, since that the three major components of cellulose, lignin and hemicellulose cross-linked closely together in original straw and different from each other on the structure. Hence to achieve high-value, it is necessary to apply appropriate pretreatment, graded separation, component separation, damage or change of the structure.

### 3.2.2. Key Issues of Plant Component Separation

Plant raw material is a complex organism, the chemical composition, internal structure and other characteristics are different in different types of plants, so component separation is a key step of raw material efficient using. Analyzing the principles and techniques of components separation in raw materials will provide theoretical and technical basis for the efficient use of plant resource [11].

In the pretreatment process of straw, there is still a bottleneck of safe, efficient and pollution. The efficiency of the pretreatment process, mainly based on the following considerations: 1) the removal of lignin and/or hemicellulose; 2) degradation and loss of cellulose; 3) enzymatic hydrolysis ratio of cellulose; 4) the amount of by-products formed. The key difficulty for biomass conversion is the high cost and pollution: 1) the pretreatment only emphasizes on cellulose, neglect the recovery of hemicellulose and lignin. 2) study on pretreatment technology is not sufficient, 3) new pollution are arisen [10].

To improve the efficiency and decrease the cost of biomass pretreatment, resolve the problems that exist in pretreatment from the principle, the following concepts have been made in the field of biomass conversion: 1) Biological refining: the chemical industry based on biomass must break the traditional concept of a single product, make the best use of the components, translate them into different products, consequently realize the full use of raw materials, highest value of products and most using efficiency of land. 2) Integrated biological processing: integrated biological processing was previously known as the direct microbial converse on, which completed the cellulase production, hydrolysis and fermentation in one step, the cellulase formation and ethanol fermentation are carried out in a bioreactor, thereby converting biomass into ethanol, so as to simplify the process and reduce costs. 3) Components separation and multi-layered transformation: Based on the concept of biomass whole utilization, we introduced ideas of components separation and multi-layered conversion, namely: through the combination of clean and highly efficient pretreatment technology, converting lignocellulose to various components, then process these components into high value-added products. [6]

Pretreatment research and practice shows that components separation is necessary for comprehensive utilization of biomass. Only stressed the use of a single component, cellulose technological development is greatly restricted. How to make use of all existing pretreatment technology to achieve the efficient use of lignocellulose? First of all, the pretreatment technology is feasible. Second, it should be energy conservation and pollution-free. Third, it should be economically viable. From the perspective of biomass full utilization, component separation technology could be evaluated as following [11]: 1) The purpose of separation and recovery of components is to be used, it is not necessary to achieve complete components separation; 2) Reduce the interaction between the various components, reduce enzyme invalid absorption and chemical usage; 3) Effective and economic recovery of all components, 4) Clean production; 5) Low operating and investment costs.

### 3.2.3 An Example for Separation of Plant-Based Biomass Components

Giant Napier (*Pennisetum sinese* Roxb), a high yield grass, is introduced to China from Colombia, belonging to Angiosperms, Monocotyledon, *Pennisetum* genus. It is a hybrid from

*Elephant Grass* and *Pennisetum americanum*. Giant Napier is an erect perennial plant with the high of 4-5 metres and internode perimeter of 5–9 cm. In China, Guangdong and Guangxi provinces are the perfect region for Giant Napier growth with annual harvest of 4 to 6 times and annual yield of up to 375 tons/hectare. The main components of Giant Napier are cellulose, hemicellulose and lignin with the contents of 50%, 20% and 20% respectively, and also 3% of ash. Giant Napier is an excellent species for producing plant-based resource because it can grow fast and has high biomass yield. Ethanol, xylo-oligosaccharide and other important fuel and chemicals can be produced from Giant Napier biomass by fermentation and the fermentation residue can be burnt for electricity generation with components separation and cascade use methods. The total utilization model of Giant Napier is shown in Fig 3.1.

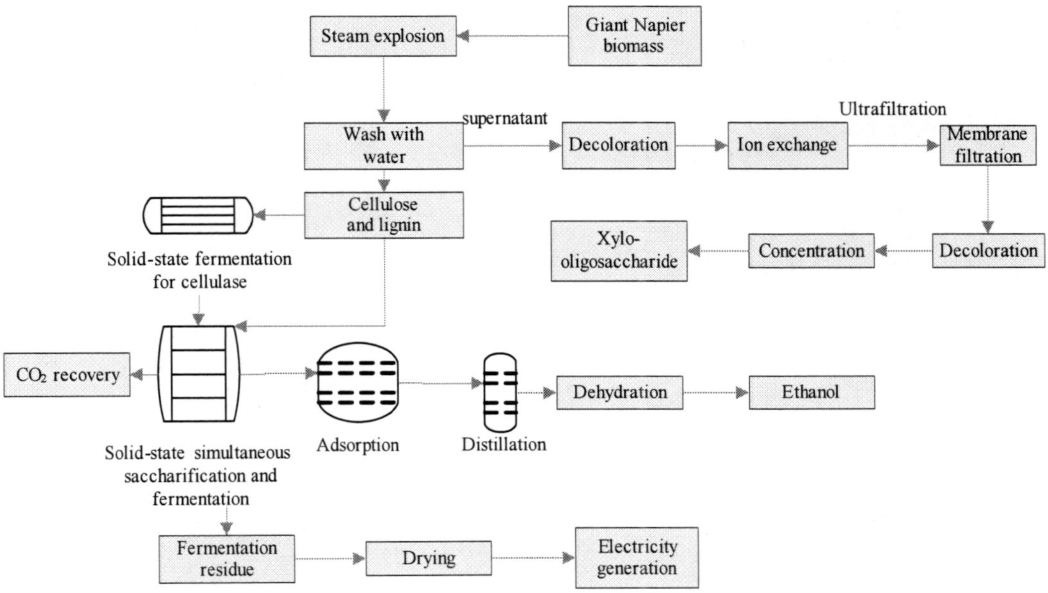


Figure 3.1. Total utilization model of Giant Napier

## 3.3 Principle of Components Separation for Plant-Based Materials

### 3.3.1. Components Separation Technology for Lignocellulose

Now commonly used methods for components separation include physical, chemical, biological and integrated method [12, 13].

#### *3.3.1.1 Physical Method*

*1) Mechanical Smash*

The chemical composition and physical characteristics of plant vary considerably. Cell toughness is related to the content of cellulose, hemicellulose and lignin, different plant usually apply different approach to smash. Commonly used smash machine includes ball milling, compressed milling, dual-roller crushing machine, fluid momentum grinding machine, and frozen grinding machine. Freeze grinding method is to smash the sample at -100 °C, which can retain lignin after crushing but destroy the combination of lignin and hemicellulose. The disadvantage of mechanical smash is energy consumption, high cost of grinding.

*2) High-Energy Radiation Treatment*

High-dose radiation treatment can reduce neutral detergent cellulose, acid detergent fiber, acid insoluble lignin and reducing sugar content in the cell wall, thereby enhancing the digestibility of straw. Low dose radiation can be used for the sterilization disinfection of agricultural by-product. As for cellulose, hemicellulose and lignin degradation, complex treatment with radiation and chemical methods is better than applied them alone. It can get same conclusion that whether using high-dose radiation and low concentrations of chemical reagents or low radiation doses and high concentrations of chemical reagents to deal with agricultural waste.

*3) Microwave*

Microwave is electromagnetic waves with frequency of 300 MHz to 300 GHz. Microwave treatments can change the hydrogen bond of cellulose and increase the availability of cellulose. Microwave has the advantages like short processing time and simple operation. The effect of microwave heating is superior to conventional heat treatment.

*4) Supercritical $CO_2$ Processing*

Supercritical $CO_2$ has the advantages of economy, clean, environmental friendliness and easy recovery, which is frequently used as solvent extraction in recent years. Study found that the enzymatic sugar yield increase when lignocellulose treated with SC-$CO_2$.

*5) Supercritical Water Treatment*

Cellulose will be degraded in the supercritical water ($P > 21.8$MPa, $T > 374$ °C), the main product are dihydroxy acetone, fructose, glucose and oligosaccharides. In supercritical water system, cellulose was firstly broken into oligosaccharides and glucose, glucose through convert to fructose through isomerization.

*6) Steam Explosion*

Steam explosion is the widely used method for lignocellulose pretreatment. The impact of steam explosion on straw structure includes cooking effect of high-temperature and the mechanical instant decompression. Lignin is melted and hemicellulose is degraded in high-temperature cooking process. Straw was torn into small fragments after explosion. After this process, the surface area and porosity of straw are greatly increased, hydrolysis rate markedly improved [14].

The advantages of steam explosion are as follows: 1) it can be applied to all kinds of plant biomass and easily controlled. 2) hemi-cellulose, lignin and cellulose can be separated in three different processes, i.e. water-soluble components, alkaline solution components and alkali insoluble components. 3) enzymatic conversion rate of cellulose can reach theoretical maximum. 4) steam-exploded lignin can also be converted into other chemical products. 5) hemicellulose can be converted into liquid fuels. 6) the fermentation inhibitors produced during steam explosion can be controlled with different conditions.

### *3.3.1.2. Chemical Method*

*1) Acid Pretreatment*

Acid pretreatment gained much attention of researchers for a long time. Such as using lignocellulose to produce sugar-formaldehyde with dilute sulfuric acid treatment, which hydrolyzes hemicellulose to xylose and other sugar be dilute mainly by sulfuric acid and then convert xylose into sugar-formaldehyde. Adding sulfuric acid was used to remove hemicellulose in lignocellulose pretreatment. Sulfuric acid can remove hemicellulose and improve the digestion of cellulose in residual solids, nitric acid, and hydrochloric acid. Phosphoric acid is also available in pretreatment study. Dilute acid hydrolysis has been successfully used for pretreatment of lignocellulose, and access to higher sugar yield.

Concentrated acid can also be used to pretreat lignocellulose, but the acid is toxic and corrosive, which needs acid-resistant equipment. Acid recycling is needed to minimize the pollution of the environment in concentrated acid pretreatment, thus increase the cost of production. The remaining acid must be recovered in the acid pretreatment for the follow-up hydrolysis and fermentation, which will produce some substances inhibiting fermentation, and removing the salt is also needed, the cost will increase.

*2) Solvent Treatment*

The organic solvents for cellulose pretreatment can be divided into organic solvent and inorganic solvent. Solvent processing can change the crystals structure of cellulose, thus hydrolysis speed and extent of hydrolysis are increased substantially. Recently, a new type of solvent - ionic liquids was invented. Ionic liquid, the liquid completely composed of ions, is a liquid salt at low-temperature (<100 °C), also known as low-temperature molten salts, which are generally formed with organic and inorganic anion cation. Compared with traditional organic solvents, ionic liquid has many prominent advantages: (1) Almost no vapor pressure, non-volatile, the pollution of volatile organic compounds is eliminated, colorless, and odorless. (2) The stabile temperature range is wilder (less than or close to room temperature to 300 °C), better chemical stability. (3) The solubility could be regulated through the designation of positive and negative ion. Because of the special nature and performance, ionic liquids, supercritical $CO_2$ and two-water state constitute three green solvent. With the proposed concept of green chemistry, the study of ionic liquid is conducted widely in the world which focused on the synthesis of new ionic liquid, the physical and chemical characteristics, and as electrolyte and solvents, and other aspects of the applied research.

*3) Alkali Pretreatment*

Some alkali can be used to lignocellulose pretreatment, of which the effect depends mainly on the content of lignin. Alkali pretreatment is based on the saponification ester of hemicellulose and other elements between the internal components, the porosity of lignocellulose increased with the reduction ester. NaOH has a stronger role in the delignification which markedly improves the hydrolytic rate of cellulose. Research shows that NaOH pretreatment has great impact on the chemical composition of lignocellulose, the cellulose material bulged after the pretreatment, the crystallization index of cellulose decreased, cellulose crystallization area is damaged, and hydrolysis is easier. In addition, calcium hydroxide and ammonia can also be used to remove lignin. Although the effects of alkali pretreatment is better, but some hemicellulose is broken down and damaged in the process. There are also the recovery of reagents, washing and other issues exit.

*4) Wet Oxidation Pretreatment*

Wet oxidation was invented in the 1980s, water and oxygen to participate in the common reaction under pressure and heating. Lignin is degradated by peroxidase with the presence of oxygen and water, increasing the hydrolytic sensitivity of materials.

*5) Ozone Pretreatment*

Ozone can be used to decompose the lignin and hemicellulose of lignocellulose. After ozone pretreatment , large degree of lignin can be degraded while hemicellulose is only minorly attacked, and cellulose is almost unaffected. The advantages of this method are that the lignin can be removed effectively which results in that there are no inhibitor produced for the latter fermentation. The reaction can be carried out under normal pressure and temperature. However, large amount of ozone is needed and the cost of process is high [14].

#### *3.3.1.3. Biological Pretreatment*

There are many micro-organisms could degrade lignin in nature. Microbial treatment is to remove lignin in lignocellulose with microorganisms. White-rot fungus and laetiporus sulphureus are often used to degrade lignin and hemicellulose, white-rot fungus is the most effective microorganism. The conditions of biological treatment are relatively moderate, and the method has advantages of less inhibit components produced, energy saving, and environmental protection. However, due to the lower enzyme activity of microbial for lignin decomposition, the processing cycle is very long, usually taking a few weeks, therefore, there is a long way for the method come into application. From a view of cost and equipment, micro-pretreatment shows unique advantage, the lignin can be degraded with enzyme, the decomposition of lignin can improve digestibility of lignin. But currently, the method is still remaining in the experimental stage, using genetic engineering technology to improve quality of white rot fungi will help expand the application of pretreatment.

#### *3.3.1.4. Comprehensive Method*

*1) Ammonia Explosion Pretreatment*

Ammonia explosion is a promising pretreatment technology. Using liquid ammonia in a relatively low pressure and temperature (60-100 °C) of raw materials will be handled a certain period of time, and then a sudden release of pressure by the bursting of raw materials, including liquid ammonia can be recovered for reuse, lower energy consumption of the whole process. We found that ammoniation steam explosion could degrade the hemicellulose in corn stalks, the enzymatic hydrolysis rate of cellulose increased 42.92%, and the organic nitrogen content was increased 1.27 times. Ammonia steam explosion will not produce microbe inhibition materials and can remove lignin. Most of the hemicellulose and cellulose are retained and can be best used, but the effective recovery of the ammonia the problem that must be properly resolved [15].

*2) $CO_2$ Explosion Pretreatment*

As ammonia and steam explosion pretreatment, $CO_2$ explosion pretreatment is also a pretreatment method for wood materials, which is just different in this approach that $CO_2$ must be converted to carbonate to increase the rate of hydrolysis.

*3) Steam Explosion and Carding*

The separation technology coupled with steam explosion and machinery carding technology is an invention patent invented by us. According to the structure and component characteristics of raw materials, the two technologies are combined together. Taking corn stalks as an example, the details of steam explosion and mechanical are as follows. Steam explosion processing was first applied, then separate fibrous tissue by mechanical means carding. Steam explosion is the main factor for successful pretreatment, of which the conditions are different from different parts of corn stalks. At the same time, the water content of materials is also an important factor for carding [16].

*4) Steam Explosion and Ethanol-Extraction Pretreatment*

We also invented the component separation technology coupled steam explosion with ethanol pretreatment. The structure of straw is complex, cellulose, hemicellulose and lignin linked with each other to form dense structure. Generally a single technology can not achieve rapid separation of the three components. In steam explosion pretreatment, hemicellulose can be effectively recycled, but lignin still remains in solid compound [16]. Ethanol can extract lignin effectively, but could cause hemicellulose loss. Therefore, the two technologies can be combined for biomass pretreatment [7].

*5) Steam Explosion for Extraction of Bast Fiber Crops Functional Fractions*

Bast fiber crops (BFC) are annual, herb and dioecism, belonging to Moraceae. China has the most abundant BFC in the world with about twenty species and more than 6500 varieties such as Hemp, Flax, Ramie, Jute, Sisal and Apocynum. The plants of BFC were once important crops of fibre, oil, and grain. The most important traditional use of the BFC is their fibre for producing rope, bag and cloth. With the development of the health care, the medicinal value of the BFC is exploited.

Apocynum fibre has medicinal value, the root, stem and leaf of it can all be used as medicine. It is used in clinical treatment for high blood pressure, high cholesterol, heart disease, asthma and chronic heart failure because it contains flavones, alkaloids, amino acids, cardiac glycosides and multi-trace elements. Apocynum is calling 'The king of wildness fibre' in China because it has not only the characteristic of fibre but also the medicinal value and antibacterial property.

Extraction with organic solvent is the most common method for obtaining the functional fractions from BFC. Traditional extraction and refining method had various steps such as solvent cooking, purification, refining et al. So much steps result in high energy consumption and complex techniques. The main components of BFC are cellulose, hemicellulose and lignin. Steam explosion pretreatment is an efficient method for increasing the yields of the functional fractions, because after steam explosion the cell wall of the plant was broken hence the compounds in the cell are easily to be released.

The hemp stem was chopped in 5cm and dripped in water for 1 h then put into the steam explosion reactor remaining for 5 min at 1.5 MPa, finally with a sudden discharge. After such treatment the steam exploded biomass was washed with water and separated to solid and liquid sections. Flavones and xylo-oilgosaccharide can be obtained from the liquid section after ion exchange, membrane filtration and micro-filtration. Compared to the hemp stem without steam explosion, the extraction yields of flavones and xylo-oilgosaccharide were increased by 60-95% and 3-8% respectively. We concluded that steam explosion pretreatment can decrease the cost of extracting the functional compounds from the BFC.

*6) Lignocellulosic Biomass Fractionation with Glycerol Pretreatment*

Lignin of the lignocellulosic biomass can be dissolved or hydrolyzed when treated in organic solvent at appropriate temperature and press. This is an effective method for separating lignin from lignocellulose with several advantages. This method can be widely applied for lignocellulosic biomass, further more the solvent can be recovered, therefore saving the cost of treatment.

The organic solvent treatment method has attracted much attention and has demonstrated the potential utilization in lignocellulosic biomass. Glycerol, as a high-boiling-point organic solvent, is presently the main byproduct of oleochemicals industry, as high as 10% of the total bio-diesel output. Since the soaring petroleum price has made the oleochemicals industry, especially biodiesel production, more flourishing, glycerol products are driven rapidly to oversupply and have created a drastic glut on the market. And the cheap glycerol, whose price even declined to about 10 cents per pound, has slaughtered on the whole EU and North America market. Therefore, utilization of glycerol to treat the lignocellulosic biomass may escape the collapse of glycerol price and further defray the cost of biodiesel production. We have performed a series of work on lignocellulosic biomass fractionation using glycerol treatment [17-20].

With preliminary optimization, the atmospheric glycerol autocatalytic organosolv pretreatment (AGAOP) of wheat straw was carried out at 240 °C for 4 h with the glycerol addition of 15 g/g dry feedstock, resulting in high cellulose recovery (95%) and removal of hemicellulose (90%) and lignin (>70%). Meantime, 48 h enzymatic hydrolysis yield of the wet pretreated fiber was high up to 90%. With the initial liquid-to-solid ratio of 20 g aqueous glycerol/g dry wheat straw and at 220 °C for 3 h, the AGAOP enabled wheat straw to remove 70% hemicelluloses and 65% lignin, with 98% cellulose retention. After pretreatment, 48 h enzymatic hydrolysis yield of the wet fiber achieved 90% (Fig. 3.2). When crude glycerol from biodiesel and oleochemical industry was directly used as pretreatment solvent in AGAOP, lipophilic compounds contained formed '*pitch deposition*' on the pretreated fiber. The pitch deposition led to '*lignin passivation*' which adverse to lignin removal, resulting in low delignification (app. 30%). Another striking finding was that some residual lignin deposited on the pretreated fiber in form of the lignin recondensation and/or lignin–carbohydrate complex (LCC). The deposition was aggravated due to fiber hornification occurring when the fiber was drying, which helped to explain why dried pretreated wheat straw had a low enzymatic hydrolysis yield, above 10% lower than that of the wet substrate. The crude glycerol was applicable for AGAOP, but it was advisable to remove lipophilic compounds from crude glycerol before its utilization.

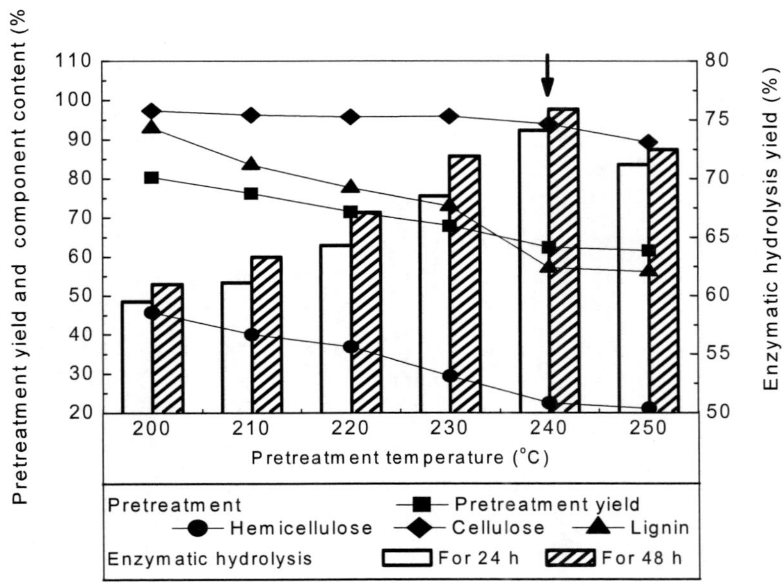


Figure 3.2. Effect of different pretreatment temperature on enzymatic hydrolysis (44 FPU g-1 dry substrate) of wet fiber from glycerol pretreated (20 g glycerol / g dry wheat straw, 2 h) wheat straw [20]

### 3.3.2. Component Separation Methods for Medicinal Plants Resources

#### *3.3.2.1. Traditional Methods*

*1) Solvent Extraction*

The principle of solvent extraction is as follows, according to the solvent character of Chinese herbal medicine, appropriate solvent was selected for active ingredients separation. When the solvent was added to Chinese herbal medicine raw materials, solvent will permeate into the biomass.

The soluble substances then infiltrate into the cell, and the concentration difference of within and outside the cells is formed, the solution with high concentration will come outward, solvents then come to the cell. With the fresh solvent added, all of the active ingredients could be obtained.

a) Soakage method: The method is putting powder or fragments of Chinese herbal medicine to appropriate containers, and adding appropriate solvent (such as ethanol, dilute alcohol or water) to dissolve the components. The method is simple, but the extraction rate is poor. When taking water as solvent, the liquid is easy to degenerate, therefore appropriate preservatives should be added.

b) Seeping method: Chinese herbal medicine was first installed in seeping utensil, then new solvent is added constantly to make it infiltrate into medicine, the liquid was leached out from top to bottom. When the solvent containing higher and higher proportion of extracted powder composition, it would move downward. The top of the solution would dilute it and replace its location, resulting in a good concentration difference, so that the spread can be better than leaching impregnation. However, the flow rate should be controlled, new solvent showed be added timely in the process. When the colors seepage drops become light or the volume of fluid seepage equivalent 10 times of original ingredients, the extraction can be treated as basically completed.

c) Boiling method: Boiling method is a traditional method for leaching first used in China. The general used containers include pottery, copper, ceramic, which not suitable for iron container for fear of decoloration. When heating directly, stirring is usually needed to avoid local high temperature. Steam-heated methods are also applied in many factories. In some factories, several containers are interconnected through the pipeline to improve efficiency.

d) Circumfluence extraction: When using organic solvent for extraction through heating, circumfluence heating device is usually applied to avoid loss of volatile solvents. The efficiency of this method is higher than cold soaking method.

e) Continuously dynamic extraction: when volatile organic solvent is used for effective extraction of components from Chinese herbal medicine, which prefer to continuous extraction method that fully extracted with less-solvent. Fat extracting utensil is often used for component extraction in laboratory. Complete extraction normally need hours for continuous extraction. As the heating time is longer, some compositions are not stable in higher temperature which limits the application of this method. Continuous extraction is often applied in mass production.

*2) Steam Distillation*

Steam distillation method is suitable for some composition of Chinese herbal medicine extracts, which can not be destroyed in steam distillation. The boiling point of such components is usually higher than 100 °C, and the vapor pressure at 100 °C is present. When heated together with water, the liquid will boiling when the sum of vapor and water vapor pressure is one atmospheric pressure, volatile substances will be brought out with vapor. Some volatile components with good solubility in the water, the distilled liquid is often re-distillated, the volatile oil is separated firstly in distillation.

*3) Sublimation*

Solid material is directly heated and gasified, then cooled and met for the solidification of solid compounds, which is known as sublimation. Some elements of Chinese herbal medicine have the nature of sublimation, which can be extracted using this method. For example, the camphor can be extracted from camphor wood using the sublimation method, which was detailedly recorded in the "Compendium of Materia Medica", the world's first application of the sublimation. Although this method is simple, there have often volatile tar bar adhering to sublimation, and is difficult to remove. Secondly, not entirely sublimated, low yield, sometimes accompanied by decomposition phenomenon.

Traditional sublimation method has some shortages such as loss of active ingredients, long cycle and complex processes. On the other hand, the crushing level of raw materials, the extraction time, extraction temperature, equipment conditions and other factors can also affect the efficiency of extraction, which must be taken into account. Therefore, the traditional extracting methods are unable to meet developing needs. We should be strengthened to new extraction technology research to the Chinese medicine industry of China's development by leaps and bounds.

#### *3.3.2.2 New Technologies and Methods*

*1) Supercritical Fluid Extraction Technology*

Supercritical fluid extraction is a new technology using supercritical fluid to replace conventional organic solvents for extracting the active ingredients of Chinese herbal medicine. Its principle is that the liquid (solvent) in a region near the critical point (supercritical District) has different abnormal behavior and transmission performance from the solute and made the dissolution of the solute change in a wide range according to change of pressure and temperature. Using supercritical fluids as solvent, some components can be extracted and separated from a variety of solid or liquid mixture. $CO_2$ is commonly used for supercritical fluid, because $CO_2$ has the advantages like non-toxic, non-explosive, cheap and low critical pressure and temperature, which is easy to separated from the mixture safely. Compared with conventional extraction methods, the biggest advantage of supercritical $CO_2$ extraction method is performing the extraction under the conditions of close normal temperature.

*2) Membrane Technology*

The principle of membrane separation technology is application of pressure as driving force for the separation of solute and solvent under normal temperature without phase transition. The use of membrane technology, such as micro-filtration, ultrafiltration, nanofiltration and reverse osmosis, etc, can achieve separation in the original biological material system, also can accumulate products and remove impurities efficiently.

*3) Ultrasonic Extraction Technology*

The principle of ultrasonic extraction technology is that the ultrasonic and air speed up the role of the dissolution of active ingredients. On the other and the secondary effects of ultrasound, such as mechanical vibrations, emulsion, spreading, crushing, chemical effects, can also accelerate the pervasion of components and fully mixture with the solvent. The technology has merits such as extraction in a short time, high yields, without heating, extraction at low temperature protecting the active ingredients. Ultrasonic extraction technology can avoid the damage of active ingredients causing by high temperature and high pressure, but requires thick wall container vessel and should be place highly to achieve perfect medicine extraction. Research and experiments are now in a very small scale. The scale-up problem of engineering equipment would be solved for industrialization.

*4) Microwave Extraction Technology*

Microwave extraction is a latest development new technology using microwave energy to increase the extraction rate and. The principle of the microwave extraction technology is that the different absorbability of microwave makes the substrate heated selectively. So that material was extracted from the matrix system and separate. The microwave extraction technology has some advantages, such as with simple equipment, application of a wide range of extraction efficiency, reproducible, save time, save reagents and little pollution. At present, except for using in environmental sample pretreatment, it was also used for biological and chemical, food, industry analysis and extraction of natural products, and other fields.

*5) High-Speed Counter-Current Chromatography Extraction Technology*

High-speed counter-current chromatography (HSCCC) extraction technology is the distribution of liquid chromatography without any carrier or solid support, having the advantages of high separation efficiency, high purity products, without absorption and contamination of samples from the carrier, a large quantity of simple can be deal with and solvent consumption less, especially fits for the extraction of preparation.

*6) Enzymatic Technology*

Most of the impurities in Chinese medicine preparations are the starch, pectin, protein, etc. The appropriate choice of enzyme can remove impurities. Enzyme reaction will decompose plant tissue moderately and can greatly improve the yield, which is one of the efficient methods for achieving maximum yield from plants. At present, the most research of extracting active components from Chinese medicine focus on cellulase, since that the cell wall of majority Chinese herbal medicines is consisted of cellulose. Plant's active ingredients are often wrapped in the cell wall. The cellulase can break β-D-glucose bond so that the plant cell wall destruction for help to the effective components of extraction. Application of enzyme can increase the effective elements of Chinese herbal medicines in the extraction rate, but the concentration of enzyme and substrate, reaction temperature, acidity, inhibitors, activator, and so on, all need further study.

*7) Semi-Bionic Extraction*

Semi-bionic extraction method is the consistency of overall drug research and molecular medicine research. From the perspective of biological pharmacy, Semi-bionic extraction method provides a new extraction process for gastrointestinal medicine formulations, which simulates the principle of oral administration and the gastrointestinal drug transshipment It refers that the drug is first extracted by acid Hydrolysis with certain pH, then extracted by a certain pH of alkali solution, finally the extracted liquid is filtrated and concentrated to make preparations. By unifying the analytic thinking and systematic thinking, it can form a new thinking that in traditional Chinese medicine extraction of adhering to "a component of, not only elements of it and by the body's response to pharmacodynamics." "Semi-bionic extraction" of Chinese medicine practitioners can use a combination of clinical features, with oral administration of gastrointestinal transit by the principle of absorption. But this method still follows the traditional high-temperature decocting, which will destroy many effective active ingredients and reduce medicinal effects under long time and high temperature decoction.

*8) Broken Extraction*

Broken method is to achieve component extraction by broken plant material through appropriate solvent extraction. Although the broken extraction method has the advantages such as simple operation, avoiding of high temperatures heating and short extraction time, the extracting rate is not high. On the other hand, it is still confined to experimental research which should be further studies for the large-scale industrialization.

*9) Air Explosion*

The principle of air explosion is similar to the steam explosion method. It use the powerful force from the compression air suddenly released after decompression to break the plant cell wall and tear plant tissue, so that the osteoporosis structure of medicine benefit the solvent infiltrating internal and increase the contact area, at the same time conducive to the movement and transportation of solvent particles in internal medicine. Air explosion has been used in the pulp and paper industry. Also it applies to the plant roots, stems, bark, leaves and other multi-fiber ingredients, but it is not appropriate for the short fiber and high-starch-containing medicines.

*10) Macroporous Resin Adsorption Separation*

Principle of macroporous resin adsorption separation technology is that special absorbent, combining it absorptivity and selectivity, selectively absorbs the active ingredients from the decocting Chinese herbal compound and remove invalid components. The process has the advantages of short time, high-yield, high purity, renewing resin easily and so on. At present still a few species of adsorption resin are reported, actually only two polar and non-polar types, that can not meet the need of extracting active ingredients from traditional Chinese medicine.

*11) Two-Phase Water Extraction Technology*

Application principle of two-phase water extraction technology is using the distribution difference of the materials in two different systems to separate components. Compared with the traditional water-cook and alcohol-deposition, two-phase of two-phase water extraction is mostly water, so that interface tension of two-phase is very small, which provided a suitable environment with many advantages such as fast mass transfer, convenient operation, time saving, mild conditions, easily scale-up and continuous operation.

It is indicated by researches that new technology and new methods used in extraction of Chinese medicine really possesses the virtues such as high extracting rate, less loss of active ingredients and short cycle, which has shown broad prospects of them in industrial production in the preparation of Chinese medicine. At present, we should strengthen these researches of new technologies and new methods in industrial production, so as to improve the quality of traditional Chinese medicine.

## 3.4. Principle and Technics of Active Components Separation from Plant Cell Wall

### 3.4.1 Extraction and Separation Principle for Effective Component

Structure and composition of plant cell wall is a complex three-dimensional structure, in addition to cellulose, hemicellulose and lignin, and other major components, also has protein, ferulic acid, fat, SiO2, and other ingredients. The cell wall of some Chinese herbal medicine plants is also rich in active ingredients. Therefore, besides taking the advantage of the main component of plant cell wall, the full development of other components of the low content of value is to raise the value of plant resources of the effective means.

Because the structure of plant cell wall is dense, the major components of the cell wall is combined with protein, ferulic acid and other ingredients by hydrogen bonds, hydrophobic, ionic bond, and so on. Therefore, it is difficult to extract various active ingredients by conventional methods. For example, the effective extraction of protein usually needs high-strength salt solution and strict regulation of the pH, or elutes various protein by high concentration surfactant. It is the key for total utilization of plant cell wall that separates and extracts each component through the combination of physical and chemical methods According to the structure and composition of their own characteristics and the characteristics of the various components.

### 3.4.2 Ferulic Acid Extraction and Separation Process

Ferulic acid, named as 4-hydroxy-3-methoxy-cinnamate, is a common existing phenolic acid with a molecular weight of 194.19 and melting point of 174 °C. Other properties of ferulic acid are cold water-soluble, insoluble in water, soluble in methanol, ethanol, and acetone, difficult to dissolve benzene, petroleum ether. Ferulic acid is sensitive to light and heat that isomers conversion or degradation would happen if illuminated in light for a long time, so it always be kept in dark. Ferulic acid has many unique features, and low toxicity. Ferulic acid is the active ingredients of many Chinese herbal medicines, such as angelica, Chuanxiong, a natural anti-oxidants, it also has many medical effects, such as anti-thrombosis, hypolipidemic, prevention and treatment of coronary heart disease, anti-inflammatory anti-bacterial, anti-mutagenic and anti-cancer effects, such as, In food, medicine, and other industries in a wide range of applications.

As oxidation resistance, ferulic acid has been widely used in the food industry as antioxidants. The United States and Japan have ratified it as a safe food additive, applicable to the processing of meat and pasta products. As ferulic acid can make the cross-linking between protein with polysaccharides and polysaccharides with the polysaccharide, it could be used for food crosslinker. Ferulic acid with UV absorption, anti-oxidation and a whitening effect, has widely used in the cosmetic industry. Ferulic acid emulsion performs antioxidant activity, which can cure skin aging and skin diseases, including multi-light skin is useful, without stimulus [21, 22].

Ferulic acid widely exist in various plants in particular herbs, such as angelica, Chuanxiong, etc., as a Chinese herbal medicine in the active ingredients. Ferulic acid in the plant rarely exists in free state, but mainly as one part of cell wall by cross-linking with the polysaccharides and lignin. In straws, ferulic acid exists in the cell wall by linking with lignin and hemicellulose with ether or ester bonds. Among them, there has part of the two elements of ferulic acid reduction of ferulic acid dimer in the form of connection between the lignin and hemicellulose.

#### *1) Solvent Separation*

Solvents for the separation of ferulic acid include petroleum ether, ethyl ether, ethyl acetate, acetone, alcohol, chloroform, n-hexane, methyl alcohol, mixed solvents (n-hexane:ether:methanol=5:3:2). The extracting effect of different solvent on ferulic acid in Chuanxiong show that methanol has the highest extracting rates, and ethanol and chloroform followed. In addition to methanol ultrasound extraction and supercritical extraction method can also extract Chuanxiong of ferulic acid, supercritical extraction ferulic acid extraction rate higher than the methanol solvent extraction and ultrasonic extraction.

### *2) Enzymatic Separation*

Enzymatic extraction of ferulic acid hydrolysis taking biological hyddrolysis as its core technology, which compared with chemical method has the advantages as follows: mild reactive conditions, relatively small toxicity and less pollution to the environment, short producing cycle, relatively simple processing. Feruloyl esterase was separated and purified recently, which is an extracellular enzyme falling into sub-category of carboxylate hydrolase. Ferulic esterase can hydrolyse polysaccharide - ferulic ester into ferulic acid and polysaccharide.

The structure of feruloyl esterase is very complex and various from different origin. Fig. 3.3 shows the three-dimensional structure of the Xyn10B ferulate esterase module. Most of the feruloyl esterases have catalytic triad of Ser-His-Asp (Fig.3.4). But the feruloyl esterase from *Mycobacterium tuberculosis* has catalytic triad of Ser-Glu-His.

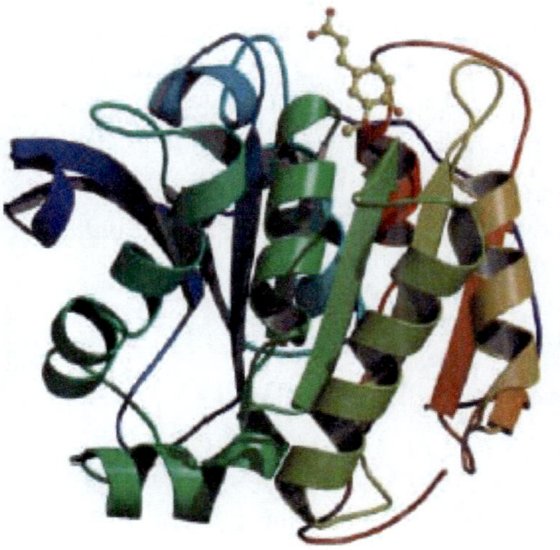

Figure 3.3. Three-dimensional structure of FAE Xyn10B [23]

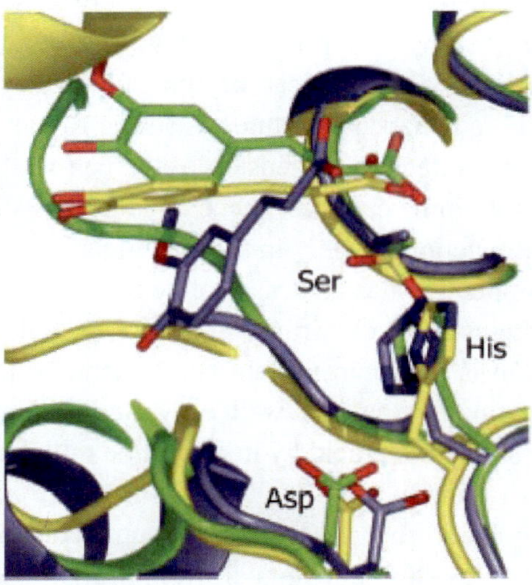


Figure 3.4. Three different active centres for FAE [24]

Feruloyl esterases can be produced by fungal, bacteria and yeast. The feruloyl esterases from different microbes show different optimum pH, even the feruloyl esterases from the same microbe in different fermentation conditions have different emzymatic characteristics.

Although feruloyl esterases can accelerate the release of ferulic acid from the cell wall, the yield of feruloyl esterases does not achieve the requirement of commercialization and should perform advanced researches about the function of these enzyme. The extract yield of feruloyl acid from plant cell wall by feruloyl esterases are essentially influenced by the pretreatment method for the substrate. So finding out an effective pretreatment method is very important to produce feruloyl acid from plant biomass by feruloyl esterases.

We had performed a series of work on extracting feruloyl acid from straw by feruloyl esterases. The effects of steam explosion pretreatment and feruloyl esterases cooperating with cellulase on the yield of feruloyl acid from straw were investigated. The releasing ferulic acid were 0.17 mg/g, 1.01 mg/g, 1.17 mg/g after enzymatic extraction by using cellulose, feruloyl esterases and mixed enzyme of the two enzymes, which showed that the double-enzymatic method was better than single-enzymatic method and feruloyl esterases was the key factor in enzymatic extraction of ferulic acid. For the pretreatment methods, steam-exploded was higher-efficient than smash. When using double-enzymatic method, the quantity of ferulic acid releasing from steam exploded rice straw was 1.17 mg/g, while it was 0.027 mg/g(<60sieve mesh), 0.069 mg/g(60-100sieve mesh) and 0.087 mg/g(>100sieve mesh) after smash.

### 3.4.3 Separation of Protein in the Cell Wall

***Existence State of Protein in Plant Cell Wall***

Straw cell wall contains about 2-8% of the protein. Researchers began the research on cell wall protein as early as 1924. According to its physiological function, the cell wall proteins can be broadly classified into three categories: structural proteins, enzymes and those have not yet been fully identified proteins. Plant cell wall, the most important structural protein for the extension of category [25], but also the most extensive and in-depth study of a class of proteins, is a kind of hydroxyproline-rich glycoprotein, and its peptide matrix contains a unique hydroxyproline-rich repeat of the unit. Most of hydroxyproline were glycosylated to form a stable rod-shaped structure, its strong role in the cell wall is protecting of cells free from the pathogen infection. Extension of the cell wall structure as a component in the defense and plays a role in disease resistance and stress resistance. Disease resistance and resilience of its defense functions mainly in the development, mechanical damage, fungal infections, plant anti-toxin-induced, such as heat treatment.

Since the cell wall proteins were confirmed in 1960, the cell wall research has become increasingly active. Expansin is the main protein component, which is a kind of sugar-rich hydroxyproline protein existing in the cell wall of various plants. Protein-rich stretch of hydroxyproline, the core protein amino acid hydroxyproline in the composition of the general content of the total number of amino acids is 30-40% (in moore). Non-sugar-protein part of the composition is mainly arab and sugar galactose, the sugar content is about 60%.

Enzyme is another protein existing in plant cell wall, nearly 30 kinds of protein have been found, mostly belonging to hydrolysis enzymes or oxidoreductases. Study found that cell growth is closely related to the activity of expansin. Protein on the cell wall of the gene expression and regulation also need further in-depth study.

***Cell Wall of Separation and Purification Technology***

*1) The Way without Damage of the Cell Wall*

Extracting protein from the plant cell suspension is one of the strategies to reduce pollution of protein from the protoplast. The method maintains the integrity of the cell, at the same time, through a specially designed solvent sequentially extracts the lipid membrane protein from the integrity of cell suspension, as much as two thirds of the plant protein can be extracted by this method.

*2) Way of Destroying the Cell Wall*

The major route for protein extraction is destruction of plant cell wall. This specific method is firstly to mix tissues homogeneously in a low salt buffer, and then using a series of low-salt buffer to remove intracellular protein pollution, finally use high concentration of salt to dissolve the protein from ions. "Grinding and separation" means is very effective for many types of protein extraction, especially those with the extracellular matrix in close connection with the protein.

As the pectin poly-galacturonic acid in cell wall is a component polyanionic matrix, so once the cell wall rupture, the positively charged protein in the cell wall will be combined, resulting in pollution of extracted protein. In some actual operation, intracellular protein can be very solidly combined with the cell wall, so high-salt buffer can not be dissociated the interactions between them. Some detergents such as the sodium dodecylsulphate (SDS) can be re-dissolved with the protein in the cell wall.

## 3.4.4 Pectin

***Pectin Components***

Pectin is a polymer of cell wall of plant roots, stems, leaves and fruit. The relative molecular weight of them is between $5\times10^5$~$3\times10^6$. It is generally believed that the D-D-galacturonic acids through α-1,4 glucoside bond links into the main chain and the side chain of miscellaneous polysaccharide is consisted of sugars including the D-galactose, L-arab sugar, D-sorbose, L-rhamnose, etc. and the main chain of carboxyl methyl usually exists as methyl esterification. Pectin material is a pure white or pale yellow powder and has some special smell, 20 times weight of the water can almost completely dissolve it and forming a negatively charged viscous colloidal solution, but it can not be dissolved in ether, acetone and other organic solvents. The solubility of pectin is relative to its molecular structure, decreases with the increase of the length of chain, and increases with the increase of esterification degree, because the pectin chain is not a straight line but a folded form so easily forming molecular hydrogen bond, and when the degree of esterification becomes higher the hydrogen bond is relatively weakened, the solubility will be increased to a certain extent. Under the condition of without any reagents the pectin solution is acidic substances, mainly pectin acid and galacturonic acid. Therefore, pectin is stabile in an appropriate degree of acid conditions, but the strong acid or alkali conditions will caused the degradation of pectin.

The molecular structure of pectin can be divided into smooth area and hair area, smooth area is composed of α-D-galacturonic acid linking by 1, 4-glucosidic bond in line correlation; hair area is composed of the high branches of α-L-rhamnose galactose acid. Therefore, classified in accordance with the structure, pectin includes three categories: galacturonan, rhamnogalacturonan-I and rhamnogalacturonan -II.

1) Galacturonan: galacturonan is composed of α-D-galacturonic acid residues by 1, 4 glucoside bond linking with a linear chain of sugar, of which the galacturonic acid carboxyl has different levels of methyl ester, and therefore natural pectin is divided into high methoxyl pectin and low methoxy pectin. The carboxyl that is not esterification exists as free acid form or K, Na, Ca salt form. Pectin can form gel if there is a sufficient concentration of sugar and acid in the presence of $Ca^{2+}$ pectin, through galacturonic acid chain and $Ca^{2+}$ cross-linking molecules in the association forms gel.

2) Rhamnogalacturonan-I

Pectic polysaccharides are usually heteropolysaccharide and homopolysaccharide with single galacturonic acid is seldom. If inserted a number of D-rhamnose-buckthorn in the middle of the α-1,4-connected multi-polygalacturonase main chain it becomes Rhamnogalacturonan-I, which is repeat units composed of the turn of rhamnose and galacturonic acid.

3) Rhamnogalacturonan-II

Rhamnogalacturonan-II was first reported in 1978, it exists in higher plants and parts of the cell wall of the lower plants. Research has been confirmed that in wine and other juices the content of rhamnogalacturonan-II is relatively high as 20-150 mg/L. Rhamnogalacturonan-II is a low-molecular-weight (5-10 kDa) pectin type polysaccharide with complex structure, including 11 kinds of different sugar residues.

Pectin, found in all higher plants, is deposited in the primary cell wall and interstratification of cell. In the wall of the newborn cross-linking with different levels of cellulose, hemicellulose and lignin and some stretching protein, making the variety of tissue structure hard and showing the inherent shape. Pectin molecular structure varies according to the plant species, location organizations, and growing conditions. Just as other polysaccharides, pectin is multi-molecular, multi-decentralized, multi-structure, senior space conformation and a relative molecular weight distribution.

### *Traditional Production Methods of Pectin*

Existing in the general nature of the pectin is a high-methoxy-(LM)-pectin (methoxy-content between 7%-14%).The natural LM-pectin is little. But HM-pectin can gel only in the condition of soluble solids higher than 55% and the pH range of 2.0-3.5. Therefore it mainly used for the production of high-carbohydrate foods. LM-pectin can create excellent gel in the presence of Ca cation and other bivalent cations. Most researchers have generally considered that the role of the introduction of amide group is to improve the gelling capacity of LM-pectin, mainly because amide may help to form hydrogen bonds. Therefore, according to the different demands pectin production can be divided into HM-pectin production, LM-pectin production and the amidated pectin production.

Fruits are usually contained in the general category of raw materials for the HM-pectin, low-sugar jams as popular in consumers, demands of low methoxyl pectin is increasing. At present, apart from a small amount of access to natural resources in the LM-pectin, the majority of LM-pectin is usually be produced by methoxy-decomposition of HM-ester pectin, which mainly includes base catalytic skim, acidification ethanol, amidation and enzymatic method, etc. (1) . alkali catalytic skim refers to the HM-pectin base catalytic skim of LM-pectin is usually strong base of pectin molecular weight will cause a substantial decline in the use of weak base such as sodium carbonate catalytic treatment, irrespective of stem samples or frozen samples can be processed after all the good performance of LM-pectin; (2) . point to acidified ethanol is the pretreatment of the HM-pectin solution to join the acidification of ethanol, that is, from ester into a LM-pectin, the law has been applied from Pingguo Pi Java and Java potatoes in the LM-pectin extracted in the production; (3) . amidating method is, in alkaline conditions, the use of ammonia treatment HM-pectin, so that part of methyl change for the primary alcohols amines, this method is also known as pectin amidation pectin, which has good dissolution and easy to form a gel; (4) . Enzymatic preparation LM-pectin, including sources within the enzyme enzymatic and plus. The former is to increase the source of activator activated pectin esterase (PE), then PE will HM-pectin into LM-pectin; plus enzyme is used first of plus or microbial fermentation plant tissue fluid extraction of PE, re-use of sources within the enzyme will HM-pectin into LM-pectin.

Pectin is the original material and does not dissolve in water, but in the acid and alkali, salt and other chemical reagents and of the role, the decomposition of water into water-soluble pectin. Extraction of pectin is insoluble pectin into soluble pectin and soluble pectin to the liquid phase transfer process. Traditional pectin extracting include acid extraction, ion exchange, microwave extraction, currently there are microbes method, ammonium oxalate perchlorate method, counter-current extraction method, and enzymatic method.

1) Acid extraction method: Traditional acid extraction method is the most used method; the principle is using acid to peel cells in the original non-water-soluble pectin into water-soluble pectin, then add alcohol or metal salts to get pectin precipitation, by washing, filtration, drying, grinding, then obtain pectin powder.

2) Ion Exchange: Ion exchange solution is to use all kinds of charged particles and ion exchange between the binding forces of the material differences between the operations of separation. By ion exchange extracting, the shortcomings of traditional acid pectin can be overcome, since the removal of $Mg^{2+}$, $Ca^{2+}$ plasma through the ionic bond with the pectin with the closure of pectin, so as to enhance the rate and to obtain product quality.

3) Microwave extraction: Microwave extraction technology is a new technology, microwave heating of raw materials will damage the parenchyma cells, the microwave processed raw materials and the permeability of the porous absorbent capacity has increased, and heat treatment can inactivate pectinase. Microwave has strong extracting selectivity of pectin, short operating time, small amount of using solvent, and heating uniform components in the target rate. It will not undermine the long-chain structure of pectin, both the yield and quality have improved, which might be an effective way.

4) Ammonium oxalate extraction method: Peel with ammonium oxalate solution processing, filtering be pectin after extract. This method can become insoluble pectin soluble calcium ammonium salt, $Ca^{2+}$ precipitation in the form of calcium oxalate to remove. Chelating agents can also use the six partial phosphates, so that the solubility of insoluble pectin increase, to obtain better results from.

5) Continuously counter-current extraction method: Continuous counter-current extraction is the extraction method in which the extraction phase and raffinate phase move continuously on the opposite directions. During this process, the solute-enriched extract

contacts with fresh extracts before emit out of the system, and the raw materials that want be completely extracted contacts with the extracting solvent before it emit out of the system, so that the extraction ratio is higher. Compared to the alcohol-extraction process, continuously counter-current extraction method increases the handling capacity. Countercurrent extraction equipment is commonly used in pot countercurrent extraction equipment, spiral drum countercurrent extraction equipment.

6) Enzymatic extraction method: Enzymatic extraction is using of enzyme to impove pectin extracting. General steps of enzymatic extraction includes as follows: add the buffer to grounded raw materials, enzymatically extracted in a heated water bath oscillator, after the end of reaction use ethanol to deposit, then filtration separation, drying, crushed and obtain in the pectin product. As enzyme extracted pectin need a long reaction time and a large amount of enzyme, it hindered the application of enzymatic extraction. However, if using acid combined with the enzyme, to extract a small amount of pectin firstly using acid, then to extract the remaining pectin using enzyme, will greatly shorten the reaction time, and reduce the amount of enzyme. With the development of industrial enzyme, the enzyme cost will be reduced. The prospect of enzymatic extraction of pectin will be great.

### *Modern Pectin Separation Methods*

It is necessary to further isolate the rough pectin extraction liquid before it can be purified. The most commonly method is alcohol precipitation and salting, there are also using membrane directly enrichment and purification of the pectin [26].

1) Alcohol precipitation method: alcohol precipitation was the first method to achieve industrialization of pectin. The basic principle of this method is that pectin does not dissolve in organic solvents, when a large number of alcohols are added to the pectin extract a mixture of water and alcohol will be formed and pectin will precipitate. This method is simple. The pectin obtained by this method has good color and less ash. But a large amount of alcohol is needed and it is difficult to recycle, also it needs huge energy consumption, so the production cost is high and it is not conducive to large-scale production.

2) Acidified ethanol method: in the process of pectin precipitation, ethanol can reduce the dielectric constant of pectin solution, at the same time pectin dehydration, therefore the pectin is precipitated from the solution. Trace electrolyte in the conditions, the precipitation of pectin will be more complete, even a very low concentration of pectin can also be precipitated from solution. This may be due to ions in the pectin molecules formed bridges, connecting the pectin molecules and polymerization, thereby speeding up the pectin precipitation.

3) Salting-out method: the principle of salting out method is that the salt ions and the carboxyl of pectin have opposite charges, which results in the precipitation of pectin by charge neutralization. Taking advantage of this characteristic, by firstly adding ammonia in pectin solution and then adding the electrolyte metal salts in, some insoluble pectate pectin salts and a small amount of salt hydroxide and other impurities will be produced. After separation, desalination, performed using acidification ethanol and replacement reaction, was taken placed between acid and metal then the pectin is produced. A small amount precipitation of salt hydroxide disappeared and the produced pectin does not dissolve in alcohol and precipitated. Now the precipitator in salting-out method is generally aluminate, molysite, calcic salt and mixed salt. Production of pectin by salting-out method has some advantages, such as high yield, less alcohol consumption, low cost, high esterification of pectin than that of alcohol precipitation and the gel is stronger.

4) Membrane separation method: membrane separation method is refers that by using the selectivity of natural or synthetic polymer membrane, with the impetus of outside energy or the chemic potential difference, to separate, class, purify and enrich solute and solvent with double or multiple compositions. Membrane separation method has some advantages such as needing simple equipment, being operate at room temperature or cold, high efficiency, energy-saving, especially fitting for the thermosensitive material separation. Membrane separations include penetration, reverse osmosis, ultrafiltration, nanofiltration, electrodialysis and dialysis. At present, the main methods of extraction of pectin are ultrafiltration and electrodialysis which simplify the production process, greatly reduced the production costs of pectin, the pectin separated by this method was beige, transparent film, the main indicators of it are in line with national standards and performance of dissolved in water is better than the pectin produced by alcohol precipitation.

## 3.5. Separation Principle of Plant-Based Active Component

China has some advantages for research and development of new drugs due to the abundant plant resources and long history of traditional Chinese medicine compared with other countries. But the resources advantages has not translated into economic advantages, compared with 12,800 kinds of natural medicinal resources, our own creation of new drugs so far not more than 200 kinds. Deep development of systematic, comprehensive study on the novel chemical structure and physical activity of the natural product, not only have important academic significance for organic chemistry, biology and other disciplines, but also have both direct or potential social, economic value for the research of China's new drug with independent intellectual property rights and the rational use of development resources development.

### 3.5.1. Flavonoids

Flavonoids are, broadly speaking, a series of compounds which mutually linked two benzene ring through three carbon atoms. Flavonoids can be divided into flavonoids, flavonoids alcohols, double flavonoids, isoflavones, flower pigments, and other categories. They are secondary metabolites widespread in nature, mainly in plant leaves, fruits, roots, bark. As of 2000 statistics, the total number of flavonoids has been over 8000, there are so many structures because of the difference of number and location for the hydroxyl and methyl substituted. Flavonoids are usually in form of glycoside, have loose bodies (aglycones), in addition to O-glycoside, natural flavonoids also have C-glycosides, such as puerarin, puerarin xylosidase.

#### *3.5.1.1 Structure and Distribution*

Flavonoids are widely distributed in higher plants, very few flavonoids also exist in algae, fungi, and other lower plants. Flavonoids are abundant in angiosperms, particularly in Legumes, Rutaceae, Umbelliferae, Labiatae, Scrophulariaceae, Rhamnaceae, Ginger and other plants. Amentoflavone is more limited distribution in the gymnosperm, is the characteristic elements of gymnosperms, particularly widespread in Ginkgopsida and Coniferopsida plants. In addition, the flavonoids are also widely distributed in the fern and moss plants.

Flavonoids are the 2-or 3-phenyl derivatives of chromogen alkane and chromogen ketone, broadly speaking, are series of compounds having two aromatic rings through three central carbon chain connected, generally has C6-C3-C6 basic skeleton characteristics. Flavonoids are mainly the basic structure of the derivatives and the common substituents of the rings are hydroxyl, methoxy, isopentenyl side chain and monoterpene-side chain. In plants, most of flavonoids are exist as glycosides forms, which can create a wide variety of flavonoid glycosides category because the sugar type, quantity, location and link the different connections, and most of them are the oxygen glycosides and also a small number of C-glycosides. The linking position of aglycones in flavonoid is relative to the structure of glycosides. The glycosides of flavonoid include single sugars, double sugars, trinal sugars and acylation sugars [27].

#### *3.5.1.2. Extract*

There are many methods for flavonoids extracting. We can select the most suitable extraction process according to the nature of extracts, cost of extraction, processing equipment and other conditions to, and increase the extraction rate, of increasing the effect of using raw materials. The traditional means, including organic solvent extraction, dilute alkaline water or alkaline alcohol extraction and system solution extraction, generally spend a longer time, low extraction rate and easy to sneak impurities leading inconvenient for the follow-up separation and purification. With the development of depth study for flavonoids, many extraction methods were emerged to extract flavonoids, such as ultrafiltration, biological enzyme, supercritical fluid extraction, macroporous resin adsorption, ultrasonic extraction, and microwave extraction technology.

*The Traditional Method for Extraction*

The traditional method for flavonoids extraction are mainly organic solvent extraction, hot water extraction, alkaline water or alkaline dilute extract, the system solvent extraction method. Ethanol and methanol is the most commonly used solvent for extraction of flavonoids. High concentration of alcohol suitable for extracting aglycones, about 60 percent concentration of ethanol or methanol-water solution is suitable for the extraction of glycoside substances. Acetone and ethyl acetate are also commonly used to extract flavonoids. The extraction process includes quench method, percolation method and reflux method. For example, quench method is used to extract total flavonoids from Puerarin, percolation method is used for hesperidin extraction, reflux method is used to extract total flavonoids from Ginkgo. Hot water is limited to extract flavonoid glycosides, such as rutin, because when using hot water extraction, impurities (such as protein, tannin, starch, sugar and more compounds, etc.) are soluble in water, leading to low extraction efficiency, therefore not commonly used. Flavonoids mostly have phenolic hydroxyl group, so they can be extracted by alkaline water or dilute alcohol, flavonoids can be precipitated from leaching solution by acidification. When using alkaline solvent extraction, the concentration should not be too high, in order to avoid damage flavonoids nucleus when heating under the strong alkali. When flavonoids contain o-phenolic hydroxyl, boric acid can be used to protect it.

*New Extraction Methods*

1) Ultrafiltration

Molecular weights of flavonoids are no more than 1 kDa, and molecular weight of polysaccharides, proteins and so on is more than 50 kDa, using the method of ultrafiltration can remove these impurities. Hence increasing solution clarification and the content of active ingredients, also in the process of analysis no phase changes, effective physical and chemical compositions are stable and high accuracy. Using ultrafiltration to extract of flavonoids can effectively separate the fat-soluble chlorophyll and tannin.

2) Enzymatic Method

Enzymatic extraction can be used for raw materials in which flavonoids were surrounded by walls. For example, in hawthorn, because flavonoids are surrounded by the cellulose-based wall and there are pectin bond between the cell wall, using the enzyme extraction will obtain higher extraction rate than the general approaches. The principle of this extraction method is pectinase fully destroyed the pectin substances of the wall reducing the resistance of extract material so that the flavonoids of flesh fully released. In addition, the cellulose also contributes to the release of flavonoids.

3) Supercritical Fluid Extraction

Supercritical extraction of flavonoids began in the 1990s, is now widely used in the extraction of flavonoids, such as extraction of rutin from *Sophora japonica* using supercritical $CO_2$.

## 3.5.2 Alkaloids

Choosing suitable method for active ingredients extraction of alkaloids is of great significance to maintain the activity. With the increasing emphasis on alkaloids, the extracted method is developed from the traditional to a large number of new extraction methods.

### *3.5.2.1 Extraction Methods*

In accordance with the classification of the used solvent, the methods of solvent extraction alkaloids include acid solution extraction, alkaline solution extraction, organic solvent extraction, etc.

a) Acid solution extraction

The basic alkaloids in the plant usually exist in the form of salt, and the weak neutral or alkaline alkaloids exit in the form of free base or unstable salt. 0.5-1% of acetic acid, hydrochloric acid is usually used to extract alkaloids.

b) Alcohol extraction

Free alkaloids and its salts are soluble in methanol and ethanol. Different parts of the alcohol are suitable for different methods, such as water ethanol extraction are suit for roots and stem of the plant.

c) Lipophilic organic solvent extraction

As alkaloid exits in the form of salt, it is necessary to convert alkaloid salt into free base with lipophilic organic solvent extraction. The method is as follows, the plant first soaked with Na2CO3 solution or dilute alkali, then extract with organic solvent. However, the organic solvent extraction method is not suitable to the systematic alkaloids separation, only suit to the extraction of medicinal alkaloid.

#### *3.5.2.2 According yo Extracts Conditions*

a) Steam explosion extraction of alkaloids

Steam explosion could loose plant material, destruct plant cells, so alkaloids can be extracted easily.

b) Ultrasound-assisted extraction

Ultrasound-assisted extraction can shorten extract time and increase extraction rate, and without heating, increased the extraction rate of thermosensitive alkaloids, and does not affect the basic physical activity, a relatively small amount of solvent used and lower costs.

c) Microwave Extraction

Compared to traditional methods, microwave extraction has the advantages of high quality and yield, high selectivity, save time and lower energy consumption. However, microwave extraction is affected by solvent, extraction time, extraction temperature and pressure.

d) Supercritical Fluid Extraction

The study on supercritical fluid separation of alkaloids has conducted for many years, Germany and the United States scholars have studied removal of caffeine from coffee beans using supercritical $CO_2$, and a commercial-scale production has been formed.

e) Enzymatic method

Professor Chen Hongzhang applied enzyme to extract alkaloid, the alkaloids and other material were released after the cell wall damage. The enzymatic extraction of ephedrine have been studied, the concentration of enzyme, substrate concentration, temperature, acidity and other parameters inhibitors were analyzed. Through the degradation of cellulose, the ephedrine was effective extracted.

### 3.5.3 Tannin Isolation

Tannin is a widespread multi-plant phenolic compound. According to the structure, tannin can be divided into three categories: hydrolysis tannin, condensed tannin and compound tannin. Tannin is widely distributed in plant leaves, paper, wood and fruit, which is one of the targets for comprehensive utilization of forest resources. The content of tannin in the plant is only low than cellulose, hemicellulose and lignin. The content of tannin in the bark of mangrove is more than 20% and even 20-40% in the bark of some conifer. In the 340 billion cubic meters forest resources of the world, the bark accounts for 10% to 15%, therefore tannins resources in bark can amount to 100 million tons.

#### *3.5.3.1 Existence State*

At present, the tannins not only closely related to health and human life, and with the development of technology, but also studied in chemical, biological, medical, agriculture and animal husbandry, and many other areas. So far, more than 400 new tannins and phenolic compounds pluralism have been found.

### *3.5.3.2 Production and Separation*

The highest purity of tannin extraction from gallnut is 78-80% using water as solvent. The method only suit for industrial demand, but can not for the demand of high-purity, high-quality tannin [28]. At present, there are several separate ways exit.

a) Impregnation

The method is putting plants into a covered container, and adding appropriate solvent (water or alcohol), soaking for some time at room temperature or heated circumstances. This method is simple and appropriate for plant material containing starch, gums, mucus quality, pectin and other ingredients. The drawback of the method is low extraction rate, using easy-mould raw material, and requirement of preservatives.

b) Seeping method

The plant powder was first installed in seeping utensil, then new solvent is added constantly making it infiltrate into medicine, the liquid was leached out from top to bottom. When the solvent dissolved the powder and increased the proportion then would move downward, the top of the solution would dilute it and replace its location, resulting in a good concentration difference, so that the spread can be better, and it is better than leaching impregnation. However, the flow rate should be controlled and new solvent should be added timely in the process. When the colors seepage drops become light or the volume of fluid seepage equivalent 10 times of original ingredients, the extraction can be treated as basically complete.

c) Decocting method

Plant materials are boiled in ceramic containers with water to extract the active ingredients. This is a traditional method that earliest used in China and is still using. This method is simple, but also dissolves the majority composition of plant material. However plant materials containing volatile components and heat-damaged active ingredients are not suitable for this method.

d) Ultrasonic extraction

Adding extraction solvent and plant material to ultrasonic generator, with the destruction of raw materials, active ingredients can be easily dispersed to the extracting solution.

Using ultrasound to enhance the leaching and chemical extraction process is a great potential direction of development in the field of ultrasound. The traditional extraction method is to select the right target solvent, using heating or stirring. The higher the temperature is conducive to target components extraction, but too high temperature often results in the degradation or structure change of active ingredients.

e) Tissue broken extraction

Tissue broken extraction is often used to animal tissue, plants seeds and tender leaf. In different experimental scale, different experimental materials and experimental requirements, the crushing methods used and conditions are also different.

### 3.5.4 Plant Oils

Plant oils are mainly derived from the seeds of plants, containing essential fatty acids for human body, such as linoleic acid, linolenic acid and oleic acid. China is the world's largest country for edible oil and oil-producing.

#### *3.5.4.1 Traditional Vegetable Oil Extraction Technologies*

Traditional vegetable oil extraction technologies include press and the leaching.

*1) Press Method [29]*

Press method is squeezing the oil with help of machinery, which is the main method for oil extraction. Press method has the advantages of adaptability, process simple, easy maintenance of production equipment, production scale of flexible, suitable for all kinds of vegetable oil extraction and production of relatively safe. Press method has many disadvantages, including low rate of oil extraction, labor-intensive, low production efficiency, and low comprehensive utilization rate of oil resources.

*2) Leaching Method [30]*

Leaching method is a more advanced method for extracting oil, which is based on the application of principle of solid-liquid extraction. The organic solvents used include paraffin and carbon cycloalkanes, propane and butane.

Leaching method has the advantages of high oil extraction rate, low residual oil, low labor intensity, easy scale-up and automation. Its shortcomings include the heavy color, poor quality, and the leaching solvents used are inflammable and explosive, but also certain toxicity.

#### *3.5.4.2 Modern Vegetable Oil Extraction Technologies*

*1) Water Substituting Method [31]*

The water substituting method is different from the ordinary method, using hot water to replace the oil. Water substituting method has a lot of advantages: good quality; less energy consumption, no danger of blasting, not polluting environment, and can separate oil and protein simultaneously. But the major drawback is low oil extraction rate than traditional method, easily be microbial contaminated in the process of extraction.

*2) Water Enzyme Method [32]*

Water enzyme method is a relatively new method for oil and protein separation, the oil extraction rate is increased through the machinery and enzymatic degradation. Compared to traditional method, water enzyme method has the advantages of moderate condition, simple process, low energy consumption and high-quality protein.

*3) Supercritical $CO_2$ Extraction Method [33]*

With the excellent solubility of supercritical fluid, supercritical $CO_2$ extraction method is to extract by adjusting the density of extract fluid. Supercritical $CO_2$ extraction of vegetable oil has many advantages, such as the simplification of process, energy conservation, lower extraction temperature, low damage of active material; abundant resources of $CO_2$, low prices, non-toxic, no environment pollution, etc. Supercritical $CO_2$ extraction has the disadvantages of expensive high-pressure equipment, high production costs and difficult to operate, the industrial production was limited in some extent limit.

*4) Ultrasound Treatment [34]*

Ultrasound is sound waves with more than 20 KHz frequency and with the dual nature of wave and energy. Its vibration can generate and deliver a powerful energy, making the great acceleration of molecular. As the role of ultrasonic energy, particles will be in a medium gravity about $10^4$ times. Ultrasonic has been widely used in the extraction of bioactive substances, but is still at exploratory stage in oil extraction area.

### 3.5.5. Organic Acid

Organic acid is the compounds containing carboxyl in the molecular structure, which is widely distributed in the leaves, roots and particularly fruit of Chinese herbal medicine, such as dark plum, schisandra, raspberry, and so on. The organic acids in common plants are aliphatic acids' Monobasic Acid, diatomic acids and multi-carboxylic acid such as tartrate, oxalic acid, malic acid and ascorbic acid, also aromatic organic acids, such as benzoic acid, salicylaldehyde acid, caffeic acid, and so on. Apart from a few of organic acids are existing in the free form, they are in the form of combined salt by generally combining with potassium, sodium and calcium, and some with alkaloid. Mostly, fatty acids are combined with glycerol to form ester or with advanced alcohol to form wax. Some organic acids are the composition of volatile oil and resin [35].

Organic acids are more soluble in water or alcohol with significantly acidic reaction, but difficult to dissolve in other organic solvents. Adding calcium chloride or lead acetate or barium hydroxide solution to the organic acids aqueous solution, it can generate the water-insoluble calcium salt, lead salt or barium salt precipitation. Aliphatic organic acid is generally believed that without special biological activity, but some organic acids such as tartaric acid and urinary citric acid can be used as medicinal. Some special acid is the active ingredients of certain Chinese herbal medicines, such as pseudolaric acid in the Pseudolarix has anti-fungal effect. Caffeic acid derivatives have some biological activity, such as chlorogenic acid is the active ingredient in many Chinese herbal medicines, having anti-bacterial, cholagogic, elevated white blood cell functions. The main methods of extracting organic acids are reflux extraction and ultrasonic extraction. The choice of extraction reagent is based on the nature of organic acids which usually using water and 50 percent methanol [36, 37].

### 3.5.6. Volatile Oil Extraction

Volatile oil, an oil-like liquid, can be volatile with the steam distillation, most of which has aroma, such as mint oil, eugenol, and so on. Chinese herbs containing volatile oil can be used as medicine.

#### *3.5.6.1. General Character*

Most of volatile oils are colorless or pale yellow, with special gas (mostly aroma) and spicy taste, and can be volatile at room temperature. The general density of volatile oil is between 0.850-1.180. Volatile oil is difficult to dissolve in water, but can be completely dissolved in ethanol, ether, chloroform and fat oil. Volatile oil can be dissolved in various concentrations of ethanol, but the solubility is decreased with the decrease of alcohol concentration.

Chemical composition: volatile oil is a mixture of multiple compounds, including aliphatic compounds, aromatic compounds and terpene [38]:

1) Aliphatic compounds: such as hydrocarbons, alcohols, aldehydes, ketones and esters, they are all widely distributed in fruits.

2) Terpenoids: The general formula of terpenoids is $(C_5H_8)n$. $C_{10}H_{16}$ called monoterpenoids, $C_5H_{24}$ called sesquiterpenes, $C_{20}H_{32}$ known as Diterpenoids, six or eight isoprene composed of triterpenes or tetraterpene, more isoprene composed of the compounds of polyterpenes. The terpenes in the volatile oil are mainly monoterpene and sesquiterpenes. Polyterpenes do not exist in the volatile oils and are the ingredients of some resin, pigment and rubber. Terpenoids in volatile oil contain oxygen or without oxygen. For most volatile oil are not oxygenated hydrocarbons. Although there has a large number of composition without oxygenated hydrocarbons, but most of aroma result no better than an important component. Oxygen derivatives are alcohols, aldehydes, ketones, ether, acid, phenol, esters, although less content but most have excellent aromatic gas, are essential elements in volatile oil [38].

#### *3.5.6.2. Extraction Methods*

*1) Supercritical $CO_2$ Extraction*

Supercritical $CO_2$ extraction technology can be applied directly to extraction of components with smaller molecular weight, fat-soluble, thermosensitive, volatility. Volatile oil is widely used in the separation of material. Supercritical fluid extraction have the advantages as: 1) at low temperature extraction and separation, reduce or prevent loss of thermal components, 2) have no solvent residue 3) selectivity, the leaching of high purity. 4) extraction, distillation step operation, high efficiency, energy conservation.

*2) Microwave Extraction*

Microwave extraction for volatile oil is one of the effective methods, which requested that medicine has good water absorption. Otherwise, it will be difficult to handle with microwave work. Microwave extraction has the following advantages: ① excellent selectivity: Molecules with high polar can get more microwave energy, which can be used to selectively extract polar molecule, thereby improving the purity of products, quality improvement. ② Rapid extraction speed: The object is heated in the microwave often transparent or translucent container and heat for the poor conductor, the material rapidly warming, which can save 50% to 90% of the extraction time. ③Good quality of products and avoid prolonged high temperature causing the decomposition of sample, which is conducive to the extraction of volatile components. ④ simple process, save energy and simplify processes, reduce the amount of solvent, reducing investment to reduce human consumption. ⑤extraction efficiency. Particularly for the organizational structure of the dense herbs, the traditional method of extraction from the low rate of herbs can significantly increase its extraction rate.

*3) Enzymatic Extraction*

Using the biological activity and specificity to decompose the pectin, crude protein, cellulose, starch, fat and other impurities in the pharmaceutical raw materials can be more effective in extracting essential oil. As the extraction of auxiliary agents, enzyme can damage plants cell wall structure, improve the efficiency of extracting enzyme. Extraction efficiency of enzyme mainly depends on the type, dosage, hydrolysis time, temperature, acidity, fine materials, mixing strength and other factors.

### 3.5.7. Glycosides Extraction

Glycosides are ring acetal derivatives formed by carbohydrate or carbohydrate derivatives and another non-carbohydrate substances through dehydration. Glycosides can be hydrolyzed to carbohydrate and non-carbohydrate substances. Non-carbohydrate substances usually include phenols, anthraquinones, flavonoids and other compounds.

Most glycosides are colorless, odorless and bitterness, but a small number of glycosides has color, such as flavones glycoside and anthraquinone glycosides. Some glycosides have sweetness, such as liquorice saponins. Most glycosides were neutral or acid and can dissolve in water, ethanol. Some glycosides can be dissolved in ethyl acetate and chloroform, but difficult to dissolve in ether, petroleum ether, benzene, and other small polar organic solvents. With more hydroxyl existing, glycosides showed strong hydrophilic character. Water is a strong polar solvent that can dissolve most glycosides. Some glycosides such as saponins, flavone glycoside can react with lead acetate to generate precipitation, and after the lead removed from precipitation glycosides can be obtained again [39].

### 3.5.8. Plant Pigment

Plant pigment is widely existed in Chinese herbal medicine, including fat-soluble and water-soluble pigments. Fat-soluble pigments mainly include chlorophyll, carotene and lutein. Pigments are insoluble in water and methanol, but soluble in high concentration ethanol, ethyl ether, chloroform and benzene. In preparation of Chinese herbal medicine and other active ingredients extraction, chlorophyll is often be removed as impurity [40, 41]. Chlorophyll has inhibitory effect, and can make drugs anti-inflammatory. Water-soluble pigments are mainly color glycosides, also known as proanthocyanidins, commonly found in flower. Water-soluble pigments are soluble in water and ethanol and insoluble in ether, chloroform and other organic solvents. In Chinese herbal medicine preparation or active ingredients extraction, water-soluble pigments are often removed as impurities [40, 42].

### 3.5.9. Resin

Resin is a class of substances which are secreted in many plants and in plants often mixed with volatile oil, rubber and organic acids. Coexisting with oil it is called oil resin, such as pine oil. Coexisting with rubber it is called plastic resin. Coexisting with a large number of aromatic organic acids it is called incense resin, such as benzoin. Co-existing with some sugar it is called glycosides resin, such as thunbergia resin glycosides [43].

Resin is usually amorphous solids, crisp, which can be melted after be softened in the case of hot tacky. Resin is insoluble in water, soluble in ethanol, ether, and other organic solvents. In alkaline solution it can be partially or completely dissolved, but insoluble in acid solution. Resin is widely distributed in the body of plant, such as frankincense, and myrrh to promoting blood circulation, pain, swelling, benzoin promoting blood circulation, anti-corrosion, Styrax aromatic fails to grasp the situation, ferulic for San lump in the abdomen, a rosin-drive style, such as the role of pain. The majority of Chinese herbal medicines contain a small amount of resin which is removed in the production of Chinese herbal medicine as impurities [44, 45].

### *3.5.10. Polysaccharide Extraction Method*

Carbohydrate, widespread in Chinese herbal medicine, can be divided into monosaccharide, oligosaccharides and polysaccharides. The chemical formula for monosaccharide is $(CH_2O)_n$, is a multi-hydroxy aldehydes or ketones, the vast majority of monosaccharide exist in nature with five, six or seven carbon [46]. Most monosaccharide are crystal, sweet, soluble in water and dilute alcohol, difficult to dissolve a high concentration alcohol, do not dissolve in ether, benzene, chloroform, and other small polar organic solvents [47]. Oligosaccharide is composed by 2-9 monosaccharide molecules from polymerization. Oligosaccharide has similar characterization with monosaccharide, such as crystalline, sweet, soluble in water, insoluble or do not dissolve in organic solvents. Polysaccharide is a polymer composing of aldose and (or) through glucoside bond, is an important macromolecular besides protein and nucleic acid.

Polysaccharide is usually composed of more than 10 single-sugar molecules, most of which are amorphous compounds, with greater molecular weight, no sweet, and difficult soluble in water, some can form paste or colloidal solution in hot water.

In recent years, polysaccharide from plants, marine organisms and fungi other sources has become an important bioactive natural products, anti-tumor, immunization, anticoagulant, hypoglycemic and anti-virus activity of polysaccharide has been found.

#### *3.5.10.1. Common Polysaccharide*

Starch, inulin, gums, mucus, cellulose are the most common polysaccharide in Chinese herbal medicine.

1) Starch, extensively stored in plant seeds, tubers and underground stems, is composed by hundreds of glucose molecules. Starch is insoluble in cold water and organic solvents.

2) Inulin is composed of fructose through polymerization, of which the molecular weight is smaller than starch. Inulin is granular crystals, soluble in hot water, micro- soluble or insoluble in cold water and organic solvents [55].

3) Rubber is a mixture of macromolecule, its chemical structure like polysaccharides, but contains carboxyl, which can combine potassium, calcium, magnesium to form salt,

hydrolyze to uronic acid and monosaccharide. Rubber could expand in the water to form colloidal solution, do not dissolve in organic solvents, and react with lead acetate or basic lead acetate to form precipitation. In addition to taking few Arab gum as excipients and suspension in medicine, the majority of gum were considered as invalid components, and removed in preparation.

4) Mucus is a normal physiological product, existing in the mucous cells, its chemical composition is similar to gum. Extracted mucus is usually amorphous solids, which can be expanded to form colloidal solution in hot water, but insoluble in organic solvents. Mucus could react with lead acetate to form precipitation, and ethanol can also make mucus precipitate. This property can be used to extract or remove mucus from Chinese herbal medicine. Except that some mucus of Chinese herbs is used for medical treatment, most mucus is used as lubricants and suspension or removed.

#### *3.5.10.2. Polysaccharide Extraction Methods*

1) Solvent Extraction [48, 49]

Solvent extraction is a commonly used method for polysaccharide extraction, which applies with the principle of similarity and intermiscibility that polar active ingredients choice polar-solvent, weak polar ingredients choose weak polar solvent. Polysaccharide is a polar macromolecule compounds, usually choose polar solvents water and alcohol. In all solvents, water is a typical strong polar solvent, with better penetrating power for plant tissue, high efficiency. Hot and cold water can also be used to polysaccharides extraction. Most polysaccharide is extracted by hot water that the insoluble matters can be removed through centrifuge, or to precipitate polysaccharide with high concentrations of ethanol.

2) Acid Extraction

Acid extraction are suitable for some types of polysaccharides, such as the HC1 solution, and can attain a higher extraction rate. However, acid extraction is only suit for some specific plant polysaccharides, the acidity should be strictly controlled, as the acidic conditions may cause the glucoside bond fracture in polysaccharides.

3) Alkali Extraction

Similar to acid, some polysaccharide has better extraction rate in alkali solution, especially polysaccharide with uronic acid and acid polysaccharide. Similar to acid extraction, the concentration of alkali should also be effectively controlled, as some polysaccharide will be hydrolyzed in alkaline solution. In addition, polysaccharide precipitation can be obtained through dialysis and concentration from acid and alkali dilute extract.

4) Enzymatic Extraction

Enzyme technology is a widely utilized method for active ingredient extraction. Polysaccharide can be extracted from in moderate conditions. In addition, enzyme can also decompose starch, pectin and protein. The commonly used enzyme include protease, cellulose and pectinase.

5) Ultrafiltration

Ultrafiltration is a membrane separation technology, the gel and macromolecules can be isolated from water and other extracted liquids. As ultrafiltration membrane with asymmetric porous structure, and the use of friction and turbulent flow structure, membrane pollution was reduced. Small molecule material and solvents can pass porous membrane, and macromolecular was obtained. Membranes used for polysaccharide bioactive substances separation has advantages of no damage to activity, high efficiency, low energy consumption, simple equipment, continuous production and pollution-free.

6) Extraction of Polysaccharide from *Polystictus Versicolor* by Superfine Comminution

Superfine comminution, also called cellular level pulverizing or critical particle comminution, is a method more aimed for crushing the cell wall than for making substrate into granule. Superfine comminution technology becomes more and more mature and economical now and was widely used in chemical industry, medicine, dyestuff, agriculture, etc. Professor Hongzhang Chen, Chinese Academy of Science, has investigated the effects of superfine comminution conditions on the extraction yield of polysaccharide from Polystictus Versicolor and optimized the superfine comminution conditions and the extraction parameters. The optimum parameters of superfine comminution were airflow pressure of 0.8 Mpa, classifier wheel frequency of 24 Hz and comminution time of 20 min. The optimum parameters of extraction process: 100 ℃, 80 min and distilled three times. By optimization, 91.96% of the total polysaccharide was harvested. The harvest rate was increased 61% comparing to the fruiting body of *Polystictus Versicolor* which was not superfine-comminuted. This suggested that superfine comminution was an effective method for extracting the polysaccharide of *Polystictus Versicolor*.

## 3.6. Concluding Remarks

Plant-based biomass is the most abundant renewable resources in the world, including agricultural products and crop stalks, wood and wood waste, energy plants, medicinal plants and aquatic plants, and other plants and their residues, animal manure, organic waste. Through industrial processing and transformation, biomass not only can be converted to bio-energy, with the principal component of carbohydrates, biomass can also be used to produce thousands of environment-friendly chemical products. Therefore biomass is becoming a new industrial raw material with broad prospect.

Plant resources are complex organism, different species with different components, besides major components of cellulose, hemicellulose, lignin and starch, are also rich in active ingredients, such as flavonoids, organic acids, volatile oil, glycosides, pigments, resins and carbohydrates. Plant-based raw materials also have complex three-dimensional structures, different plants or even different parts of the same plants possess distinctive structures and components. Therefore exploiting plant resources is a systematic project, it not only calls for the completely understanding of the plants' compositions and structures, effective measures also need to be adopted to exploit the active ingredients.

The purpose of components separation is to separate the plant-based raw materials with the largest rate, maintain the integrity of elements as far as possible, and achieve the greatest value. Systematic analysis of components separated from plant-based raw materials from the principles and technologies, not only has great significance for active ingredients, but also have indirection for the rational development of natural resources.

Though the total utilization of biomass is focus on the major components of cellulose, hemicellulose and lignin, exploiting the active ingredient of plant-based materials is an effective way of raising the value of biomass industry, as well as an effective way to increase their economic viability.

## REFERENCES


[1] Xu, F.; Chen, H., New Developments of solid - state fermentation in the applied aspects of resource andenvironment. *Biotechnology Bulletin* 2002, (3), 27-30.

[2] Yang, S.; Ding, W.; Chen, H., Enzymatic hydrolysis of steam-exploded rice straw in membrane bioreactor. *Environment Science* 2005, 26, (5), 161-163.

[3] Zhai, W.; Chen, H.; Ma, R., Structural characteristics of cellulose after dissolution and regeneration from the ionic liquid [bmim]Cl. *Journal of Beijing Unilversity of Chemical Technology (National Science Edition)* 2007, 34, (2), 138-141.

[4] Chen, H.; Li, Z., Biological reaction reactor engineering. *Progress in Biotechnology* 1998, 18, (4), 46-49.

[5] Jin, S.; Chen, H., Structural properties and enzymatic hydrolysis of rice straw. *Process Biochemistry* 2006, 41, (6), 1261-1264.

[6] Chen, H.; Xu, J.; Yao, J.; Li, Z., Studies on multi-level conversion of liquid fuel with straw. *Chinese Journal of Bioprocess Engineering* 2005, 3, (1), 9-13.

[7] Chen, H.; Liu, L., Unpolluted fractionation of wheat straw by steam explosion and ethanol extraction. *Bioresource Technology* 2007, 98, (3), 666-676.

[8] Chen, H.; Li, Z., No-contaminative steam explosion and its application. *Journal of Cellulose Science and Technology* 2002, 10, (3), 47-52.

[9] Yuan, Y.; Chen, H., Application of steam explosion in ephedrine extraction. *Journal of China Pharmaceutical University* 2005, (5), 5.

[10] Chen, H.; Qiu, W., The crucial problems and recent advance on producing fuel alcohol by fermentation of straw. *Progress in Chemistry* 2007, 8, 1116-1121.

[11] Chen, H.; Li, Z., Key technologies of straw eco-industrial building. *Transactions of the Chinese Society of Agricultural Engineering* 2001, 17, (3), 1-4.

[12] Sun, Y.; Cheng, J., Hydrolysis of lignocellulosic materials for ethanol production: a review. *Bioresource Technology* 2002, 83, (1), 1-11.

[13] J, Z.; J, N.; L, O., Fuel ethanol production from lignocellulose: a challenge for metabolic engineering and process integration. *Applied Microbiology and Biotechnology* 2001, 56, (1), 17-34.

[14] Chen, H.; Liu, L.; Yang, X.; Li, Z., New process of maize stalk amination treatment by steam explosion. *Biomass and Bioenergy* 2005, 28, (4), 411-417.

[15] D, M. J., Pretreatment of lignocellulosic biomass. *ACS Symp. Ser* 1994, 566, 292-324.

[16] Jin, S.; Chen, H., Superfine grinding of steam-exploded rice straw and its enzymatic hydrolysis. *Biochemical Engineering Journal* 2006, 30, (3), 225-230.

[17] Sun, F.; Chen, H., Organosolv pretreatment by crude glycerol from oleochemicals industry for enzymatic hydrolysis of wheat straw. *Bioresour Technology* 2008, 99, (1), 5474-5479.

[18] Sun, F.; Chen, H., Evaluation of enzymatic hydrolysis of wheat straw pretreated by atmospheric glycerol autocatalysis. *Journal of Chemical Technology & Biotechnology* 2007, 82, (11), 1039-1044.

[19] Sun, F.; Chen, H., Comparison of atmospheric aqueous glycerol and steam explosion pretreatments of wheat straw for enhanced enzymatic hydrolysis. *Journal of Chemical Technology & Biotechnology* 2008, 83, (5), 707-714.

[20] Sun, F.; Chen, H., Enhanced enzymatic hydrolysis of wheat straw by aqueous glycerol pretreatment. *Bioresource Technology* 2008, 99, (14), 6156-6161.

[21] Ding, M.; Ma, S.; Liu, D., Stability of ferulic acid and its existing form in Ligusticum chuanxiong and Angelica sinensis. *Chinese Traditional and Herbal Drugs* 2004, 35, (1), 28-30.

[22] Song, H.; Ruo, H.; Guan, H., Ferulic acid and its agents Stability. *Anhui Medical and Pharmaceutical Journal* 2003, 7, (4), 252-253.

[23] J, P.; N, T.; J, C. S.; C, F.; L, F.; J, D. G., The Structure of the Feruloyl Esterase Module of Xylanase 10 B from Clostridium thermocellum Provides Insights into Substrate Recognition. *Structure* 2001, 9, (12), 1183-1190.

[24] E, M. K.; A, S.; A, P. S.; S, W. K., Structure of a feruloyl esterase from Aspergillus niger. *Journal logo* 2004, 60, (5), 878-887.

[25] Liu, J.; Deng, Z., Study Progress of the Extensin. *Food Science and Technology* 2006, 31, (1), 12-15.

[26] Chen, H.; Li, Z., Study on the dealing with steam-explosion wheat straw by ethanol. *Chemistry & Industry of Forest Products* 2000, 20, (3), 33-39.

[27] Pan, F.; Feng, Y., Saponins component analysis in Astragalus medicinal plants. *China Journal of Chinese Materia Medica* 1995, 20, (7), 391-393.

[28] Wong, Y.; Fu, J.; Xu, A., Plant essential oils Progress. *Journal of Henan Forestry Science and Technology* 1999, 19, (4), 23-26.

[29] Yang, L.; Hao, L., Vegetable protein and vegetable oil separation technology. *Food Research and Develop*e*nt* 2003, 4, 5-8.

[30] Zhao, G.; Liu, X.; Liu, Z., Leaching technology development trend of oil. *Cereals & Oils* 2007, (4), 1-7.

[31] Li, X.; Dong, W., Vegetable oil extraction technology progress. *Modern industry and trade* 2007, 19, (8), 201-202.

[32] Du, Y.; Zhang, L., Preliminary study on aqueous enzymatic extraction oil. *Cereals & Oils* 2005, 6, 10-12.

[33] Wang, C.; Liu, Z., Supercritical $CO_2$ extraction of Vegetable oil. *China Oils and Fats* 1997, 22, 3-6.

[34] Guo, X., Application of ultrasound technology in the extraction of oils and fats processing. *China Oils and Fats* 1996, 21, (5), 36-37.

[35] Wang, H.; Ji, Q., Eucommia leaf of chlorogenic acid substances extraction research. *Journal of Jiangxi Normal University (Natural Sciences Edition)* 1997, 21, (4), 339-341.

[36] Wang, M.; Wu, H., Organic acid component of traditional Chinese medicine extraction and separation method Progress. *Journal of Nanjing University of Traditional Chinese Medicine(Natural Science)* 2004, 20, 190-192.

[37] Ma, L.; Tang, J.; Li, Z.; Liao, Q.; Xiao, X.; Zhao, X.; Jin, C., Determination of total organic acids and salicylic acid in extract of Radix Isatidis. *China Journal of Chinese Materia Medica* 2006, 31, (10), 804-806.

[38] Sun, L., Plant essential oils and biological activity of terpene composition. *Journal of Jiangxi Normal University (Natural Sciences Edition)* 2000, 24, (2), 159-163.

[39] Jia, Q.; Zhang, R., Saponin chemicals in Bupleurum plant research progress. *Acta Pharmaceutica Sinica* 1989, 24, (12), 961-971.

[40] Wen, G.; Zhu, W., The extraction and exploitation of natural pigment. *Journal of Henan University of Science and Technology (Agricultural Science)* 2003, 23, (2), 70-78.

[41] Chen, B., Edible natural pigment production and research progress. *Guangzhou Food Industry Science and Technology* 1998, (3), 35-37.

[42] Chen, J., China's national plant pigment applications. *Wild plant resources in China* 2000, 19, (2), 21-24.

[43] Yang, H.; Zhen, M., Resin traditional Chinese medicines and their chemical identification. *Nei Mongol Journal of Traditional Chinese Medicine* 2001, 20, 38-38.

[44] Wu, X.; Liu, M., Resin and aromatic series tanning agents and their application. *China Leather* 2001, 30, 38.

[45] Li, B.; Guo, J., Liquidambar two types of resin incense drug – styrax and resina liquidamberis research. *Natural Product Research and Development* 1995, 7, (3), 53-59.

[46] Tian, G.; Feng, Y., Research Progress of Polysaccharides in Plant. *China Journal of Chinese Materia Medica* 1995, 20, (7), 441-444.

[47] Yu, Z.; Zhou, X.; Guo, H.; Pu, X.; Sun, L.; Wang, D., Comparison of the monosaccharides in tea polysaccharides. *Journal of Anhui Agricultural University* 2007, 9, (4), 547-550.

[48] B, S. A.; S, P. B.; K, W. J.; H, O.; H, Y.; T, E., Isolation and partial characterization of biologically active polysaccharides from Plantago major L. *Phytother Research* 1995, 9, (3), 211.

[49] B, S. P., Biologically active polysaccharides as possible lead compounds. Phytochemistry Reviews 2002, 1, (3), 379-387.

*Chapter 4*

# PROCESSING ENGINEERING AND APPLICATION OF PLANT-BASED CHEMICALS

## 4.1. OVERVIEW

Progress of petrochemical technology was used to play a most significant effect on the rapid development of world economy. Petrochemicals are exist in everywhere of our life, such as synthetic fiber, plastic equipment and resin packing filmy, which bring serious white pollution to the earth because of the undegradability. Owing to the unrenewablility and the increase of exploitation cost, compete advantages of synthetic products from petroleum are losing gradually. According to the requirement of reducing greenhouse gas emission and the target of achieving sustainable development, higher value-adding outputs from renewable plant-based resource, especially lignocellulose, is becoming important in many countries development strategies and the hot field of scientific research.

Lignocellulose biomass is in the form of plant and composes of about 20% of lignin, 40% of cellulose and 25% of hemicellulose. Lignocellulose biomass, generated total over 200 billion tons on Earth each year by photosynthesis, is the most abundant renewable resources. In the 1990s, developed countries have successfully developed oil-instead renewable chemicals, and further accelerate the research and development process of the utilization of non-edible renewable resources, such as agricultural plant waste straw and city fiber.

America plans that at least 10% of chemicals in 2020 from lignocellulose biomass and enhance to 50% in 2050, by 2030 above 20% of the liquid fuel and more than 25% of the organic chemicals will obtain from biomass, declares that the fuel and chemicals industrial transit from the era of non-renewable 'hydrocarbons' to renewable 'carbohydrates'[1, 2]. Research report in September 2004 of World Organization for Economic Cooperation and Development (OCED) indicated that governments should vigorously support and encourage technological innovation in the field of high value-added chemicals from biomass in order to reduce the price gap between traditional fossil raw materials and ultimately replace traditional fossil materials [3, 4].

Human has strong demand that produce organic compounds from biomass, which cannot act as the direct raw material because of the unhomogeneity. So, how to transform biomass efficiently into energy and chemicals becoming a potential research field which include thermochemistry transformation, biology conversion, chemistry transformation and the combine of these methods. In thermochemical reaction, biomass was directly used as fuel without fractionation after being gasified or liquefied. However, the fractionation was the foundation for biochemical conversion of biomass.

Judging from the current system of products, biomass processing technology has reached a level that products could be prepared from each one of the major carbon chain systems. The basic idea of these technology was that biomass firstly transformed into fermentable sugar and then into the major carbon chain systems. Products from C1 system include methane, methanol etc. Products from C2 system mainly include ethanol, acetic acid, ethylene, glycol etc. Products from C3 system mainly include lactic acid, acrylics acid, propanediol etc. Products from C4 system mainly include butanedioic acid, fumaric acid, butylenes glycol etc. Products from C5 system mainly include itaconic acid, xylitol etc. Products from C6 system mainly include citric acid sorbitolum etc. Some of these chemicals had been widely used. The industrialization of agrochemicals, fine chemicals, commodity chemicals, medication and macromolecule materials are presenting a rapidly upgrowth [5].

In conclusion, the multifariousness, environment-friendliness and biodegradability of fine-chemicals make circulate economy and sustainable development feasible, which present a favorable developing opportunity for plant-based chemicals. The most representative fine biomass-based chemicals, large scaly produced and widely applied, basically prepared from the monosaccharides, disaccharides, the starch, the celluloses and the hemicelluloses, the lignin and the lipids, etc [5].

# 4.2. Category of Plant-Based Chemicals

## 4.2.1. Classified by Raw Materials

### *4.2.1.1. Chemicals using Sugar as Platform*

Biomass degrades into fermentable sugar through hydrolysis and then transformed into several of chemicals by fermentation. The main approaches and classes of chemicals are shown in figure 4.1

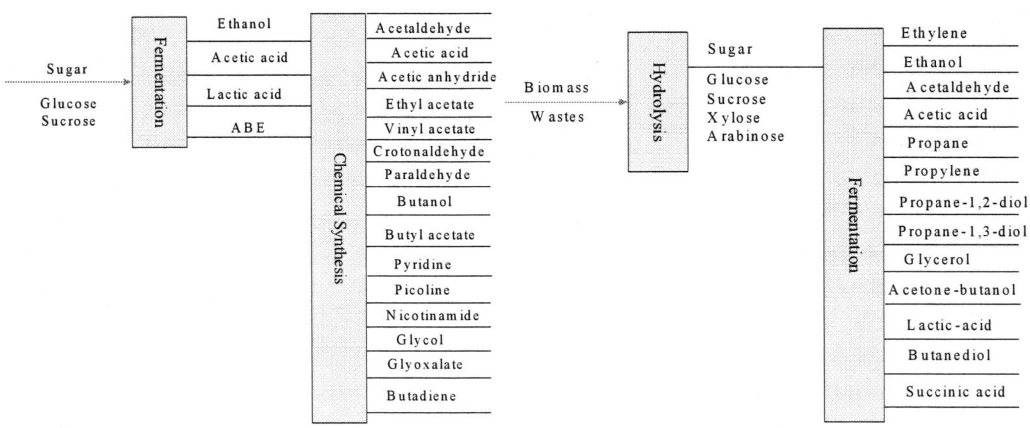


Figure 4.1. Chemicals from biomass [6]

#### *4.2.1.2. Chemicals using Syngas as Platform*

Chemicals using syngas as platform mainly refers to the thermochemical conversion process of biomass into a variety of small molecule synthesis gas. During this process, biomass is treated as a whole to deal with, which neglect the comprehensive utilization of protein, lignin, oils, and other lipid components, making it difficult to get high value-added chemicals [7].

### 4.2.2. Classified by Production Scale

#### *4.2.2.1. Commodity Chemicals*

Since the early 19th century, the beginning of industrial modernization, chemical industry considered the production of commodity chemicals as its main task. The production of commodity chemicals very large (at least an annual output of 1 000 t), the production process control and cost account for a large proportion. Benzene, hydrogen, polyester and titanium dioxide, etc, are typical commodity chemicals. The same kinds of chemicals provide by different manufacturers has the same quality and composition, the difference is reflected in product prices.

#### *4.2.2.2. Fine Chemicals*

Fine chemicals refer originally to the products with low yield, high purity, high prices, such as medicine, dyes, paints, and so on. In recent years, experts have some new insights on the definition of fine chemicals. For example, some Europe and the United States countries define small yield products produce and sale according to different chemical structure as the fine chemicals, and define small yield products as specialty chemicals, which, after the preparation process, have special function or final usable performance. China and Japan name define these two types of products as fine chemicals.

## 4.3. TRANSFORMATION METHODS OF PLANT-BASED CHEMICALS

### 4.3.1. Acid Hydrolysis

Acid hydrolysis of biomass includes strong acid hydrolysis, dilute acid hydrolysis and non-acid hydrolysis. Through acid hydrolysis, Furfural, levulinic acid, xylose etc can be gained from lignocellulose. For example, cellulosic and starch biomass direct hydrolysis into mono-sugar (such as glucose and fructose) catalyzed by inorganic acid (usually vitriol or hydrochloric acid) at high temperature. The mono-sugar dehydrated into hydroxymethylfurfural by heating which further decarboxylated into levulinic acid.

Hydrolysis of biomass arouse great interests since that use abundant and renewable biomass or waste residue rich in biomass as material to produce levulinic acid by one-step hydrolysis which reduce the cost obviously. Presently, inside and outside of country had used various materials to study the production of levulinic acid by acid hydrolysis, such as waste residual or liquid of xylose, glucose solution, starch, waste residue of pulp, jowar starch, etc.

### 4.3.2. Biotransformation

Biomass is degraded into low molecular chemicals by enzymes and microbial fermentation. For example, cellulose is degraded into glucose by cellulase, hemicellulose is degraded into xylose by xylanase, xylitol and lactic acid etc, can be produced by fermentation. The first important thing of biotransformation is to degrade biomass into fermentable sugar, second is having high efficient microbe or enzyme, and finally is to establish adapted separation and purification system of outputs. Biotransformation technology has widened into yield of commodity chemicals, polymer and special chemicals. Presently, almost 5% chemicals base on biotechnology, which possibly enhanced to 10%~20% at year of 2010.

### 4.3.3. Thermochemical Transformation

Thermochemical transformation of biomass includes direct combustion, liquefaction, pyrolysis and gasification, etc. Compared with other technology, chemicals gained by thermochemical transformation are mainly come from fuel gas, tar and waste residue etc which are low molecular degradation products of cellulose, hemicellulose and lignin under high temperature. For example, under higher temperature and pressure the degradation products of cellulose and hemicellulose including sugar (Oligosaccharides, mono-sugar and laevoglucose, etc), organic acid (Formic acid, acetic acid, lactic acid, 2-hydroxy-butyric acid, 2-hydroxy valeric acid, levulinic acid, Benzoic acid, etc), furfural, furan ramification, aromatic materials, etc, the degradation products of lignin including hydroxybenzene, naphthol, Guaiacyl, aromatic oxy-compounds （such as methoxybenzene etc.） . The degradation products of cellulose under pyrolysis including ketone, phenol, aldehyde, mellow, etc. Tar includes benzene, naphethalene, toluol, dimethylbenzene, styrene, hydroxybenzene and indene, etc.

### 4.3.4. Extraction

Besides, above biochemical and chemical transformation methods, plant-based chemicals also can be gained by extraction. More and more high value-added chemicals have been produced by extraction, which exist on the cell wall of plant or in the cell of plant (mainly effective ingredients of Chinese traditional medicine). Chemicals on cell wall of plant include asafetida acid, pectin, protein, and so on. Chemicals in the cell of plant are much more abundant, as mentioned effective ingredients of Chinese traditional medicine e.g. flavone, alkaloid, anthraquinones, saponins, coumarin, lignans, terpene, polysaccharides, which all can be extracted and isolated from plant by appropriate method. Recently, along with the development of modern technology, supercritical fluid extraction, ultrasonic extraction technology, microwave extraction technology, enzyme engineering technology, semi-bionic extraction technology, high-speed counter current chromatography (HSCCC) extraction technology, macroporous resin adsorption separation technology are introduced into the extraction of chemicals which enhanced the quality and production efficiency of effective ingredients [8, 9].

## 4.4. Transformation Research Thoughts and Methods for Plant-Based Chemicals

Since the plant biomass is composed of numerous macromolecules, existed with complex and compact structure, the utilization of it faces an ambivalent thoughts that to derive and/or degrade after been separating into individual chemical compounds or to degrade and/or converse directly. Many components with higher relative molecular mass are gained by the former method, which could degrade further into fine compounds. However, a difficult problem must be dissolved before this goal to be realized that how to separate each high molecular component efficiently and cleanly. In the latter method, firstly decomposed biomass directly into a mixture with many lower relative molecular mass compounds through the thermochemical degradation, and then use them, or gasifying them into small molecules gas, or separating them into various compounds. Although this method avoids the problems of separating macromolecules, but have to face the problems of separating small molecule complex mixture. Which is the better route? The former might be a good one, because the direct use of these natural macromolecular materials will be able to avoid many polymer polymerization reactions, besides, the product of further degradation after the separation of macromolecules will be much simpler. But to achieve this goal, we must firstly resolve the clean separation problem of major components (cellulose, hemicellulose and lignin) and other macromolecules of raw materials. Therefore, in this area needs a great deal in-depth and meticulous researches, such as: (1) from the molecular-supramolecular-macrostructure - microstructure point to recognize the complex structure of macromolecules in biomass, to define the interaction of these macromolecules, (2) based on the careful study of structure, using chemical, physical and biological approaches to investigate the clean and efficient separating course of macromolecules. If we can properly solve these two problems, then the first line on the success of the half. Before the bottleneck in the first line did not effectively dissolved, most of the researchers chose the second route that through the degradation of biomass into a mixture of small molecules firstly, and then use and transformation.

### 4.4.1. Research and Development of Lignocellulose Transformation into Chemicals

#### *4.4.1.1. Chemical Composition Analysis*

*Chemical Composition of Cellulose*

Glucose is the monomers of cellulose which connect each other through β-1, 4 - glycoside bond to form macromolecular cellulose. Cellulose has complex crystal structure and amorphous structure. Lignocellulosic biomass was degraded into glucose by strong acid or dilutes acid Hydrolysis. Cellulose also be degraded by cellulose enzymes (Fig 4.2), including endo-1, 4-β-D-glucanase (EG), exo-1, 4-β-D-glucanase (CBH) and β-1, 4-glucosidase (BG), into glucose.

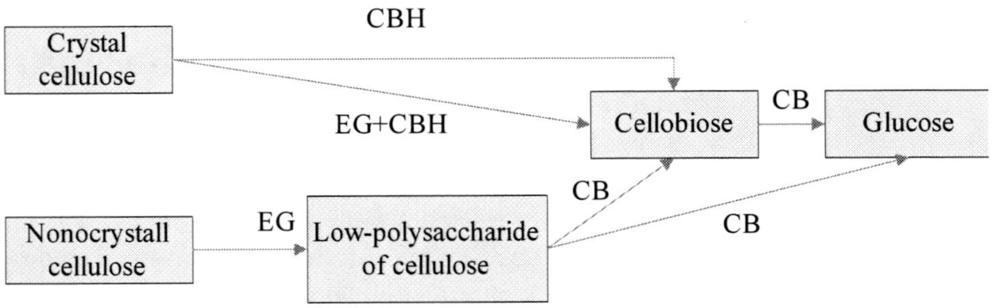


Figure 4.2. Enzymatic hydrolysis of cellulose to glucose [3]

*Chemical Composition of Hemicellulose*

Sugar constitution of hemicellulose main chain includes D-xylose, D-glucose, D-mannose etc, of side chains includes D-xylose, D-galactose, D-glucose, L-Arab sugar, D-galacturonic acid, D-glucuronide, etc. Xylose is the major component of hemicellulose. Xylose can be get through acid hydrolysis of hemicellulose, of which the condition is much tender than acid hydrolysis of cellulose. Enzymatic hydrolysis of hemicellulose also is a main process to get xylose. Holocellulose of Spruce and Fagus, Arab sugar-based -4 - O-methyl glucose acid and 4-O-methyl glucuronic acid -based xylan all can be hydrolyzed into xylose in addition to generate Arab sugar and oligosaccharides by β- 1, 4 - xylanase endonuclease (or Xylanase) isolated from *Trichoderma viride*.

*Chemical Composition of Lignin*

Lignin separated from organic solvents pulping, acidic sulphite pulping or alkaline sulphite pulping was oxidized by potassium permanganate to generate propane benzene compounds, which is the monomer of lignin. Lignin can also be decomposed by lignin peroxidase (LiP), manganese peroxidase (MnP) and laccase to generate some propane benzene compounds or its dimer. LiP and MnP can catalyze the benzene circle of lignin to generate phenoxy radical through single-electron oxidation, then degraded through a series of non-enzymatic catalyst radical reaction, including the removal of methoxy, the fracture of C α - C β, C β-C γ and alkyl-butyl ether bond on the side-chain of lignin, into monomer or dimer (Fig 4.3).

In addition, lignin can be degraded through hydrogenation, catalytic oxidation, heating and several chemical reactions to generate phenol, benzene, vanillin, benzaldehyde, benzoic acid, dimethyl sulfoxide and different aromatic and heterocycle compound with low-molecular-weight. Therefore, lignin may become an important organic industrial resource. For example, phenolic material is typical bulk chemical products, which shows that transformation of lignin has a vast potentiality in industrial applications.

Liu and Xu [10] detected many high value chemicals from the degraded products of lignin, oxidized by photocatalysis, including vanillin, guaiacyl-acetone, guaiacyl acetic acid, 1 - hydroxy -3 - guaiacyl acetone, etc. Yang et al [11] degraded lignosulfonate using high efficient PbO2 membrane electrode. The high-performance liquid chromatography analysis indicated six kinds of low molecular weight phenolic compounds: vanillin, Lilac aldehyde, syringic acid, hydroxybenzaldehyde, acetophenone, and benzoic acid.

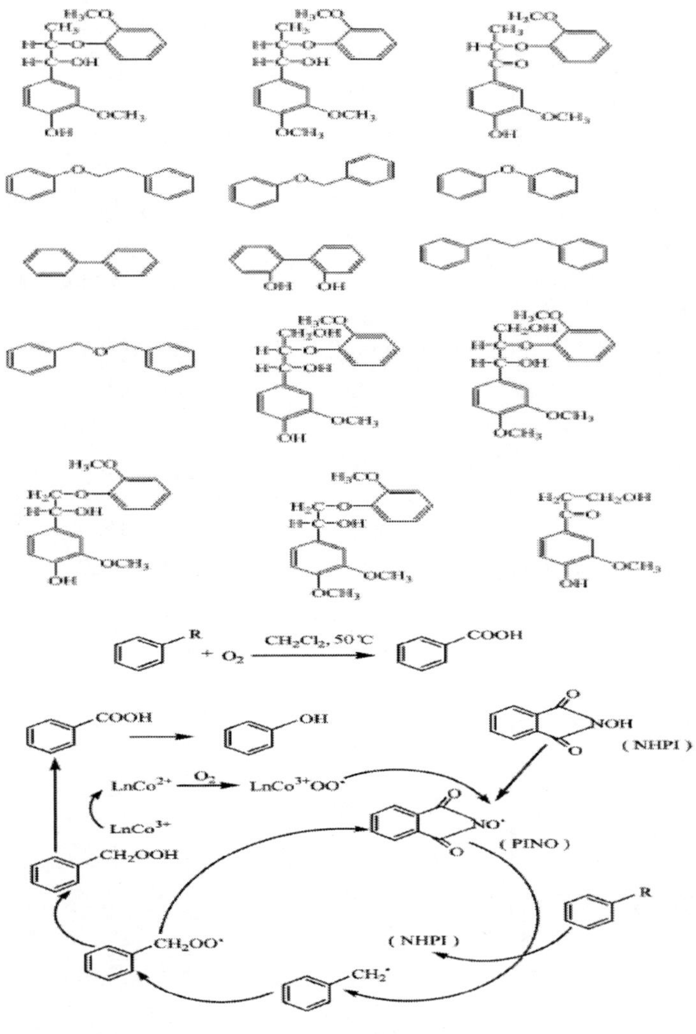

Figure 4.3. Products from degradation of lignin and phenols from conversion of lignin-based monomers [3]

### *4.4.1.2. Direct Conversion*

The direct conversion courses of lignocellulose mainly include acid hydrolysis and therochemical transformation (combustion, gasification, pyrolysis, liquefaction, and so on). In thermal conversion, combustion is already widely applied, gasification attracts a great of interest since it can offer higher efficiencies compared with combustion; and fast pyrolysis arouses much interest because liquid has the advantages of convenient for storage and transport and far-ranging in application [12].

*Gasification*

Biomass gasification is a thermal decomposition process in the presence of limited supply of air or oxygen to produce combustible gases thus converting calorific value of organic material into a gaseous energy carrier [13]. The basic principle of biomass gasification is in certain thermodynamic conditions by means of air (or oxygen) and vapors the macromolecular in biomass takes pyrolysis, oxidation, reduction and reforming reactions, then the tar formed company with pyrolysis further thermical cracking or catalytic cracking into low molecular hydrocarbons such as gases including CO, H2 and CH4. Upon heating the biomass gets dried up to 120℃, devolatalized the volatile matter up to 350℃ and gasified the char above 350℃. Therefore, it is customary to classify the entire gasifier process as drying, devolatiliation and gasification.

A biomass gasifier-based power generation system normally consists of a biomass gasifier, a gas cooling and cleaning unit, and a dual-fuel-engine-electrical generator as shown in Fig 4.4. Design of gasifier depends upon type of fuel used and whether gasifier is portable or stationary [14].

Tar is the inevitable product of biomass gasification. The composition of tar is very complicated and most are benzene derivatives. Precipitation of tar have more than 100 kinds of components, the main component of is not less than 20 kinds, including seven kinds more than five percent, which are benzene, naphthalene, toluene, xylene, styrene, phenol and indinavir. The gasified tar content of biomass is related with the reaction temperature, heating speed and retention time, etc. For example, tar content is decreased obviously with the temperature increase and usually the maximum yield is gained at 500℃.

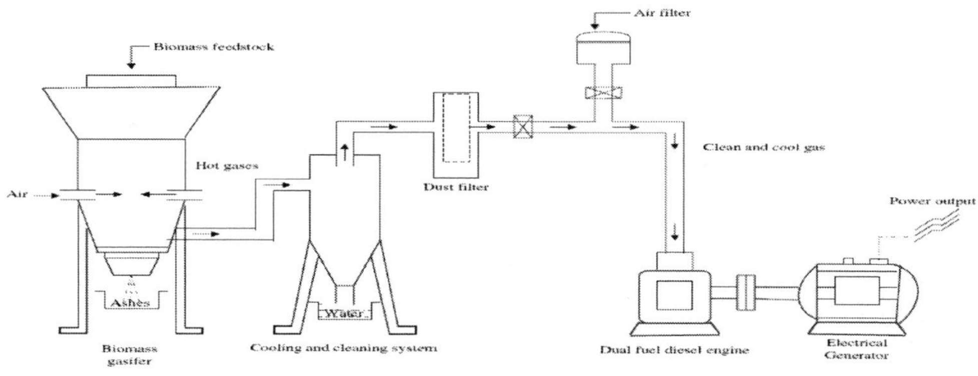


Figure 4.4. Schematic diagram of a biomass gasifier-based power generation system

*Pyrolysis*

The pyrolysis is degradation of biomass by heat in the absence of oxygen including the reactions of bond rupture of macromolecules, isomerization reaction and polymerization of small molecules which converting biomass into liquid product often called 'bio-oil'. Bio-oil is a complex mixture of several different chemical species which can be broadly classified into hydroxyaldehydes and ketones, carboxylic acids, anhydrose sugars and phenolic compounds [5].

The pyrolysis process can be divided into three subclasses: conventional pyrolysis, fast pyrolysis, and flash pyrolysis. The hemicelluloses break down first at temperatures of 470 to 530 K, cellulose follows in the temperature range 510 to 620 K, and lignin is the last component to pyrolyze at temperatures of 550 to 770 K. By pyrolysis, High value-added light aromatics (benzene, toluene and xylene and naphthalene) will be produced.

Phenol-rich tars formed during lignin pyrolysis, causes by decomposition of benzene propane of lignin, can be used to distill commercial value phenol compounds. Klein and Virk gave a mathematical lignin pyrolysis model which simulated the temporal evolution of lignin thermolysis products in each of four categories: namely, (i) gas, comprising methane and carbon monoxide; (ii) aqueous liquid, consisting of water and methanol; (iii) tar, containing single-ring aromatic liquids, the majority related to guaiacol, catechol, and phenol; and (iv) a carbonaceous residue, of multiple-ring aromatic products [15].

Through the GC-MS analysis, the major products of cellulose pyrolysis are ketone, phenol, aldehyde, and alcohols compounds containing methyl, ethyl, methoxy, hydroxyl etc. In these complex mixture L-glucan, ethanol aldehyde and 1-hydroxy-2-acetone are account for about 80 percent. Alcohol aldehyde and 1-hydroxy-2-acetone as a product of cellulose pyrolysis directly generated in the process of competition with L-glucan [16].

Generally speaking, if the purpose is to maximize the yield of liquid products resulting from biomass pyrolysis, a low temperature, high heating rate, short gas residence time process would be required. For a high char production, a low temperature, low heating rate process would be chosen. If the purpose was to maximize the yield of fuel gas resulting from pyrolysis, a high temperature, low heating rate, long gas residence time process would be preferred [17].

Pyrolysis has the advantage of being able to recover the storable fuel, directly recover fuel gas and fuel oil from organic waste under lower temperatures than the burning of the conditions. From the point of resources recovery, pyrolysis has a key role in the conversion of plant-based raw materials into high value-added chemical products.

*Liquefaction*

Liquefaction is another thermochemical conversion process of biomass which sometimes confused with pyrolysis. Unlike pyrolysis, liquefaction decomposed feedstock into fragments in the presence of a suitable catalyst and these fragments are unstable and reactive, repolymerize into oily compounds having appropriate molecular weights. A simple comparison of the operating conditions for liquefaction and pyrolysis is shown in table 4.1.

**Table 4.1. Comparison of liquefaction and pyrolysis [18]**

| **Process** | **Temperature** | **Pressure (MPa)** | **Drying** |
|---|---|---|---|
| Process | 525-600 | 5-20 | Unnecessary |
| Pyrolysis | 650-800 | 0.1-0.5 | Necessary |

Biomass liquefaction is a chemical process to convert biomass into liquid products which can be divided into indirect liquefaction and direct liquefaction. Indirect liquefaction refers to the process that after biomass gasification other liquid products are synthesized such as methanol and dimethylether. This method has the advantages of high-quality products, but the fixed investment too high and the total thermal efficiency very low. Refers to the direct liquefaction, used water or organic solvents as media, biomass was catalyzed in a relatively low temperature to produce a small amount of gas, a large number of liquid products and a small amount of solid residue. According to the different using temperature and pressure, direct liquefaction can be divided into high temperature and pressure directly liquefaction and moderate liquefaction. High temperature and pressure directly liquefaction as a traditional method, its products usually used as fuel oil but before use needs necessary modification. Recently, low pressure even moderate liquefaction gets more and more interest, the products of which are generally used as raw materials for macromolecule(such as adhesives, phenolic resin, polyurethane foam), or as a fuel additive.

Plant-based liquefied low molecule organic compounds cover the considerable part of petrochemical products, such as acids, alcohols, aldehydes and phenol, etc. However, due to the highly complexity, how to separate and purify industrialized scale chemicals from these compounds and gradually replace the current petroleum-based chemicals still under further research and development.

1) High Temperature and Pressure Direct Liquefaction

Liquefaction techniques that converts lignocellulosic materials into fuels and chemicals have been studied since Fierz-David (1925), who converted cellulose directly by simulating coal liquefaction conditions into tarry product quantitatively by applying hydrogen pressure and high temperature in the presence of $Ni(OH)_2$.

In 1971, Apell et al under the conditions of 28 MPa and 350~400℃, using Na2CO3 as catalyst and anthracene oil as solvent, liquefied wood in the medium of $H_2$ and (or) CO and eventually received 53 percent of the crude fuel oil. The study was done in Pittsburgh Energy Research Centre and named PERC method. Lawrence -Beckley Laboratory (LBL) proposed that pretreated lignocellulose at 180℃ and weak acid (pH = 1.8) for 45 min to prepare aqueous pulp, and then adjusted pH to 8 with Na2CO3 and liquefied the pulp with a similar conditions of PERC (using water as the solvent). This process could receive 33 percent of crude oil and been entitled as LBL method.

1990, Shell's Goudriaan and Peferoen proposed the HTU (Hydro Thermal Upgrading) process that converted wood into crude oil, containing 5~15% oxygen, under 300~350℃ for 5~15 min by using aqueous water as solvent.

The product of high temperature and pressure liquefaction is very complicated. The products from cellulose and hemicellulose include: carbohydrate (oligosaccharides, single-sugar, L-glucose, etc.), organic acids (formic acid, acetic acid, lactic acid, 2-Hydroxybutyric acid, 2 -hydroxyisobutyric acid, 2-hydroxyvaleric acid, levulinic acid, benzoic acid, etc.), furfural, furan derivatives, and aromatic compounds. The products from lignin include: phenols, naphthalene phenols, guaiacol, syringin phenols, oxygenated aromatic compounds (such as methoxy-benzene). Other products include: long-chain fatty acids, benzofuran (such as methyl benzofuran, hydroxide benzofuran, methyl-2- benzofuran), a ring compounds (such as 2-Methylcyclopentanone). Overall, the product of high temperature and pressure liquefaction mainly including low molecular weight compounds such as sugar, acid, ketones, phenol and other kinds of carbon hydroxide compounds.

Thermal gravitational analysis indicates that as the temperature increase lignocellulose degradated into aqueous products firstly, then further into gas, oil and tar, and finally repolymerized into high molecular products. In the case of cellulose, it firstly degrades into oligosaccharides, then into glucose which dehydrates to form L-glucose and 5-hydroxy furfural, and finally furfural reacts with solvent (such as phenol) or homopolymerized to form compound containing furan ring structure.

In the catalysis of alkaline salts (such as sodium carbonate and potassium carbonate), macromolecular compounds in the products of liquefaction such as cellulose and hemicellulose are decomposed firstly into micelle-like debris which are degraded into small molecular compounds through dehydration, dehydrogenation, deoxy and decarboxylation. Once these compounds formed, new compounds are further repolymerize through condensation, cyclopolymerization and polymerization reactions.

Yu et al studied the solvent decomposition process of cellulose and reported that at 233.9 ℃ there has phenoxy intermediates in the liquefying products with phenol as solvent and alkoxy intermediates in the liquefying products with alcohol as solvent. If keep heating, phenoxy intermediates started to decompose and produce HMF and phenol, and HMF will further polymerize into high molecular weight compounds if still keep heating.

Due to high energy consumption, high-pressure demand for equipment, complication of equipment because the need for continuous feedstock, and harsh operating conditions, the cost of traditional high temperature and pressure liquefaction process is much high. On the other hand, the main products of high temperature and pressure liquefaction process are crude fuel oil which has the problems similar to fuel oil from pyrolysis that it needs further distillation and modification before application and difficult for chemicals extraction because of the complicated composition. Account for these problems, many researchers put the interest on the lower pressure and temperature direct liquefaction.

2) Moderate Liquefaction

The study of moderate liquefaction began in the 1980s, Shiroishi et al of Kyoto University made a series wood liquefaction tests for the purpose of useing chemical raw materials which made wood directly liquefied in atmospheric pressure with acid as catalyst. This research caused widespread concern in the world because of the simple operating condition, easy to control, low-cost, significant economic benefits and little pollution, and promoted the rapid development of biomass raw material liquefying technology.

Compared with high temperature and high pressure liquefaction, high boiling point solvent is prefer to use in moderate liquefaction since that moderate liquefaction is usually operated under mild condition. Liquefying solvent of moderate liquefaction includes water or aqueous solutions, low alcohol (such as ethanol, butanol, cyclohexanol, etc.), multiple alcohol and its derivatives (such as ethylene glycol, polyethylene glycol, glycerine, etc.), phenols (such as phenol). It was generally believed that under the conditions of same temperature and pressure phenols and multiple alcohols have good liquefing effects than water and simple alcohol. So, the researches on moderate liquefaction are focused on phenols and multiple alcohol liquefaction.

① phenols liquefaction

Phenol exhibits excellent liquefying effect under normal pressure or low-pressure when using acid catalyst, while, certain pressure and high temperature is necessary when using alkali or alkali metal salts as catalyst. In the phenol liquefaction, a considerable part of the phenol will be consumed by turn to bound phenol which account for 50% to 89% of raw material. Residual phenol not involved in the reaction is free phenol. Yucang Zhang et al examined the average molecular weight and amount of bound phenol of liquefied cellulose and the substitution pattern of bound phenol as well as newly formed residue in the late stage of liquefaction in various liquefaction conditions. The results were shown that the linkage fashion of bound phenol was governed by the liquefaction condition. Phenol extended molecular weight by the reaction with decomposed fragments from cellulose and formed new residue when insufficient amount was charged. On the other hand, phenol played a role of an end-cap reagent when sufficient amount was charged. Bound phenol with o-substitution increased with increasing liquefaction temperature [19].

In the system of phenol liquefaction, lignin most vulnerable to liquefy, followed by semi-cellulose, and cellulose is most difficult to liquefy. The phenol liquefied products of cellulose are 5-HMF, oligosaccharides, glucose, etc; lignin phenol liquefied product is mainly lubanol, guaiacol and phenol compounds [20, 21].

Characterization of the residue from wood liquefaction provides a new approach to understand some fundamental aspects of the liquefaction reaction. The Klason lignin content of the liquefying residues decreased, while the holocellulose and a-cellulose contents increased as the phenol to wood ratio (P/W) increased. The SEM images of the residues showed that the fiber bundles were reduced to small-sized bundles or even single fibers as the P/W ratio increased from 1/1 to 3/1, which indicated that the lignin in the middle lamella had been dissolved prior to the cellulose during liquefaction [22].

② polyhydric alcohols liquefaction

Polyhydric alcohol solvent includes ethylene glycol, polyethylene glycol, glycerin and polyhydric alcohol derivatives (such as esters). Shiraishi had reported a almost complete liquefaction of wood using polyhydric alcohol as a solvent which was liquefied by acidic catalyst under the conditions of 150 ℃ and atmosphere pressure. Under similar conditions, the liquefied efficiency of ethylene carbonate ester solvent is 27.9 times of glycol and 10 times of polyethylene glycol (PEG 400)-ethylene glycol (EG) mixture (V PEG400/VEG = 8:2). However, the problems of ethylene carbonate ester solvent include the high cost, the decomposition under acidic conditions and the almost completely lost (for example, ethylene carbonate decomposed into ethylene glycol and carbon dioxide).

Kobayashi et al analyzed the reaction process of wood using polyhydric alcohol through the liquefactions of cellulose powder, steamed lignin, alkali lignin, and their mixtures which indicated that the liquefaction of wood proceeded immediately and wood components were converted to *N*, *N*-dimethylformamide (DMF)-soluble components. In the initial stage of wood liquefaction lignin was converted to DMF-soluble compounds, followed by cellulose gradually being converted into soluble compounds. After that, condensation reactions took place among some parts of depolymerized and degraded compounds from cellulose and lignin, and were converted into DMF-insoluble compounds. Kobayashi et al gave a conclusion that the rate-determining step of wood liquefaction was the depolymerization of cellulose. Furthermore, it was suggested that the condensation reaction was due to the mutual reaction among depolymerized cellulose and degraded aromatic derivatives from lignin or due to the nucleophilic displacement reaction of cellulose by phenoxide ion [23].

Yamada reported that in the liquefied products of cellulose liquefied by ethylene glycol (EG) EG-glycosides were detected as the only saccharides and 2-hydroxyethyl levulinate as the highly decomposed compound derived from cellulose. The mechanism in the EG-liquefaction of cellulose proposed by Yamada was described as follows: First, cellulose is degraded and produces considerable amounts of EG-glycosides during the early stage of liquefaction. Then, when liquefaction is prolonged, the glycosides are decomposed, leading to a large quantity of levulinates [24].

#### *4.4.1.3. Transformed by Catalysis*

Presently, acid catalytic hydrolysis is still the main ways of small molecule chemicals production. A number of platforms compound such as furfural and hydroxymethylfurfural (HMF), levulinate, and xylitol.

*(1) Furfural and Hydroxymethylfurfural (HMF)*

The acid hydrolysis of lignocellulose to produce furfural and HMF includes hydrochloric acid and sulfuric acid hydrolysis.

Sulfuric acid hydrolysis is more extensive application of one of the ways, it using 3% - 6% of dilute sulfuric acid as the catalyst will transform pentosan (or hexosan)into pentose(or hexose) and then dehydrate to get furfural (or HMF). This method has many advantages, such as mature technical line, process simple, easy to operate and control, and lower investment, but also has the disadvantages like low utilization rate of raw materials, high energy consumption of intermittent operation, difficult by-products recovery, and so on.

Hydrochloric acid hydrolysis has the similar technical process and reactive principle with sulfuric acid hydrolysis, but has higher catalytic activity, lower usage of acid, faster hydrolysis speed, shorter reaction time, without using high-pressure steam, higher capacity of hydrolysis reactor, and higher effluent rate of aldehyde [6, 25].

Biomass is also available through direct acid-free hydrolysis process in which it is hydrolyzed under high temperature and pressure to produce furfural and acetic acid, the latter also acts as catalyst. The most significant characteristics of this method are without additional catalyst and higher effluent rate of aldehyde. The process has advantages over other technologies, such as simplification of production technological process, continuous process production, less investment in construction, lower equipment corrosive and easier to deal with furfural [6, 26].

*(2) Levulinic Acid*

Recently, the aggravation of energy crisis and the increasing petroleum oil prices are forcing the chemical industry to find alternative raw materials for the production of basic chemicals. Levulinic acid has the potential as an important basic chemical material. Levulinic acid can be produced by heating hexose, or any carbohydrate containing hexose with a dilute mineral acid for an extended time. Many materials such as glucose, sucrose, fructose and biomass materials including wood, starch, cane sugar, grain sorghum and agricultural wasters have been used to produce levulinic acid [27].

Biomass levulinic acid is prepared by two consecutive autoclave using ligoncellulose as raw materials and dilutes sulfuric acid as the catalyst. Biomass containing cellulose and starch is broken into a single sugar, catalyzed by inorganic acid under high temperature, and then dehydrated to form 5-HMF which further decarboxylated to form Levulinic acid [28]. Levulinic acid is an important platform compounds, it can happen a variety of chemical reactions, such as salt and esterification reaction, and so on.

Levulinic acid has obvious weak acid, while slightly stronger than acetic acid in aqueous solution, and therefore easy to react with solution of alkali (hydroxide and carbonate) to produce levulinate. For example, in boiling water, levulinic acid solution reacts with excessive calcium carbonate, magnesium carbonate, copper carbonate, manganese carbonate or nickel carbonate to produce corresponding levulinate. levulinic acid reacts with excessive copper carbonate to form levulinate copper at 60℃ and with excessive calcium hydroxide to generate levulinate calcium at room temperature [29].

#### *4.4.1.4. Multi-Level Conversion*

The key obstacle of biomass used for industrial raw materials is lack of efficient fractionation technology. The characteristic and quality of final products are impossible to control if the components of biomass cannot be effectively isolated, however, because of the complex composition and high intensity, it is very difficult to separate useful component from biomass. Therefore, the existing industrial production of biomass raw materials are only emphasized the utilization of cellulose. The neglect of other components of biomass, such as lignin, not only caused a great deal of environmental pollution problems, but also the enormous waste of resources.

This single utilization model of biomass component, lack of fundamentality, integrity and systematism, can not achieve the simultaneous utilization of cellulose, hemicellulose and lignin. In order to solve this problem essentially, we must first of all fully aware of that other components of biomass is not waste, but resources, is potential products needs to be developed. It is indicating that fractionation is very important to the utilization and development of bio-based materials, and therefore more and more fractionation method were exploited.

Hongzhang Chen, from the point of total utilization, proposed the fractionation and muti-level conversion thoughts of biomass, which treat biomass not only as cellulose resources, but regards it as a multi-component material. By fractionation, lignocellulosic biomass is refined into various pure components, and these components can be processed into valuable products. It is necessary to develop new utilization methods for bio-base raw materials since that the transformation of macromolecule components, separated from these materials, with physical, chemical or biological approach into chemicals has the disadvantages of high cost and complex operation. The multi-level conversion means that biomass is firstly pretreated, and then produces bio-based products according to the composition and nature of raw materials, using biological transformation and thermal chemical conversion as the theme line, through biotechnology, physical and chemical technology. The ultimate aim of multi-level conversion of biomass is to achieve its total utilization of biomass, including the utilization of cellulose, lignin and hemicellulose, ash and other substances also is included.

During the process of multi-level conversion, various technologies are integrated to achieve the complete utilization of cellulose, hemicellulose and lignin. Steam explosion technology is a low-cost, effective method for hemicellulose recycling but lignin still remains in solid residues. On the other hand, ethanol can effectively extract lignin but destroys hemicellulose obviously for the intense condition. The combination of steam explosion and ethanol extraction can achieve the separation of cellulose, hemicellulose and lignin. Steam explosion firstly reclaim hemicellulose and split propylene ether bond, which made the ethanol extraction of lignin much easier. This chapter introduces the muti-level conversion approaches of bio-based raw materials into chemicals.

### 4.4.2. Process Research and Development of Plant-Based Chemicals by Extraction

Plant chemicals can be divided into chemicals in cell and on cell wall.

The component of plant cell wall includes cellulose, hemicellulose and lignin, and other major components such as protein, ferulic acid, fat, $SiO_2$, etc. In the cell wall of some Chinese herbal medicine are also rich in active ingredients. Therefore, it is the effective means to develop other high value but low content chemicals at the same time of using plant cell wall.

Components in plant cell are complex, including chlorophyll, gums, tannin, inorganic salts, alkaloids, flavonoids, amino acids, and so on. Chinese traditional medicine riches in a large number of active ingredients and the extraction of active ingredients are significant to the development of Chinese traditional medicine research and modernization. The extraction and separation of these effective components have a direct relation to the quality, efficiency and yield of Chinese traditional medicine. Therefore, it is necessary to strengthen the extraction process research of Chinese traditional medicine and develop the modern extracting method of active ingredients.

Traditional plant-based compounds extracting methods include soaking, decocting, infiltration, steam distillation, expressing, etc. Recently, along with the development of modern technology, supercritical fluid extraction, ultrasonic extraction technology, microwave extraction technology, enzyme engineering technology, semi-bionic extraction technology, high-speed counter current chromatography (HSCCC) extraction technology, macroporous resin adsorption separation technology are introduced into the extraction of chemicals which enhanced the quality and production efficiency of effective ingredients[8].

However, methods mentioned above are only the development of unit operations that mostly are manual and static extraction. With regard to these methods, it is necessary to avoid saturation by updating extracting solvent to maintain certain efficiency and extracting rate, which increases the elute efficiency and batch difference. These methods not only enhance the labor intensity and costs, but also have the disadvantages such as low extracting rate of active ingredients, long extracting period, high-energy consumption, poor product quality, etc, all of which make these method not suitable for the needs of industrial modernization [30].

#### *4.4.2.1. Demands for Modern Extracting Process of Plant-Based Chemicals*

With the progress of scientific and technological and mutual penetration of multidisciplinary, on the basis of in-depth study of the natural product extraction process and principle, a great of new technologies and approaches are appeared, on the other hand put forward some new requirements on the extracting technology and equipment of natural products:

[1] Rapid, high extraction efficiency, sufficient active ingredients extraction, high yield of effective components, low waste of raw materials;

[2] Less solvent consumption, lower elution coefficient, high lixivium concentrations, low production cost and solvent saving;

[3] Extraction temperature is adjustable. Temperature can not be too high and can be selected in 50℃~100℃. Suitable for the extraction of not only thermalstabile

materials, but also thermosensitive materials and can reduce the damage to the active ingredients;

[4] Transfer speed of raw material in extractor is adjustable. Extraction time can adjusted according to characteristics of different raw materials. Wild applicability and suitable for the extraction of different materials;

[5] Supporting the development of extraction equipment for its production of continuity, and should be suitable for modern industrialization continuous production.

#### *4.4.2.2 Extraction Method Base on Steam-Exploded Technology*

*1) Extraction and Purification of Puerariae Flavones*

*Puerariae* is an important medicinal plant resource, and its main bioactive components include puerarin, daidzin, daidzein, 3'-methoxypuerarin, and other isoflavones [31]. At present, the comprehensive utilization of *Puerariae* is absent, for example, the pharmaceutical industry often treats *Puerariae* starch as a waste, and *Puerariae* starch manufacturers are mostly ignore the recovery of isoflavones. *Puerariae* starch manufacturers and isoflavones pharmaceutical industries both elute large amount of organic wastewater, which brings serious pollution of the environment. Therefore, the first important problems of Puerariae industry are to realize the adequate and reasonable comprehensive utilization of *Puerariae* resources, to enhance the economic value of *puerarin*, and to achieve pollution-free cleaner production of *Puerariae*.

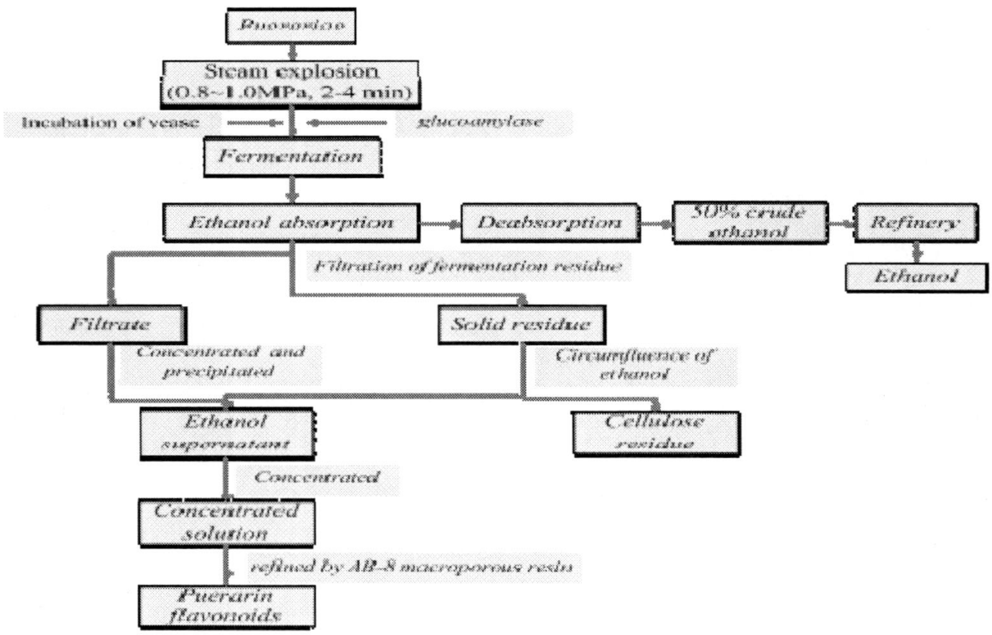


Figure 4.5. Recovery of Puerarin flavonoids from fermentation wastewater of Puerariae fermentation industry

To achieve the comprehensive, clean and highly efficient use of *Puerariae*, we put forward pretreating *puerariae* with steam explosion technology and extracting *Pueraria* flavonoids from the residue of *puerariae* after enzymatic hydrolysis and fermentation. During the fermenting process of *Puerariae* to produce ethanol, steam explosion and fermentation is conducive to the dissolution of *Puerarin* flavonoids. Thermochemical machinery effect of steam-explosion process broken the cell wall of plant tissue and destroyed the structure of solid *Puerariae*. The fermentation process utilized the starch of *Puerariae*. The steam explosion and fermentation significantly reduce the extraction solvent amount of puerarin flavonoids.

Figure 4.5 shows the process of *Puerarin* flavonoids refined by AB-8 macroporous resin. The filtrate of alcohol fermentation residue is used to extract total flavonoids by alcohol-water solution. The adding cellulase before fermentation degraded the cellulose of *Puerariae* epidermis and released the *Puerariae* starch, which changed the structure during fermentation process and produce ethanol. Ethanol is conducive to the slow dissolution of flavonoids. Extraction rate of Puerarin flavonoids from fermentation residue is 12.15 percent increase than not be fermented.

*2) Extraction and Purification of Ephedrine*

Plant-based ephedrine distributed mainly in the medial cortex, which must be overcome the mass transfer resistance of epidermis, fiber layer, in particular cell wall. Compact cell wall structure is the main obstacle of active ingredient extraction, resulting in the low extraction rate of existing industrial ephedrine extraction methods [32-34].

We make full use of steam cooking and shearing effect in the process of steam explosion to enhance extraction rate of active ingredient by plant cell walls broken. Under appropriate conditions the ephedrine extraction rate of steam-exploded ephedra up to 0.939 percent, which attribute to the structural changes of ephedra before and after steam explosion. Through the electron microscope, pores on the surface of steam-exploded ephedra increased obviously, some cell has apparent split, surficial fibers and fibril bundle crimped and folded, some fiber was ruptured and the cell wall was destroyed.

By analyzing the mechanism we believe that cooking effect of steam explosion on plant with cell structure under certain pressure and temperature makes hemicellulose degraded into soluble sugar and lignin at compound middle lamella softened and partial degraded, which weakened the bond between fibers. Then media and materials complete the physical energy release process together under a sudden decompression which releasing powerful force to break the plant cell wall. At the same time, vapor-phase materials rapidly boiled in the ejecting instant and acted on materials in the form of shock wave. Moreover, the producing shearing effect made materials deformed under softening conditions resulted in cells rupture and the formation of porosity (see Figure 2.8 B and D). In follow-up extraction process, because of the organizational structure damage of steam-exploded ephedra grass, cell wall, as the main obstacle of extraction process, is ruptured, which makes the mass transfer resistance greatly reduced. Therefore, ephedrine is much easier to contact with solvent rapidly, dissolve and diffuse completely and speed up mass transfer rate, which are conduce to enhance extraction rate in a short period of time [35, 36].

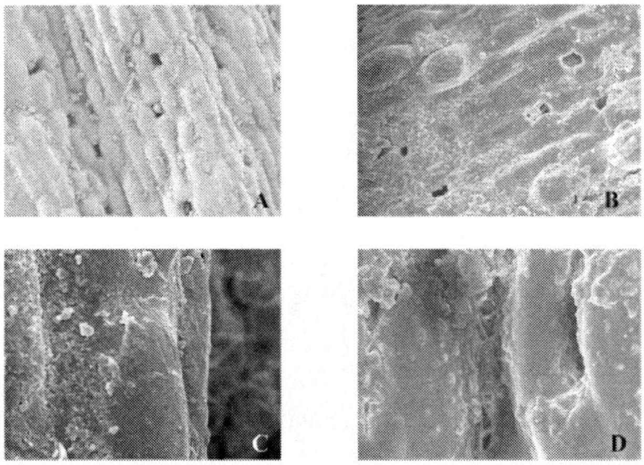


Figure 4.6. Fiber structure of ephedra grass before and after steam explosion scanned by electron microscope

#### *4.4.2.3. Dynamic Countercurrent Extraction*

*Principle*

The components in plant cell are very complexity. It is considered that invalid ingredient includes cellulose, chlorophyll, gums, tannin, inorganic salt, and so on, active ingredient includes alkaloids, flavonoids, organic acids, amino acids, coumarin [8]. The molecular weight of active ingredients in natural products generally smaller than invalid ingredients, which conduce to the permeation of active ingredients through the cell membrane and keep invalid ingredients remain in cells. Extracting process is a basic union of some operation unit such as soak, permeation, precipitation, dissolution, diffusion, replacement, etc. Extraction is a mass transfer process of solutes in materials from solid-phase to liquid-phase, explained with the theories of diffusion, which means that solutes transfer from high concentration to low concentration [37]. The concentration diffusion of active ingredients in liquid extracting solvent and solid materials may serve as motive power of diffusion. The greater the concentration difference between the liquid extracting solvent and solid materials, the larger the diffusing motive power of extracting process and the larger the leaching rate of active ingredients. Therefore, in order to extract the active ingredients of raw materials rapidly and fully, it is necessary to update frequently the interface of solid-liquid phase so that the concentration difference of two-phase maintained at a higher level. The principle of dynamic countercurrent extraction is just the method that utilizes mechanical compulsive means make relative movement of materials and solvent, so that the active ingredients in natural products will always in a higher concentration gradient.

1) Equipment

Dynamic countercurrent extraction equipment (designed by ke et al) mainly includes the power plant, feed device, extraction device, microwave-generating device, slag removing devices, solvent recovery unit, and temperature control devices [38]. Figure 4.7 shows the structural diagram of the entire dynamic countercurrent equipment.

This equipment utilizes the microwave absorbing difference of materials with different structure to heat component of materials selectively, so that active ingredient isolated from the matrix into extracting solvent with smaller dielectric constant and poorer microwave absorption capacity. As a result of integration of dynamic countercurrent and microwave radiation, it shorten the time of active ingredients transfer from inside to surface and conduce to rapid balance of active ingredients in the material and solvent, consequently shorten the extracting time of active ingredients and enhance the extraction rate of active ingredients.

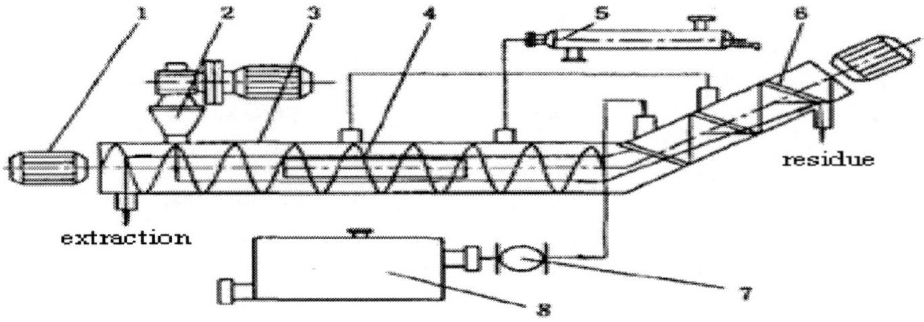


Figure 4.7. Schematic diagram of dynamic countercurrent extraction equipment [39] 1. Power plant, 2. Feed device, 3. Extraction device, 4. Microwave-generating device, 5. Solvent recovery unit, 6. Slag removing devices, 7. Liquid solvent supplying device, 8. Temperature control devices

2) Technological Process [39]

At the early stage of extraction, pretreated plants is input equably to one side of the extractor through the batchmeter controlled by feed device and pushed by the spiral-shipped to the other side. During the pushing process plants is always in a state of dynamics by stirring and rolling to make full contact with extracting solvent. Because of the initial low temperature when plants contacted with countercurrent extracting solvent, high concentration and low temperature, active ingredients such as aroma components, pigments, amino acids, and ketones, are firstly been extracted and then effuse from the exit to the next device.

At the stage of deep extraction, the contact solvent temperature keep rising and the solvent concentration becomes lower and lower, on the other hand, as the effect of microwave, the cell of materials are fully broken which made components, extracting difficultly, gradually leaching, such as soluble protein, pectin, and some polyphenols, etc.

At the stage of drip washing, materials are extracted fully and the solid residue is pushed to solvent-adding opening gradually, where has the lowest concentration of extracting solvent and highest extracting temperature. Extracting solvent washed the surface of residues, so that the soluble components on the surface of residue are washed into the extraction solution. Residues after be washed are pushed into the slag removing device and be discharged.

#### *4.4.2.4. Refinery Production*

Traditionally, fine chemicals mainly refer to the intermediate-functional and special chemicals produced from oil and coal. Recently the utilization of many natural products, especially renewable resources, also faces with the important subject of refinery producing process and products [40]. In this part refinery production of plant polyphenols used as an example to be introduced in detail.

More than 70% of products from tannin extract have been used for leather production and the rest mainly used for production of mud thinner and adhesives, however, few for the preparation of other high value-added fine chemicals which results in the low value of tannin extract. The reasons of this phenomenon are largely due to the current single production mode and coarseness of products. As for plant polyphenols, it is a natural polymer with molecular dispersion and wide molecular weight distribution. Plant polyphenols can be used for many fine chemicals areas, however, there has a certain choice for the chemical structure and molecular weight of polyphenol for different use and the appropriate molecular weight range is the most basic requirement. Therefore, molecular weight classification of polyphenol is very important.

At present, there has a significant difference between the fractionations situation of polyphenol, gained by mass-production, and applications requirements of the ideal state, making it in all aspects of the application are subject to a certain degree of restriction. Each fractionation of Polyphenol extracts has its appropriate application field, but when they exist in a form of mixture will always led to mutual hindered which has been the dominant reason that limit the application value and scope of plant-based polyphenols. Therefore, refinement and specialization of products are the efficient approaches for the application of this kind of natural products.

*1) Solvent Fractionation*

One approach of solvent fractionation is to prepare water extractions firstly and then extract by solvent with different polarity, another approach is to extract directly by different solvents. Figure 4.8 shows the typical example of first method.

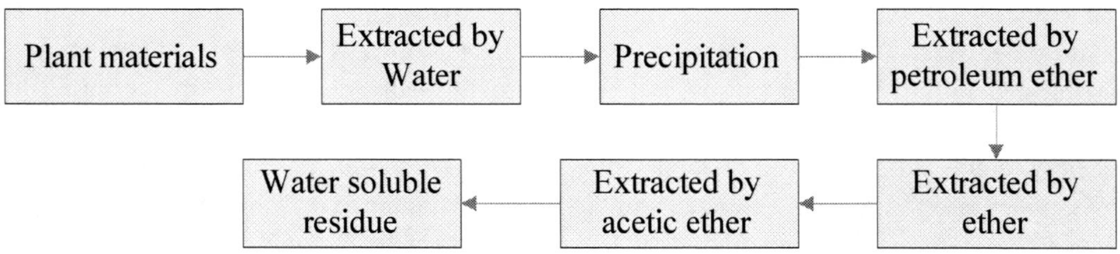


Figure 4.8. Typical example of first method [40]

The fractionation results of tannin extract indicated that petroleum ether extraction contains 5% gum, ether extraction mainly contains low molecular weight polyphenols such as flavonoids, catechins and catechin dimer, acetic ether extraction mainly contains tannin compounds with molecular weight around 1000, water soluble residue mainly contains high molecular weight polyphenols. For example, ether extraction contains the fraction with number average molecular weight of 350 and 860 for acetic ether extraction and 1550 for water-soluble residue. Figure 4.9 shows the typical example of second method.

Acetone and acetone-water system have very good solubility to the tannic component in plant material. Molecular weight of tannic components in acetone extraction is relatively small and which is relatively high in acetone-water extraction. Components in water extraction are much more complexity which including tannin and water-soluble non-tannin.

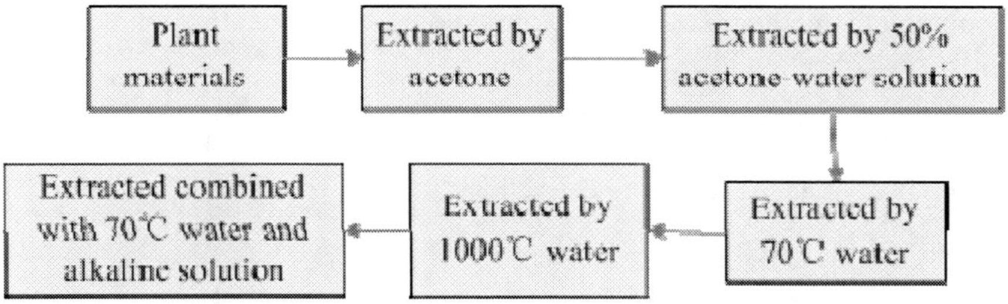


Figure 4.9 Typical example of second method [40]

*2) Precipitation Fractionation*

Precipitation fractionation mainly includes cooling precipitation and metal ions precipitation. Cooling precipitation is that a part of high molecular weight polyphenols in plant materials precipitated firstly when the hot-water extraction is been cooling which means that the extraction can be roughly fractionated by gradually piecewise cooling. Metal ions precipitation means that tannic with molecular weight higher than 500 is precipitated by metal complex effect happened under certain pH and the precipitation can redissolve by acidification. Therefore, polyphenols and other non-tannin compounds can be separate by this method. For example, after precipitated by lead acetate and dissolved by acid or H2S, the purity of tannic from polyphenols extraction reached 95%.

*3) Supercitical Fluid Extraction Fractionation*

In the near critical point, the small changes of temperature and pressure often lead to several magnitude changes of the solubility of solutes. Using this characteristic supercritical fluid has perfect extracting effect and also low pollution and energy consumption. Supercritical $CO_2$ has excellent extracting effect on small non-polar or lower polar compounds. At the same time, by adding cosolvent to increase the dissolvability of polar compounds in extracting system can enhance the extraction rate of plant polyphenols.

*4) Purification Method Combination and Condition Control*

Purification method combination and condition control means that combine the above-mentioned methods to get better fractionation effect. For example, after fractionated by water the ether extraction, acetic ether extraction and water-soluble residue a precipitated by metal ion respectively, or precipitate the water extraction by metal ion firstly and then extract fractionatedly.

In addition, condition control of various fractionation methods also has an impact on the fractionation effect. For example, the structure and content of material is relation to the pH value of extracting solvent when extracted using organic solvent which means that can control the fractions by adjusting the pH of water extraction [40].

#### *4.4.2.5. Dynamic Recycling Leaching*

Dynamic recycling leaching, condensation, extraction, column chromatography, and so on are the main section of alkaloid production [41]. Dynamic recycling leaching is a series combination system of some extracting column. Each column is pre-encased the needed materials, then chooses a column as start column randomly, after adding certain amount solvent starts the circulating pump. After dynamic recycling for a certain time gives off extracting solution to another extracting column to operate same extraction until last column, and then fulfill the collection of extracting solution.

The advantage of dynamic recycling leaching includes high extraction rate, less extraction solvent, energy saving and short production period. Figure 4.10 shows the technique process of dynamic recycling extraction.

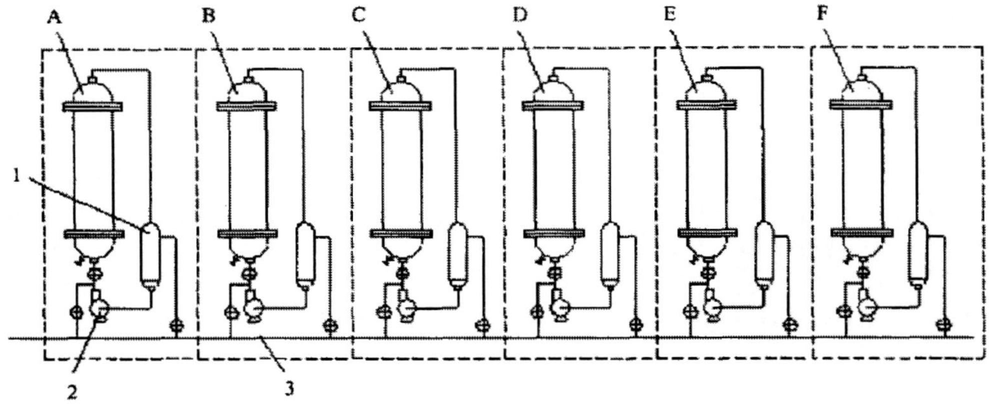


Fig 4.10 The flowchart of the dynamic recycling leaching [41] 1、storage tank 2、circulating pump 3、main pipe A, B, C, D, E, F the first to the sixth leaching columns, respectively

## 4.5. Process Development of Plant-Based Chemicals

With the gradual depletion of fossil fuels and the needs of the protection of the environment and achieve sustainable human development goals, production of high value-added chemical products from renewable plant resources, especially wood biomass, becomes an important development strategies and hot scientific research field of many countries. In the 1990s, developed countries have successfully developed renewable chemicals instead of oil, and further accelerate the research and development process of non-renewable resources such as agriculture and urban fiber plant stalk. America plan to 2030 more than 25 percent of the organic chemicals will be produce from biomass, which means that fuel and chemical industry is at transition stage from the era of non-renewable 'hydrocarbons' to the era of renewable 'carbohydrates'[1, 2].

With the development of chemical industry, the development of the chemicals also has experienced the variation from coal chemical industry to oil chemical industry. In the 19th century and early 20th century, the basis of organic chemistry industry is coal chemical industry, many platform compounds (that is some basic organic compounds have characteristics of abundant resources, low prices, widely usage) derived from coal such as ethylene, benzene, naphthalene, methane etc. Middle of the 20th century petrochemical industry replaced coal chemical industry to become the main provider of platform compounds, such as ethylene, propylene, benzene, etc. Nowadays, the green chemical industry requests the source of chemicals should derive from renewable resources, which makes new platform compounds from abundant renewable biomass are increasingly becoming the research focus of chemist.

### 4.5.1. Furfural

Pentosan is hydrolyzed to generate pentose under acid condition and then dehydrated to generate furfural by acid, which can be presented as the following reaction formula:

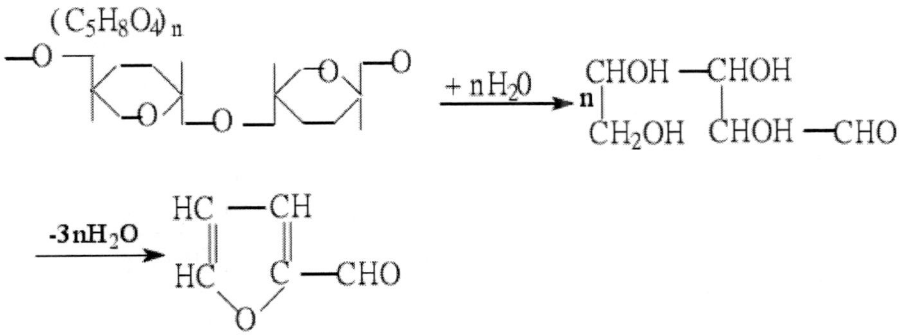

Presently, raw materials for furfural production are mainly lignocellulose of agriculture and forestry by-products, such as corncob, sunflower shells, cotton shell, rice husks, bagasse, hemp stalk, broadleaved forest, etc. The production of furfural mainly uses pentosan of raw materials. The content of pentosan in corncob as higher as 38% and the potential furfural content is 25%. In addition to xylose, Arabinose (generated from gum or derived from the degradation of L-Gluconate calcium and peroxide), D-Ribose, D-Lysose, and so on can also generate furfural, but the furfural production capacity are different. Ascorbic acid and some uronicacids, such as glucuronide, also can produce furfural [42].

#### *4.5.1.1. General Situation Industry Progress of Furfural*

Furfural production methods include into one-step and two-step. One-step is that the hydrolysis of pentosan and dehydration of pentose achieve one off in one hydrolysis-pot. In two-step, pentosan is hydrolyzed around 100 ℃ to generate pentose firstly, and then dehydrating cyclization of pentose at much higher temperature results in furfural.

1. One-step technique

One-step has been widely used in furfural industry because of low investment and simple operation. After decades of development, production techniques of furfural gained greatly improvement, developed from the initial one-pot cooking to the serial multiple-pot as well as continuous production. In the techniques have been reported, Quaker Oats technique, Agrifuran technique, Petrole-chimie technique, Escher Wyss technique, Rosenlew technique, RRL-J technique, and so on were widely used [43, 44].

2. Two-step technique

Two-step has not realized industrialization because complex technique and high investment. However, with the development of furfural industry, as well as the requirement of environmental protection and raw material comprehensive utilization, two-step technique of furfural will be an inevitable trend for furfural industry. Two-step method has the advantage that cellulose and lignin in raw materials does not react during the hydrolysis process, which after the separation can be used to produce other chemical products. In addition, alone dehydration of separated xylose solution not only reduces the steam consumption, also resolves the furfural refining problem which happened in one-step technique that decomposition of cellulose and lignin forms impurities [45, 46].

3. Acid free catalysis technique

The most prominent feature of acid free catalysis is that raw materials hydrolyzed under high temperature and high pressure to produce furfural and generate acetic acid, the latter act as catalyst and no need added catalyst. It has the advantage that the production process is relatively simple and continuous, less investment and less equipment corrosion, easy to deal with furfural residue, etc. In recent years, this technique has aroused widespread attention.

4. Steam-exploded technique

In essence, steam explosion technology is a hydrolytic process of biomass took place under high temperature and high pressure. Cellulose and hemicellulose in lignocellulosic biomass partially degraded and generated a number of acids incidentally, and thus no additional acid is need for furfural production from biomass. Taking steam explosion pretreated cornstalk as an example, the furfural preparation technique of acid-free catalysis includes following technical points.

1) Crushed cornstalk pretreated by steam explosion.

Usually through steam explosion processing, about 80% hemicellulose is degraded and generate acetic acid. Steam permeated into fiber cells soften the cellulose and degraded lignin partially, which is conducive to the latter utilization of various components.

2) Wash steam-exploded corn stalks with water.

Xylo-oligosaccharide and some residues are separated from bath lotion. The washing process requires controlling the ratio of solid to liquid, which facilitates to the treatment of bath lotion. In order to make the most of the xylose able to wash out usually washed as a batch.

3) Filtration and concentration of washing water.

4) Preparation of furfural.

Washing water pretreated as procedure 3) reacts at 140℃~160℃ for 3~4 h and condensed steam at set value to get crude furfural.

#### *4.5.1.2. Problems*

*1. Small-Scale and Low-Yield*

In theory, any material containing pentosans can be used for the production of furfural. Technically furfural is produced by acid hydrolysis of the pentosan contained in woody biomass. Almost all furfural plants employing the batch process use the Quaker Oats technology developed in the 1920's [47]. They all operate at less than 50% yield. It is difficult to enhance furfural yield through direct hydrolysis of biomass. However, the most important requirements of being platform compound is large output and low prices that only large-scale production can be a benefit.

*2. Serious Wastewater Pollution*

Production of furfural from biomass has the disadvantages of complex production, difficult separation, high extraction costs, and serious wastewater pollution. Acid catalyst of furfural production from biomass usually are hydrochloric acid or sulfuric acid which have strong corrosive and toxic and always produce large number of acidic residue and solution bringing serious environment problems. According to statistics, high concentration acidic wastewater emission from a plant with 1000t/a production capacity of furfural is 80 $m^3$/d and COD concentration reaches to 15000~23000 mg/L. Because of strong acidity, high pollutant loading, and high temperature (70 ℃), it is difficult to realize the standard discharge [48].

*3. High Energy Consumption*

Take furfural production of corncob as an example, producing each ton of furfural consumes 11~12 tons of corncob (about 18% water content) and 35~ 40 tons of saturated steam (0.6 MPa). Calculated according to the average production of 150 Kilotons furfural, consumption of steam is 6000 Kilotons, equivalent to nearly 900 Kilotons standard coal (per ton of coal steam 6 Kilotons dollars) [49].

*4. Low Level of Comprehensive Utilization*

Low level of comprehensive utilization mainly presents in the recycling of chemicals in wastewater and comprehensive utilization of waste residue.

In the producing process of furfural will produce a large number of acid-bearing wastewater. For example, a furfural plant with capacity of annual 1000 t emits about $3\times10^4$ t acid-bearing wastewater containing about 1% to 2.5% (mass fraction) of acetic acid. Though some wastewater recycling research has been reported, due to the low acetic acid content, traditional separation methods such as distillation, neutralization, electrodialysis etc has the problems like high energy consumption and low recovery. Secondary wastewater of furfural industry is still very difficult to achieve the national emission standards [50].

Another pollutant of furfural industry is solid residue. Take Chinese furfural industry as an example, China has more than 140 furfural production plant, calculating as annual furfural production capacity is about 10 million tons, the waste residue discharges 130 million tons. Most of this accumulation is disposed by deposition or dumping, which caused serious environmental pollution and take up a lot of land [51]. With the rapid development of furfural industry, the increasing discharge of waste residue has become a bottleneck to furfural industry.

#### ***4.5.1.3. Process Development***

*Multi-Level Conversion of Materials*

The multi-level conversion of biomass refers to chemicals production process that pretreating biomass by fractionation technology and preparing chemicals through biotechnology, physical and chemical methods taking biological and thermochemical conversion as the dominating route according to the composition and characteristics of raw materials. Nowadays, most of the biomass acid hydrolysis processes are directly acid hydrolysis of raw materials, lack of the idea of recycling multi-level conversion of natural organic solid-phase materials, which not only result in waste of resources, but also cause environmental pollution. The exertion of total utilization theory and multi-level recycling conversion model could increase the utilization rate raw materials in the production of furfural, achieve sustainability of raw materials and reduce the production cost of furfural. Figure 4.11 shows the total utilization of cornstalk to produce furfural.

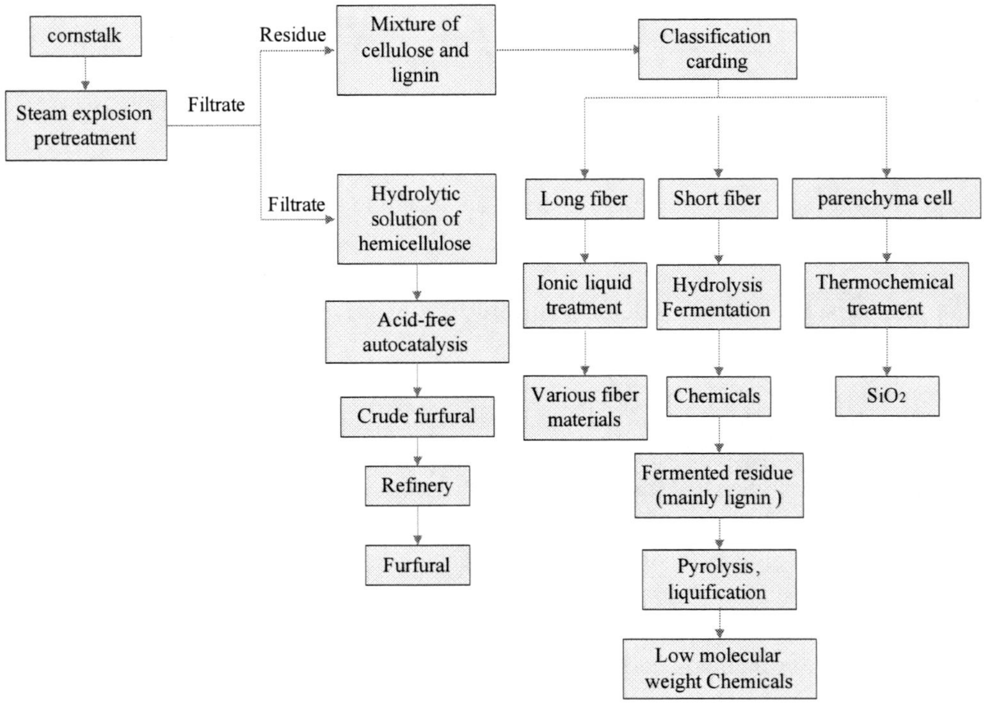


Figure 4.11. Total-utilization pathway of furfural production from Corn stalks

Multi-level utilization not only can enhance the hydrolysis efficiency of furfural and value-adding of biomass also conduce to establish an eco-industrial mode with furfural as dominant production. Furfural residue, with heating amount of about 7524~10032 kl/kg, is an excellent energy resource, which can burn easily and completely without addition fuel after been dried. The comprehensive utilization of furfural residue can save about 1.8~2.6 t coal for each ton of furfural. Zhang et al. proposed an integrated mode of furfural production combined cycle for cogeneration of heat and power to deal with the waste furfural residue efficiently, which achieve the optimized allocation and effective utilization of materials and energies. Furfural plant provides waster residues as raw materials for thermal power plant to save some coal, at the same time dry the residues of furfural by waste heat to realize cascaded utilization of energy, maximized utilization of resources and minimized emissions of by-products. It is indicated that by improving the thermal power plant equipment, water content of waste residue will be further reduced to 10% and the combustion rate of furfural waste increases up to 90%. The eco-industrial system taking furfural chemical plant and thermal power plant as centre will form a renewable resource-based sustainable eco-industrial system with solar as initial energy, planting as source, waster of agricultural products as industrial production materials.

*Integration of Unit Operation*

Energy consumption of furfural production mainly includes hydrolysis, distillation and refining section. Series-connected measure is used to reduce steam consumption and increase solution concentration of hydrolysis section, which could be divided into double-series connection and treble-series connection. Treble-series connection hydrolysis technique means that three pots all in hydrolytic status. However, if any one of these pots in residue-discharging, feedstock or pressure-boosting status it will form double-series connection with other two pots and return to form treble-series connection when it finishes the residue-discharging, feedstock or pressure-boosting unit operation. Ren Hongjun compared the steam consumption of one-pot hydrolysis, double-series connection hydrolysis and treble-series connection hydrolysis that shown in table 4.2.

**Table 4.2. Steam and heat consumption of three kinds of hydrolysis sections [49]**

| Operation Technology | Steam consumption(t)/ furfural (t) | Heat consumption (kJ)/ furfural (t) |
|---|---|---|
| Treble-series connection | 20.314 | $57.7\times10^6$ |
| double-series connection | 26.854 | $74.5\times10^6$ |
| One-pot | 46.754 | $129.7\times10^6$ |

*Comprehensive Utilization of Waste Water*

Discharge of furfural wastewater is not only a waste of resources also caused serious environmental pollution. Environmental monitoring results show that the furfural wastewater contains 1.0~2.5% acetic acid, 0.2~0.5% furfural and various organic compounds. The pH of furfural wastewater is below 2 and COD value is about 15000 mg / L, which is150 times of the wastewater discharge standards. Although there has a large number of management and research about furfural industrial wastewater, but these methods have the disadvantages like low quality of recovery sodium acetate and high operating costs. Shui Zhiliang proposed a technique that recycled butyl acetate, ammonium sulfate, ammonium acetate, acetonitrile and acetamide from furfural wastewater, and the treated water can be used directly to boiler inlet water. By the improvement of furfural production technique recycling use of furfural wastewater, closed circulation treatment of furfural wastewater not only achieves the zero-emission of wastewater also can utilize the combustible acetic acid and oil in wastewater efficiently. Closed circulation treatment does not produce secondary pollution and solves the pollution problems of furfural industry, so that the production of furfural really takes the recycling and cleaning production course. The cost of closed circulation treatment is only 0.06 $\$/m^3$ which is 20% of initial method. Take Chinese furfural industry as an example, calculated according to the production capacity of furfural is $4\times10^5$ t, each ton of furfural produces 15 $m^3$ wastewater, and COD of treated wastewater is 120 mg/L, every year can save 5.71 million dollars, reduce the discharge of COD 720 t and save water $6\times10^6$ $m^3$[52].

## 4.5.2. Levulinic Acid

Levulinic acid, also name as 4-oxopentanoic acid, is an important chemical product having a ketone carbonyl group and an acidic carboxyl group. Recently, the aggravation of energy crisis and the increasing petroleum oil prices are forcing the chemical industry to find alternative raw materials for the production of basic chemicals. Levulinic acid has the potential as an important basic chemical material. As a major source of carbohydrates, including cellulose, hemicellulose and lignin, if wheat straw can be directly used for levulinic acid production, the production cost of levulinic acid will be lower greatly [53]. Under the catalysis of inorganic acid such as sulfuric acid and hydrochloric acid, starch or cellulose materials are hydrolyzed into monosaccharide (glucose and fructose), which dehydrated into 5-HMF by heating, and then further decarboxylated into levulinic acid. The raw material of LA include paper mill sludge, municipal solid waste, unrecyclable waste paper, waste wood, and agricultural residues.

However, the traditional levulinic acid production from cellulosic biomass are using acid as catalyst which results in low yields and significant amounts of unwanted side products. In 1997, Biofine Company of United States proposed a cost-effective LA production process, which favors high-yield production of levulinic acid for as low as $0.04-0.10/lb at large scales. Due to this contribution, Biofine Company owned the "Presidential Green Chemistry Challenge Award".

#### *4.5.2.1. General Situation of LA Industry Progress*

At present, preparation methods of LA include two main ways: furfuryl alcohol catalytic hydrolysis and biomass direct hydrolysis. Furfuryl alcohol catalytic hydrolysis method refers to LA production by acid catalytic synthesis using furfuryl alcohol as raw materials. Although, furfuryl alcohol catalytic hydrolysis has the advantages like simple operation, stabile products quality, this method has the disadvantages like high cost and limited resource due to using furfuryl alcohol as raw material. Therefore, furfuryl alcohol catalytic hydrolysis process was gradually been eliminated [54].

Biomass direct hydrolysis refers to prepare LA by acid hydrolysis of starch or cellulose materials. Acid hydrolysis of biomass is a multi-phase reaction which occurred between solid-state cellulose and liquid-state acid solution, cellulose and hemicellulose acidly hydrolyzed into sugar under high temperature and high pressure. Soluble ingredients in hydrolysate include saccharide (such as xylose, glucose, and cellobiose), furfural, HMF and organic acid (such as acid, formic acid, and acetic acid), etc. Glucose, hydrolyzed from cellulose, can dehydrate into 5-HMF and then further decarboxylated into levulinic acid, which produces many byproducts [53].

In the further degradation process of pentose, hydrolyzed from hemicellulose, pentose takes place intermolecular dehydration to produce furfural by taking off three molecules of water simultaneously. Then furfural further oxidized into furfural alcohol, which could generate small amounts of LA.

The main existing processes include intermittent acid hydrolysis of biomass and continuous acid hydrolysis of biomass.

*1) Intermittent Catalytic Hydrolysis*

Intermittent catalytic hydrolysis refers to inputting biomass material and catalyst one-time into reactor and hydrolyzes until the end of the reaction, and then gets LA by separation and purification. This process of this method is relatively simple which just include the high-temperature reaction of biomass and inorganic acid (such as hydrochloric acid, sulfuric acid, phosphoric acid, nitric acid, and so on) and the separation of LA. However, the production and purity have difference according to different types and complex components of biomass.

*2) Continuous Acid Hydrolysis of Biomass*

The current representative continuous catalytic hydrolysis processes of biomass are the process proposed by Biofine Company and Ghorpade and Hanna. Biofine Company using cellulose waste as raw materials and dilute sulfuric acid as catalyst produces LA in two series of reactor. This process has the advantages like high LA yield (could reach to 70%), little content of byproducts, easy to separation, etc. At present, Biofine Company is trying to use different biomass materials, such as municipal solid waste, waste paper, wood waste, agricultural waste, as raw materials for the pilot production. Ghorpade and Hanna [55, 56] patented a process for producing levulinic acid using an extrusion process, which is advantageous in that it is continuous, requires fewer steps and reduces reaction time compared with a batch reactor process.

#### *4.5.2.2. Problems*

*1) Small Production Scale*

At present, the production of LA is still difficult to meet the large demand of fine chemical products development. With the continuous promotion of application field, the demand of LA is bound to be increase. The first of all requirements of LA as a platform chemical is large and cheap. Therefore, reducing production cost, enhancing productivity and quality is becoming a very urgent assignment.

*2) Low Yield*

As for China, the yield LA produced by direct hydrolysis of biomass is only 40% to 50%, which is about 70% for Europe and the United States. However, even the yield of 70% remains to be further enhanced. For example the continuous production process proposed by Ghorpade and Hanna could be improved to achieve higher yields at a reduced cost by removing the limitation of gas losses [55].

*3) Serious Environment Pollution*

In the preparation of LA from biomass, hydrochloric acid and sulfuric acid are commonly used as acid catalyst. As the inorganic acid has strong corrosive and toxic, the large number of acidic residue and liquid waste bring serious environmental problems.

#### *4.5.2.3 Key Technology and Process Development*

*Multi-Level Conversion of LA from Straw Catalyzed by Solid Acid Catalyst*

In order to resolve the problems of LA production from biomass as mentioned above, our group put forward a multi-level conversion thoughts of LA production from Lignocellulose, with the guidance of which can realize multi-production of LA, furfural, low molecular organic acids, etc (Figure 4.12).

This multi-level conversion process can reduce the production cost of LA obviously. The catalysis by using solid acid as catalysts overcomes the pollution problems of traditional acid production process using liquid acid as catalyst. The residue of LA production can be extracted to prepare lignin, which can decrease the environmental problems caused by large number of acidic residues in the catalysis of liquid acid.

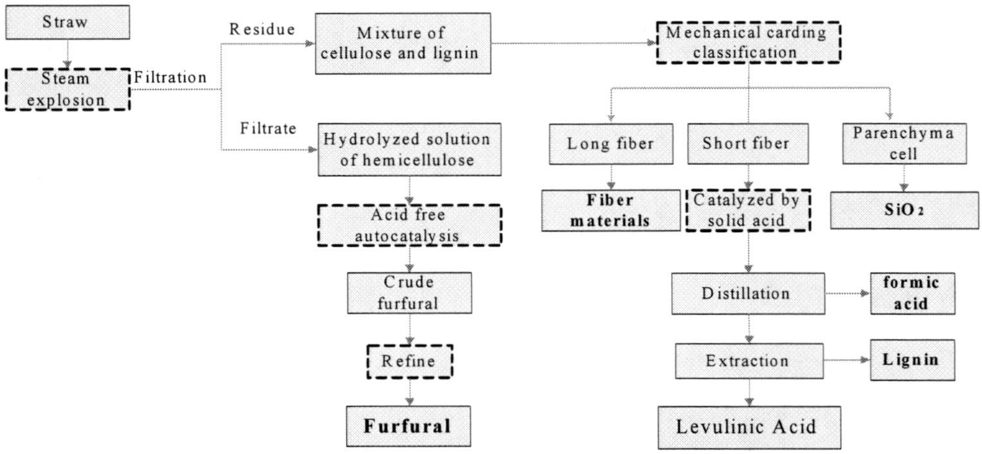


Figure 4.12. Multi-production of LA using straw as raw material through multi-level conversion

*Green Catalytic Process*

(1) LA from Straw Catalyzed By Sold Acid

Our group proposed a cleaner LA production process from carbohydrate-containing straw, pretreated by steam-explosion and fractionation, using solid super acid as catalyst. By using different amount of $S_2O_8^{2-}/ZrO_2$-$SiO_2$-$Sm_2O_3$ as catalyst to investigate the feasibility of LA production from steam-exploded straw ($lgR_0$ of steam explosion intensity is 3.5), which mechanically ground to below 60 mesh. The HPLC analysis of reaction solution is shown in Figure 4.13, and peak of 16.074 refers to the absorption peak of LA. With the increase of solid super acid concentration, the conversion rate of steam-exploded straws into LA is gradual enhanced. When the concentration of $S_2O_8^{2-}/ZrO_2$-$SiO_2$-$Sm_2O_3$ is 200%, the yield of LA reaches to 42.3%, which is higher than that of 30.5% by suing wheat straw as raw materials and 3.5% sulfuric acid as catalyst.

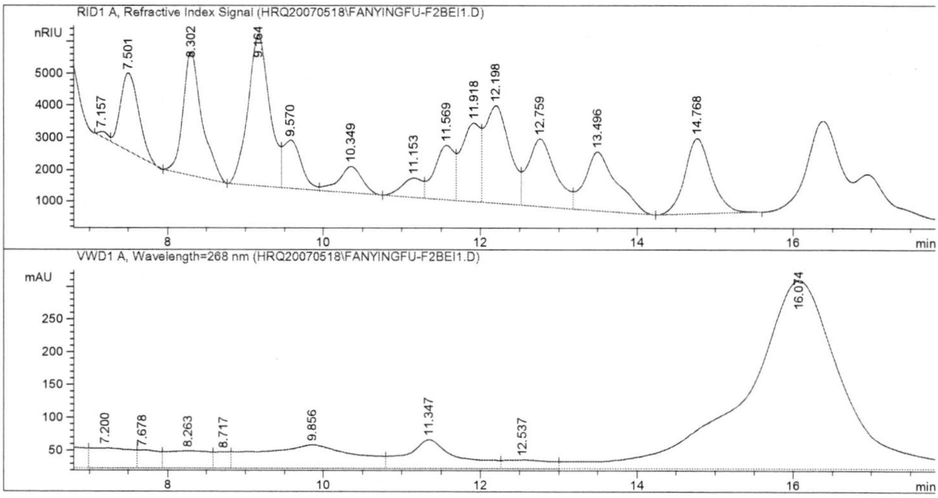


Figure 4.13. HPLC for levulinic acid yield from steam exploded rice straw catalyzed by solid super acid of $S_2O_8^{2-}/ZrO_2$-$SiO_2$-$Sm_2O_3$ in the pressurized reactor

Steam exploded-superfine ground rice straw was used as raw material to investigate the effect of steam explosion and superfine ground pretreatment on the LA yield catalyzed by solid acid catalyst $S_2O_8^{2-}/ZrO_2-SiO_2-Sm_2O_3$, catalytic reaction of rice straw and steam-exploded rice straw mechanically smashed to 60 mesh used as control group. The reaction was catalyzed in reactor, the reaction condition was as follows: reaction temperature enhancing from room temperature (26 ℃) to 200 ℃ within 40 min and then keeping at 200 ℃ for 10 min, the amount of solid super acid $S_2O_8^{2-}/ZrO_2-SiO_2-Sm_2O_3$ (presented as the mass ratio of solid acid to straw materials) is 2:1 (or 200%), concentration of substrate is 0.8:10 (g / g, adjusted by tap water), stirring speed is 198 rpm, pressure is 1.2 MPa. The result was shown in Figure 4.14.

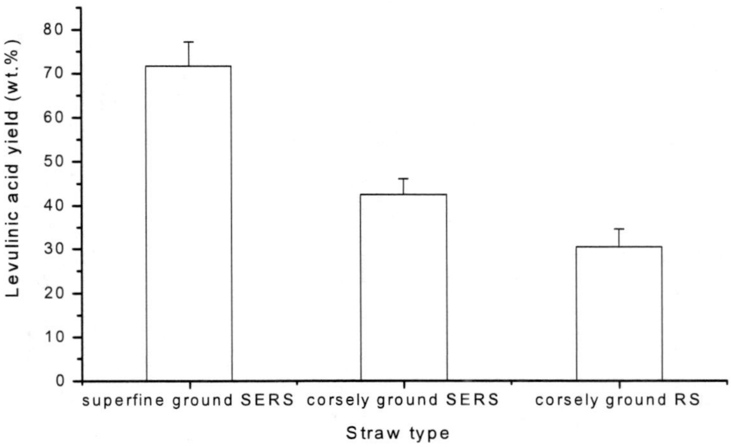


Figure 4.14. Levulinic acid production from superfine ground steam exploded rice straw (SERS) catalyzed by solid superacid of $S_2O_8^{2-}/ZrO_2-SiO_2-Sm_2O_3$ in the pressurized reactor (Solid superacid concentration: 200%, reaction temperature: from room temperature 26℃ to 200℃ in 40min, retaining time: 5min at 200℃)

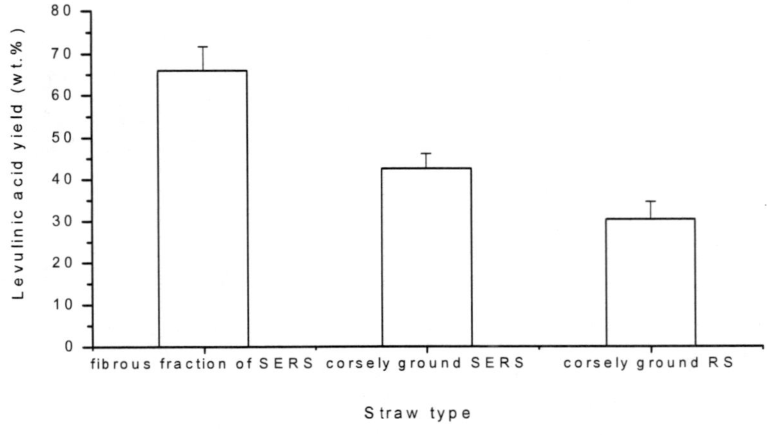


Figure 4.15. Levulinic acid production from fibrous fraction of steam exploded rice straw (SERS) catalyzed by solid superacid of $S_2O_8^{2-}/ZrO_2-SiO_2-Sm_2O_3$ in the pressurized reactor (Solid superacid concentration: 200%, reaction temperature: from room temperature 26℃ to 200℃ in 40min, retaining time: 5min at 200℃)

Fibrous fraction of steam-exploded rice straw was fractionated by steam explosion-wet superfine grinding and catalyzed by solid acid catalyst $S_2O_8^{2-}/ZrO_2$-$SiO_2$-$Sm_2O_3$ to generate LA under the same reaction condition as mentioned above, catalytic reaction of rice straw and steam-exploded rice straw mechanically smashed to 60 mesh used as control group. Figure 4.4 illustrates that under the catalysis of solid superacid catalyst $S_2O_8^{2-}/ZrO_2$-$SiO_2$-$Sm_2O_3$ steam exploded-superfine ground rice straw produce highest LA yield among three materials, reach to 65.9%, which is 1.55 times and 2.17 times of LA yield by using steam-exploded rice straw and unpretreated rice straw as raw material, respectively. This result indicated that the high fiber cell and cellulose content of fibrous fraction increase the efficient reaction catalytic activity site of solid rice straw and solid superacid catalyst thus increase the production of LA. In addition, at the end of the catalytic reaction, reactant and product will release from the catalytic site to ensure that other catalytic reaction to be continue.

2) LA Production Catalyzed by High-Temperature Liquid Water

Among multitudinous conversion technology of biomass resources, high-temperature liquid water due to its fast response, good performance solution, intrinsic acid and alkali catalysts catalytic function, selectively degradation, etc, attracted the recognition of researches.

Production of chemicals catalyzed by high-temperature liquid water has the advantages like with no catalyst recycling, reutilization of water, less pollution, etc. High-temperature liquid water catalytic technology has shown a very broad application prospects in the biomass utilization field. Cai Lei, et al [57] studied the non-catalytic decomposition mechanism of glucose and fructose during the LA production process catalyzed by high-temperature liquid water. Decomposition products of fructose catalyzed by high-temperature liquid water mainly include acid (formic acid, acetic acid and LA), glucose, 5-HMF and furfural. Hydrolytic reaction of fructose can be effectively adjusted by a certain pathway through the regulation and control of temperature, retention time, as well as suitable catalyst, so as to selectively adjust the distribution of final products.

The decomposition of glucose in high-temperature liquid water is more difficult than fructose due to the higher apparent activation energy of glucose. The decomposition reaction of fructose and glucose in high-temperature liquid water is first order reaction and second order reaction, respectively, which due to the different structure of glucose (aldose) and fructose (ketose). Although the difference of structures, glucose and fructose basically has the same hydrolytic pathway in the high-temperature liquid water, which are mainly the dehydration (hydrolysis) reaction and isomerization reaction.

## 4.5.3. Xylitol

### *4.5.3.1. General Situation of Xylitol Industry Progress*

Xylitol is produced using agriculture and forestry wastes as raw materials such as corn cob, cottonseed skins, sugar cane bagasse, birch, and other debris by processes such as hydrolysis, purification, concentration, decolorization, hydrogenation, concentration, crystallization, separation and so on. Preparation of xylitol has main methods: extraction, chemical synthesis and biosynthesis, which processes as shown in Figure 4.16.

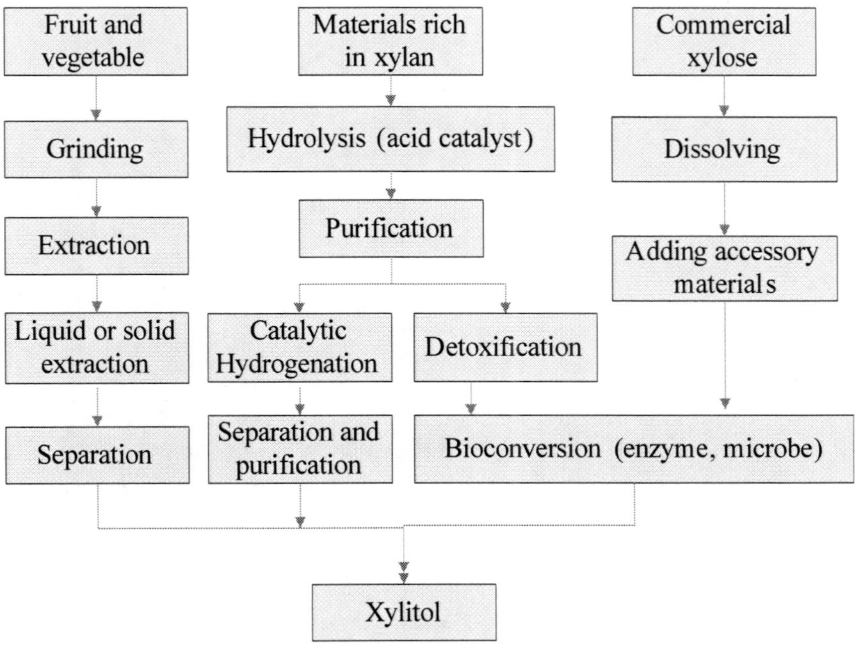


Figure 4.16. Xylitol production process [58]

The traditional production method for xylitol is the chemical process. Agricultural, forestry and industrial scrap materials containing rich condensed pentose such as corncobs, cotton seed hulls, sugar cane bagasse and birch strips are processed into xylose through hydrolysis, and xylose is further processed into xylitol through hydrogenation, refining and crystallization. Because of low xylitol concentration of natural fruits and vegetables, it is very difficult and diseconomic to produce xylitol in large scale. Xylitol is now mostly produced using chemical and biological technology. The traditional preparation method of xylitol mainly includes the extraction of xylose from hemicellulose hydrolyzate and further hydrogenation.

*Chemical Hydrogenation*

Chemical method to produce xylitol is means that xylan in plant fiber raw materials (such as straw, corn cob, and so on) is acid hydrolyzed to xylose, then xylose is hydrogenated using nickel as catalyst, under a certain pressure, to produce xylitol. The reaction equation is shown as formula (4.1)

$$(C_5H_8O_4)\,n + n\,H_2O \xrightarrow{H^+} n\,C_5H_{10}O_5$$

$$C_5H_{10}O_5 + H_2 \xrightarrow{\text{Ni catalyst}} C_5H_{12}O_5 \tag{4.1}$$

As for chemical hydrogenation method, it is necessary to multi-step purify xylose hydrolyzate before hydrogenation which increases the cost of xylitol production obviously. Chemical hydrogenated production of xylitol consumes a large amount of steam. In theory, 3-

5 t raw materials are needed to produce 1 t crystal commercial xylitol. The chemical reaction needs high temperature, high pressure. The nickel catalyst not only will pollute environment also forbidden high concentration in edible-grade standard (no more than 0.0003 percent), while the recovery rate of nickel catalysts in catalytic hydrogenation method is just about 60%. Catalytic hydrogenation reaction has no selectivity, which produces large number of impurities and makes the extraction and refinery much more difficulty. All these above-mentioned disadvantages increase the production cost of xylitol and restricted the expansion of market. Therefore, the biotransformation, not need high purity xylose solution as raw material and carried out under in moderate condition, is becoming a research focus.

*Biotransformation*

At present, biotransformation method mainly converts xylose to xylitol through the xylose reductase in microorganisms. Biotransformation method to produce xylitol overcomes the shortcomings of chemical synthesis and has many advantages like mild reaction conditions, no need high temperature and pressure, sample technique, easy control, low cost, and so on.

Although xylitol production of microbial fermentation using xylose as raw material made great progress, the resource of xylose is subject to greater restrictions. Glucose is abundant in nature. So many researchers have started on the research of conversion of glucose to xylitol by microbial fermentation. However, due to the metabolism of glucose to xylitol is very complicated and xylitol usually metabolized by microbe, it also required a substantial amount of research.

In biotransformation of xylitol, as long as the xylose plant adds ventilation fermentation equipment other units can operate in accordance with general xylose production process, which significant saves investment. It is precisely these characteristics of biotransformation attracts scientists to conduct in-depth study expecting to get a breakthrough and bringing increasing economic benefits.

#### *4.5.3.2 Problems*

*1. Cost Advantage is not Obvious*

Due to the presence of inhibitors (such as acetic acid, furfural and polyphenols) in hydrolysis solution of xylose, inhibiting yeast to produce enzyme, the process of xylitol production is much more complicated compared with the fermentation using pure xylose as substrate. Although pretreating hydrolyzate using ion exchange could increase the conversion rate, the production cost compared with the chemical method has no advantage. Therefore, still need to explore cost-effective detoxification method of hydrolyzate. In addition, to improve the adaptability of cell to hydrolyzate makes cell can effectively resist the impact of harmful substances by just dealt with the hydrolyzate simply, which is a prerequisite for rapidly and highly efficient conversion of xylose to xylitol.

*2. Low Fermentation Efficiency*

Fermentation efficiency of the existing strain is not high, and the major high-yielding strains focus on a number of strains of yeast. Therefore, it is very important to isolate new strains and broaden the microbial sources. According to strain breeding, there has not found suitable high-yield xylitol strain for hemicellulose hydrolyzate. According to gene modification and engineering strain construction, the current study mostly remain at the laboratory stage and there is still some way to go for the large-scale production. The changes of metabolic pathway caused by microbial metabolism disrupt the original physiological and metabolic state of receptor, which makes the construction of engineering strains much more complicated.

### *4.5.3.3. Key Technology and Process Development*

*Gene Modification and Engineering Strain Construction*

The xylose metabolism is started by xylose reductase (XR) and then enters the pentose phosphate pathway or generates 3-glyceraldehyde phosphate and phosphoric acid acetyl by means of xylulose-5-phosphate. Since that the participation of the main means of xylose metabolism enzyme genes have been found and cloned, the construction of high-yield xylitol engineered bacteria to become a hot research.

There are two main pathways of xylose metabolism. One pathway is called the "Xylose Reductase-Xylitol Dehydrogenase" or XR-XDH pathway. Xylose reductase (XR) and xylitol dehydrogenase (XDH) are the two main enzymes used in this method of xylose degradation. XR is responsible for the reduction of xylose to xylitol and is aided by cofactors NADH or NADPH. Xylitol is then oxidized to xylulose by XDH and accomplished exclusively with the cofactor NAD+. The other pathway for xylose metabolism is called the "Xylose Isomerase" (XI) pathway. Enzyme XI is responsible for direct conversion of xylose into xylulose, and does not proceed via a xylitol intermediate. Both pathways create xylulose, although the enzymes utilized are different. After production of xylulose both the XR-XDH and XI pathways proceed through enzyme xylulokinase (XK) to further modify xylulose into xylulose-5-P where it then enters the pentose phosphate pathway for further catabolism [59].

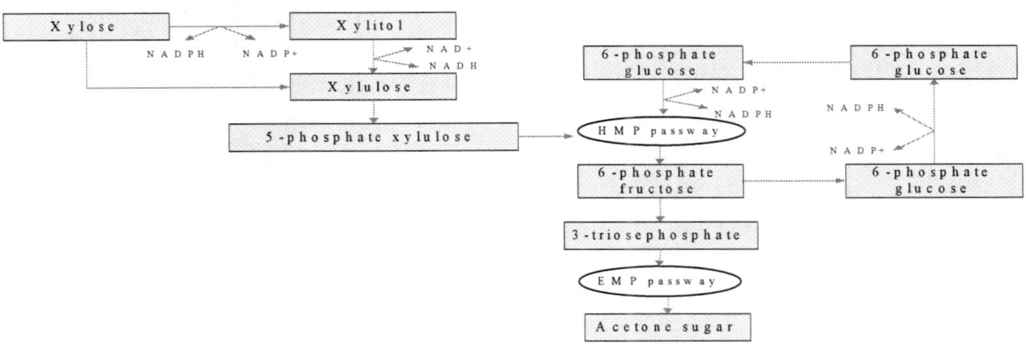


Figure 4.17. Xylose metabolism pathway

D-Glucose, on the other hand, is a common substrate in the food industry and is, compared to d-xylose, cheap and readily available. Thus, it would be attractive if xylitol could be produced from d-glucose. However, few if any natural microorganisms that produce xylitol from d-glucose are known. Although there has not found microorganisms can ferment glucose directly into xylitol, there has two metabolic pathway of known microbial can convert glucose into xylitol. One pathway is that D-Glucose converted into D-ribulose 5-phosphate by transketolse of yeast through pentose phosphate pathway, then isomerized D-ribulose 5-phosphate into D-xylulose 5-phosphate by 5-phosphate-ketose-3-epimerase in bacteria, and subsequently converts D-xylulose 5-phosphate into xylitol by kinase and xylitol dehydrogenase of yeast. Another pathway is that D-glucose converted to D-arabitol by Osmophilic yeast (such as *Debaryomyces hansenii*), then oxidized to D-xylulose by bacteria with D-arabitol dehydrogenase (such as *Acetobacter suboxydans*), and subsequently reduced to xylitol by xylitol dehydrogenase of yeast (such as *Candida guilliermondii*) [60].

*Theoretic Yield of Xylitol*

In the process of xylitol, production from xylose xylitol susceptibly to be oxidized to generate xylulose under aerobic conditions this is not conducive to the accumulation of xylitol. Under microaerobic conditions and anaerobic conditions, the xylitol yield is different.

Total metabolism equation of xylitol generated from xylose under microaerobic conditions is shown in (4.2). Calculated by this equation, the theory conversion of xylitol is 0.905 mol xylitol / mol xylose, that is 0.917 g / g.

$$126\ C_5H_{10}O_5 + 3\ O_2 + 6\ ADP + 6\ Pi + 48\ H_2O \rightarrow 114\ C_5H_{12}O_5 + 6\ ATP + 6\ O_2 \qquad (4.2)$$

Total metabolism equation of xylitol generated from xylose under anaerobic conditions is shown in (4.3). Calculated by this equation, the theory conversion of xylitol is 0.875 mol xylitol / mol xylose, that is 0.887 g / g.

$$48\ C_5H_{10}O_5 + 15\ H_2O \rightarrow 42\ C_5H_{12}O_5 + 3\ C_2H_5OH + 24\ CO_2 \qquad (4.3)$$

The xylose reductase of yeast in general uses NADPH as the main coenzyme. In the restricted oxygen conditions, the imbalance of NAD+ supply directly inhibits the activity of xylitol dehydrogenase and results in the accumulation of xylitol. And the natural assimilation of xylose in yeast can consume part of generated xylitol through D-xylulose pathway, which makes the conversion rate of xylitol only reach 65%~85% of the theory conversion rate [61]. However, the main accumulated production of yeasts with NADH2XR as dominant coenzyme is alcohol instead of xylitol because of the redox equilibrium of coenzyme.

Transfering xylose reductase gene (XYL1) of Pichia stipitis into S.cerevisiae and cultivating with some kind of hexose or ethanol as assistant substrate, the engineering S.cerevisiae strain expressed with XYL1 gene can convert more than 95% xylose into xylitol. However, these kind of metabolic engineering strains absorb and convert xylose only when hexose has consumed completely. Although the conversion rate of xylitol is high, the generation rate is very low [62].

*Cell Immobilization*

Immobilized cell systems have been traditionally considered as an alternative to increase the process productivity and minimize the production costs, while offering advantages over free cell fermentation systems [63-65]. In general, cell immobilization allows obtaining high cell concentration and productivity as a consequence of effective retention of cells within the bioreactor, thus avoiding or at least remarkably reducing the washout phenomenon. Besides, the possibility of recycling cells for inoculum permits the fermentation to be profitably carried out in continuous or repeated-batch mode [66].

Numerous biotechnological processes were shown to be favored by cell immobilization; therefore, different immobilization techniques and support materials have been proposed.

These techniques can be divided into four major categories based on the physical mechanism employed: (a) attachment or adsorption on solid carrier surfaces, (b) entrapment within a porous matrix, (c) self-aggregation by flocculation (natural) or with cross-linking agents (artificially induced), and (d) cell containment behind barriers [67].

For example, Santos et al immobilized *Candida guilliermondii* cell onto natural sugarcane bagasse fibers. Under the selected cell immobilization conditions, the yields of xylitol and biomass on consumed xylose were 0.65 and 0.18 g/g, respectively, xylitol and biomass productivities 0.66 and 0.13 g $L^{-1}$ $h^{-1}$, and the efficiency of xylose-to-xylitol bioconversion was 70.8%[67]. Carvalho et al entrapped *Candidia guilliermondii* cells in Ca-alginate beads and cultivated cells using five-fold concentrated hydrolysate to produce xylitol. The Xylitol production reached 47.5 g $l^{-1}$ within 120 h of fermentation, resulting in a bioconversion yield of 0.81 g $g^{-1}$ and a productivity of 0.40 g $l^{-1}$ $h^{-1}$ [64].

*Cell-Recycling Fermentation*

For the industrial microbiological production of xylitol, it is very important to increase volumetric productivity for a long period. Volumetric productivity can be improved by increasing both specific productivity and cell concentration, because specific productivity multiplied by cell concentration is volumetric productivity [68]. The increase in specific productivity is in the domain of microbial genetics, such as the screening and mutant selection of a high producing strain. The increase in cell concentration can be rapidly achieved using cell-recycle fermentation. Therefore, many researches are focused on cell recycling to prolong the production phase and thus increase the xylitol productivity or screening for new strains. In order to increase xylitol productivity, Kwon et al recycled *Candida tropicalis* cell in a submerged membrane bioreactor with suction pressure and air sparging. The productivity and total amount of xylitol in cell-recycle fermentation was 3.4 and 11.0 fold higher than those in batch fermentation, respectively [69].

*Hydrolysate Detoxification*

Production of hemicellulose hydrolysates by any of the currently known lignocellulosic pretreatment methods introduces byproducts, which are toxic to fermenting microorganisms. The toxic compounds include furans, aromatic aldehydes, aromatic carboxylic acids, aliphatic acids, macromolecules, and other unidentified degradation products. The degree of toxicity of the compounds in the hydrolysates varies for different microorganisms. Thus in order to decrease the toxicity caused by the inhibitory compounds, considerable efforts have been focused on detoxification procedures prior to fermentation, including physical, chemical and biological methods.

Physical detoxification methods include neutralization, extraction, over-liming, evaporation, ion exchange resins and charcoal adsorption etc. Evaporation method can remove volatile compounds in Hydrolysate. Solvent extraction method can remove toxic compounds selectively based on the nature of the toxic compounds. For example, ether extraction can remove acid, formic acid, furfural, 5-CM-furfural and benzene compounds, and so on. Extracted with acetate the removal rate of acetic acid is about 56% and furfural, vanillin and 4 -hydroxyl benzoic acid are removed completely.

Biological detoxification methods include enzymatic and microbiological method. Enzymes are mainly peroxidase and laccase. Laccase has selectivity, which is able to remove phenol monomer (usually the content in liquid Hydrolysis is 2.6 g / L or so), and carbolic acid. Microbe, which can be used to absorb benzene compounds, for example *Trichoderma reesei* can remove benzoic acid, furfural and benzoic acid derivatives from steam-exploded hydrolysate.

## 4.5.4. Xanthan Gum

Xanthan gum is a heteropolysaccharede produced by *Xanthomonas campestris* with a primary structure consisting of repeated pentasaccharide units formed by two glucose units, two mannose units, and one glucuronic acid unit, in the molar ratio 2.8:2.0:2.0. The repeated pentasaccharide units of xanthan gum consists of two β-D-glucose units linked by β-1,4 glycosidic bond. The chemical structure of the main chain is identical to that of cellulose. Trisaccharide side chains contain α-D-glucuronic acid unit between two D-mannose units linked at the O-3 position of every other glucose residue in the main chain [70, 71]. Owing to its unusual stability to temperature, pH, shear, enzyme degradation and excellent compatibility with salts over a wide pH range, etc, xanthan has widely applied in various industries field. In addition, xanthan gum has become a firstly and widely used bacterial polysaccharide in the food industry because of its best performance and repeatedly confirmed green and safety.

### *4.5.4.1. General Situation of Xylitol Industry Progress*

Since 1950s, production of microbial polysaccharides by fermentation technology has become an international hot topic, and several microbial polysaccharides has provided with considerable industrial capacity. Xanthan gum is one kind of the most representative and maximum commercialized microbial polysaccharides. The process for making xanthan is shown in Fig. 4.14 [71, 72]. The key steps of a typical xanthan production process include the selectivity and modification of strain, control of fermentation process, separation and purification of products, etc.

*Production Strain and Modification*

Xanthan gum is produced by a number of species of *Xanthomonas* like *X. campestris, X. phaseoli and X. malvacearum. X. campestris* Among all these species, xanthan gum produced by *X. campestris* has commercial and industrial significance since that xanthan gum produced by *X. campestris* using cabbage extraction as culture medium has the same chemical composition with synthetic medium [73].

Excellent genus of producing strain is very important to the yield and quality of products. It has been reported that by using recombinant gene technology sequence, separate, screen, and express gum-producing gene in bacteria, but not yet realize the industrial production of recombinant strain. If the gum-producing gene can be isolated by using genetic engineering technology and then to have double or more gum-producing genes and stably transmitted transferred into a new integral strain, xanthan gum yield will be increase by several times.

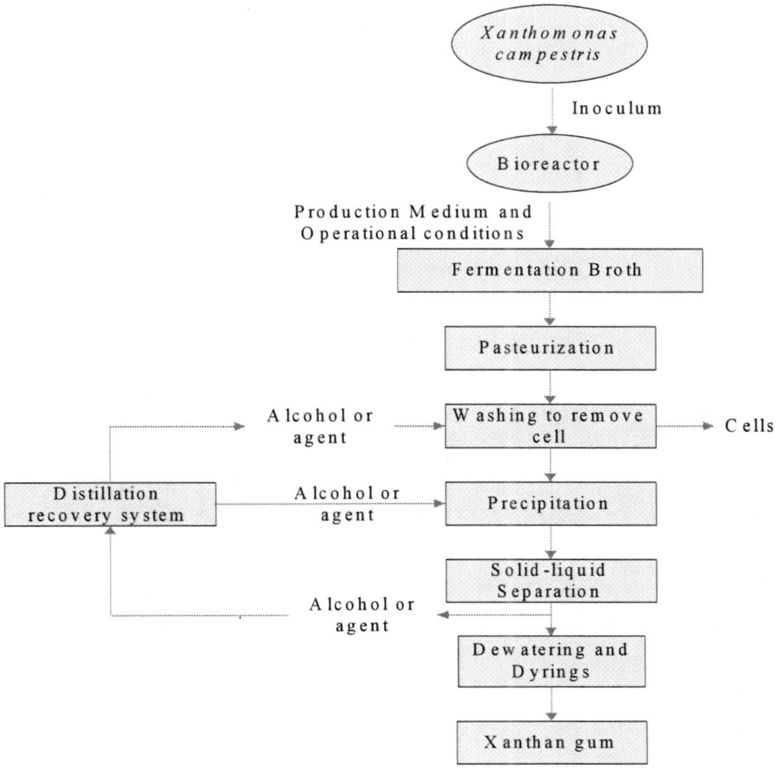


Figure 4.18. Outline of the xanthan gum production process [71, 72]

*Technology and Control of Fermentation Process*

Xanthan gum can be produced by batch fermentation, semi-batch fermentation, or continuous fermentation [74-76]. In industrial production, mixing ventilation reactor is usually used in batch fermentation process of xanthan gum. The actual production process includes seed cultivation, seed enlarge cultivation and fermentation. For a specific fermentation process, the selectivity of producing strain is decided by the metabolic features and optimum technology condition of the used strain. In batch fermentation, the substrate concentration in whole fermentation process decreased along with fermentation time.

In continuous fermentation, under normal circumstances, the substrate concentration is always enough for the xanthan gum production since that substrate concentration of the whole process of is remained stably. On the other hand, the stable substrate concentration may lead to the substrate inhibition and eventually results in the decrease of yield. However, the fed-batch cultivation adjust the additive type and content according to the growth characteristics of cells on different growth stages, which conducive to create a suitable physical growth environment for cell. Many studies have shown that fed-batch fermentation is a satisfactory way for xanthan gum fermentation [76].

As the high viscosity of xanthan gum solution and high need of adequate oxygen, some reactors not suitable for the production of xanthan gum. The airlift reactor, which is more suitable for continuous cultivation, can reduce the energy consumption since that it efficiently dissolved the problem of heat caused by ventilation and agitation. Tower reactor with fluidized bed also applied in batch-fed fermentation because of its higher air bubbles transferring coefficient. Static pump reactor can enhance the transport rate of oxygen in high viscosity broth and thus increase xanthan gum production. It is difficult achieve the demand of special characteristics of xanthan gum fermentation broth using ordinary agitator blade such as disc turbine stirring-paddle and blade turbine stirring-paddle. However, these agitators consumed low energy. In recent years, in order to adapt to the high viscosity and high ventilation rates of the fermentation process, it has developed a grid stirring-paddle and window stirring-paddle.

*Extraction and Separation of Xanthan Gum*

The purpose of extraction and purification of xanthan gum is to remove the impurity according to the end used of the gum and the economics.The final fermentation broth contains 10±30 g $L^{-1}$ xanthan, 1±10 g $L^{-1}$ cells, and 3±10 g $L^{-1}$ residual nutrients, and other metabolites[71]. Because of a high xanthan concentration, the broth is highly viscous and difficult to handle which made the recovery of the xanthan from the fermentation broth becomes generally difficult and expensive. The main steps of the recovery process (Fig. 2) are deactivation and removal of the microbial cells, precipitation of the biopolymer, dewatering, drying, and milling.

Numerous methods have been developed to deactivate, lyse, or remove cells from the broth such as chemicals treatment (e.g. alkali, hypochlorite, and enzymes), mechanical means, and thermal treatment. However, the pH elevation of chemical treatment causes depyruvylation of the product, enzymes treatment adds the costs. Therefore, generally, the fermentation broth is pasteurized or sterilized to kill the cells [77] and enhance xanthan removal from the cells. Pasteurization of the fermentation broth at a high temperature often causes thermal degradation of the xanthan gum, which conceives to the further centrifugation or filtration.

For highly viscous xanthan broths, before filtration, the fermentation broth must be diluted using water, alcohol, or mixtures of alcohol and salts as diluents. The removal the cells and solids usually include centrifugation, filtration, enzymatic degradation, hypochlorite oxidation, ultrafiltration concentration, etc. The most common technique used for the primary isolation and purification of polysaccharides is precipitation using water miscible non-solvents such as such as ethanol [78] and isopropyl alcohol (IPA) [79]. Both the cost of alcohol for recovery and the inevitable losses contribute significantly to the total cost of production.

### *4.5.4.2. Problems*

*1) High Material Cost*

Presently, raw materials for xanthan gum fermentation mainly include sucrose, glucose and starch, which results in high raw material cost. Therefore, it is urgent to develop an alternative material to produce xanthan gum.

*2) Behindhand Fermentation Technology*

Owing to the growing viscosity during fermentation, suitable aeration and mixing of the culture have to be guaranteed to avoid stagnant areas [80]. Therefore, mechanically agitated fermentation tanks instead of air-agitated tanks are usually used for the industrial production of xanthan [81]. However, because of the unique physical properties of xanthan gum, although the oxygen supply has improved, the high-energy consumption is still inevitable. At the same time, the general mechanically agitated fermentation tank has the shear force damage on cell.

*3) High Energy Consumption and Production Cost*

In the production of xanthan gum, the high viscidity makes the broth outside of paddle almost at a standstill state. In order to improve the mixing effect, must to consume large amount power at the expense of high cost. Large consumption of alcohol in the extraction of xanthan gum is another major cause for high cost, which accounts more than 50% of the total production cost. In order to reduce the usage of alcohol, it is the best by increase the concentration of xanthan gum with vacuum evaporation method, which has higher energy consumption. In addition, drying of xanthan gum and recovery of alcohol also consume high energy.

*4) Serious Pollution*

The further extraction and recovery of xanthan gum fermentation broth needs ethanol, hydrochloric acid, calcium chloride, sodium hydroxide, and other chemical raw materials. Therefore, in the fermentation production of xanthan gum will bring large amount of industrial wastewater, which will pollute the environment because of high viscosity. On the other hand, the direct emissions of high content of ammonia, glucose, acetic acid, pyruvic acid also will cause great environmental pollution.

#### *4.5.4.3. Key Technology and Process Development*

*New Fermentation Technology*

As the high viscosity of xanthan gum, the fermentation broth has the problems like difficult material mixture and insufficient oxygen supply, which result in the specificity of xanthan gum fermentation process. Therefore, it is necessary to develop new fermentation technology or equipment to improve the situation.

1) Solid-state fermentation technology by using polyurethane as inert supporter

Our group solves the problems of dissolving oxygen during the fermentation process by solid-state fermentation technology using polyurethane as inert supporter. The difference between solid-state fermentation using inert supporter (SSF-IS) and traditional solid-state fermentation (SSF) is that the substrate of SSF is not only act as a medium of solid-state also provide nutrition for the growth of micro-organisms, however, the substrate of SSF-IS is absorbed on the surface of supporter as nutrition of microorganisms. SSF-IS refers to the fermentation technology that liquid substrate is absorbed on the surface of inert supporter matrix (polyurethane foam), and microorganism grows on the supporter in the absence of water or free water. The fermentation broth was obtained by squeezing the supporter. The looseness and porosity of supporter provide microorganism a great growing interface. Meanwhile, because of no free water, the problems of dissolving oxygen can be solved under the situation of no stirring.

Fermentation steps of xanthan gum production by solid-state fermentation using polyurethane foam as inter supporter include:

a) Cut the washed polyurethane foam into small pat with the size of 0.3~0.9 $cm^3$ and spread shallowly with a thickness of 1~5cm;

b) *X. campestris* 1.1781 inoculated in fermenting medium according to the ratio of $1.4\times10^8$/ml medium and then fermented after the mixing of supporter and medium in ratio of 10~15 ml/g supporter;
c) After the end of fermentation, the fermentation broth is squeezed to separate xanthan gum.

The squeezed supporter can be reutilized after been washed and dried. The yield of xanthan gum fermented by SSF-IS using polyurethane foam as supporter is about 69.1%.

Using polyurethane foam as supporter to ferment xanthan gum has obvious advantages, which have shown as below.

a) Solve the dissolved oxygen issue without stirring using air as the continuous phase. In the latter part General liquid fermentation of xanthan gum the broth, presented as pseudoplastic fluid, often has dead zones that dissolved oxygen difficult to achieve even with high-power stirring. In this fermentation process, the loose and porous characteristics of supporter solve the dissolved oxygen issue without stirring.

b) Polyurethane foam is easy to get, low-cost, and can be reused. After the end of batch fermentation, as long as washes with water repeatedly the supporter can be reused after being squeezed and air-dried naturally. So, the use of supporter will not increase the cost of xanthan gum.

c) Microorganisms grow as film on the surface of supporter, only few cells in the outflow of fermentation broth, so the separation of xanthan gum is very simple. However, in the general xanthan gum production process, large number of water is needs to dilute broth and needs further centrifugation and precipitation.

d) After the end of the fermentation, xanthan gum can be gotten by squeezing the supporter, which is a simple solid-liquid separation process.

2) Two-step fermentation

To solve the oxygen-limit problem in the final phase of Xanthan gum fermentation and to obtain cell-free broth, Haifeng et al [82] designed a new way of producing Xanthan gum by cell-free fermentation named as "Two-Staged Fermentation". The new fermentation process included two stages. First, cells cultured in the later period of logarithmic phase were removed from the fermentation broth by the microfilering pipe. Then, non-nitrogenous medium with high carbon source was added to the fermentation broth in the second reactor. Xanthan gum was thus synthesized by the exocellular enzymes produced in the first reactor. The experimental results show that the Xanthan gum production was 24.5% higher in the two-staged fermentation than that in the normal one and alcohol consumption reduced 40%. Moreover, the final fermentation broth was cell-free and do not need to remove the cells. Therefore, it is possible to simplify the separation process and to cut the production cost. Chang et al [83] proposed a two-stage fed-batch fermentation technology for xanthan gum production by adding sugar according to the fermenting characters of xanthan gum. The new method would benefit for the yield and fluid viscosity, which was testified in 50 L airlift fermentor.

*Multi-Level Conversation of Lignocellulose with Xanthan Gum as Main Product*

In order to reduce the production costs, agricultural waste straw was used as raw materials for the production of xanthan gum. The main component of lignocellulose includes cellulose, lignin and hemicellulose. Hemicellulose can be degraded into single sugar by steam explosion and cellulose generates glucose and oligosaccharides, which all can be used to produce xanthan gum.

Our group put forward a technology that using the glucose and oligosaccharides generated from wash solution of steam-exploded straw to produce xanthan gum. Furthermore, the solid residue (mainly is cellulose and lignin) can be converted into other products. For example, cellulose can be used to produce bio-ethanol, furfural, LA, and so on; fermentation residue can used to prepare organic fertilizers or protein feed. This thought (fig 4.19) conceives to the achieving multi-level conversion of raw material.

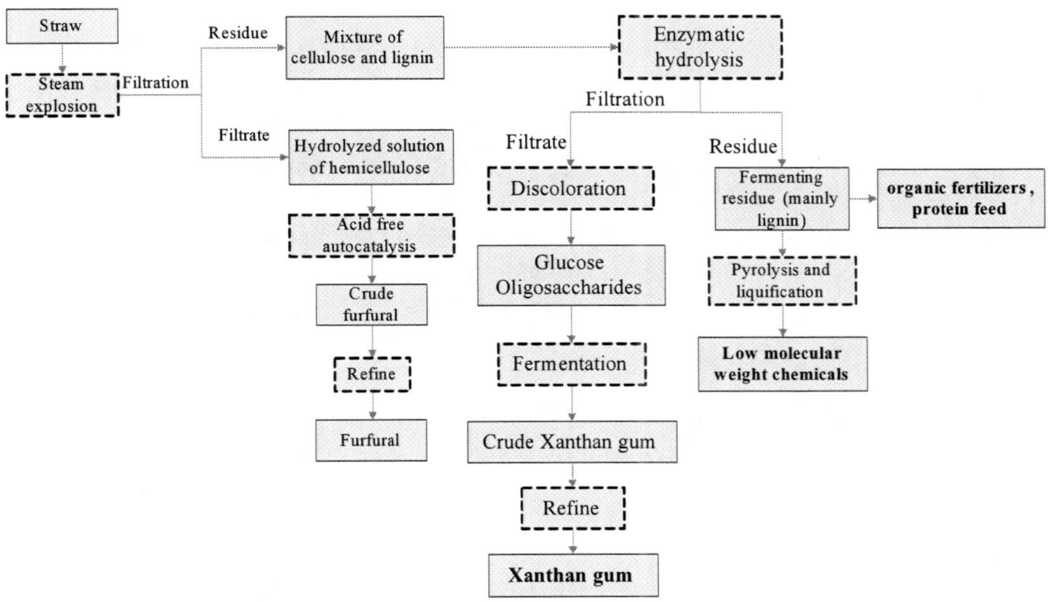


Figure 4.19. Xanthan gum production by multi-level conversation of steam-exploded straw

Xanthan gum production using steam-exploded straw mainly includes enzymatic hydrolysis of steam-exploded straw, decolorization of hydrolytic solution, cultivation of seeds and fermentation of xanthan gum using hydrolyzate, and so on. The specific operation as follows: a) adds water to steam-exploded wheat straw, smashed to 60~ 80 meshes, adjusts pH to 4.5~5.5 using hydrochloric acid, and then hydrolyzed by adding cellulase; b) after the end of enzymatic hydrolysis, the reduce sugar concentration of collected filtrate is about 40~50 g/l; c) adds certain concentration of activated carbon and boiled for hydrolyzate decolorization; d) the decolorized hydrolyzate is used for xanthan gum fermentation; e) after fermentation separate xanthan gum by precipitation or extraction.

*Cell Immobilization Technology*

There have been only a few studies of xanthan fermentation using immobilized cells. Robinson and Wang (1988) used porous Celite beads to immobilize cells in xanthan fermentation. It is not clear, however, if the xanthan broth so produced was free of cells. Furthermore, a large portion of the xanthan product was trapped in the beads and could not be easily separated from the cells. Lebrun et al. (1994) studied polysaccharide production by cells immobilized in composite agar layer/ microporous membrane structures but concluded that the immobilized-cell system was not appropriate for xanthan gum production. It is clear that cell entrapment is not an appropriate cell immobilization method for xanthan gum fermentation because of the high viscosity of xanthan solution [84].

*Energy Saving and Consumption Reduction*

In the production of xanthan gum, the energy saving and consumption reduction technologies mainly include high efficient low pressure air purification technology, air-lift fermenter instead of traditional mechanical stirring fermenter, high efficient xanthan gum enrichment technology, continuous drying technology and alcohol recovery technology, etc [72].

## 4.5.5 Oxalic Acid

Oxalic acid, scientific name adipic acid, is an important organic chemical raw materials, which used largely in separation and purification of rare metals, refining of pharmaceuticals and bleaching of textiles, but also widely used in pesticides, ceramics, plastics, leather, metal smelting, and so on. In recent years, along with the rapid development of rare earth metals refining and increasing demand for oxalic acid, which provide a broad market prospect for the development of oxalic acid industry.

### *4.5.5.1. General Situation of Lactic Acid Industry Progress*

The main commercialized methods are oxidation of carbohydrate, CO coupling, sodium formate method, oxidation of glycol, oxidation of propylene, etc.

*1) Oxidation of Carbohydrate*

Production of oxalic acid by carbohydrate oxidation is that oxalic acid is produced by oxidation of sugar and starch carbohydrates (such as glucose, sucrose, starch, dextrin, syrup preparation, etc) mainly catalyzed by nitric acid under 73℃~80℃. Choice of raw material depends on the availability, economization, and characteristics of the operation. Brazil use cassava starch as raw material for oxalic acid production, which is corn starch for United States and sugar for India and Russia. The main reaction of this technology is carbohydrate firstly hydrolyzed into glucose, glucose catalyzed by nitric acid to produce oxalic acid in the presence of $Fe^{3+}$, $V_2O_5$ as catalyst. The reaction is as follows:

$$(C_6H_{10}O_5)n+nH_2O \rightarrow nC_6H_{12}O_6 \tag{4.4}$$

$$C_6H_{12}O_6+12HNO_3 \rightarrow 3H_2C_2O_4 \cdot 2H_2O+3H_2O+3NO+9NO_{25} \tag{4.5}$$

$$4C_6H_{12}O_6+18HNO_3+3H_2O \rightarrow 12H_2C_2O_4 \cdot 2H_2O+9N_2O \tag{4.6}$$

The dominant shortcomings of oxalic acid carbohydrate oxidation technology include low-yield, high production costs, high toxicity due to large number of nitrogen oxides, and serious environmental pollution.

*2) Sodium Formate Method*

Production of oxalic acid by sodium formate method can be divided into the lead method and calcification method. In this technology, carbon monoxide reacts with sodium hydroxide solution to prepare sodium formate, and then sodium formate dehydrated to generate sodium oxalate, which reacts with sulfuric acid lead (lead method) or calcium hydroxide (calcification method) to generate lead oxalate or calcium oxalate respectively and then acidified by sulfuric acid to produce oxalic acid.

Main reaction of lead method to produce oxalic acid includes:

$$NaOH+CO \rightarrow HCOONa \quad (4.7)$$

$$2HCOONa \rightarrow (COONa)_2+H_2 \quad (4.8)$$

$$(COONa)_2+PbSO_4 \rightarrow (COO)_2Pb+Na_2SO_4 \quad (4.9)$$

$$(COO)_2Pb+H_2SO_4 \rightarrow H_2C_2O_4+PbSO_4\downarrow \quad (4.10)$$

Main reaction of calcification method to produce oxalic acid includes:

$$Ca(OH)_2+H_2O_4Na_2 \rightarrow H_2O_4Ca\downarrow+2NaOH \quad (4.12)$$

$$H_2O_4Ca+H_2SO_4 \rightarrow CaSO_4\downarrow+H_2C_2O \quad (4.13)$$

Lead method of sodium formate to produce oxalic acid has the problem of pollution and large consumption of sulfuric acid and caustic soda due to the reacted sulfuric sodium can not return to the process. Calcification method of sodium formate can eliminate the pollution of underground water caused by lead salt and sodium sulfate, and can reduce the production costs by replacing lime with caustic soda.

*3) CO Coupling Method*

Ube Kosan Co., Ltd. Of Japan on the basis of the liquid phase method jointed with Union Carbide Corporation of United States to develop the gas coupling method. Using Pd/C as a catalyst, carbon monoxide synthesized with sub-nitrate in 130℃ and atmospheric pressure to generate oxalate, and company with nitric oxide as by-product. Then, the oxalate hydrolyzed into oxalic acid and corresponding alcohol. After been separated, alcohol liquid can further oxidized to generate sub-nitrate, which can be recycled. The main reaction shown as following:

$$2\ CO+2\ RONO_2 \rightarrow C_2O_4R_2+2\ NO_2 \quad (4.14)$$

$$C_2O_4R_2+2\ H_2O \rightarrow H_2C_2O_4+2\ ROH \quad (4.15)$$

$$4\ NO+4\ ROH+O_2 \rightarrow 4\ RONO+2\ H_2O \quad (4.16)$$

4) Melting alkali oxidation of cellulose

Cellulose-rich materials (such as wood waste, rice husk, etc.) co-melt with inorganic alkali under the temperature of 513 K~558 K to generate more than 50% of oxalate, the oxalate separates out by crystallization, and then produces oxalic acid by sulfuric acid.

Preparation of oxalic acid by melting alkali oxidation of cellulose has the advantages like simple operation, but also has the shortcomings like low conversion (less than 35%), high alkali consumption and serious environmental pollution. However, facing the worldly problems of population, energy, food and other issues, it is increasingly significant to explore an efficient oxalic acid production way from lignocellulose.

#### *4.5.5.2. Problems*

At present, there have various issues of the main production methods of oxalic acid, for example, the complex technique, high investment and serious pollution for sodium formate method. Although oxidation method has the advantages like availability of raw materials, simple process and less investment, there have problems of technique and product quality stability, low yield, severe equipment corrosion and serious environmental pollution due to the large number of nitrogen oxides in the exhaust.

Another important problem of oxalic acid production is the high production cost, such as the higher catalyst cost and CO cost of CO-coupling method, higher raw material cost of glycol-nitric acid oxidation method.

Therefore, it is essentially to develop new techniques with the characteristics like availability of raw materials, simple process, low-cost and no pollution, which demands vigorous researches on the key technologies and processes of oxalic acid industrialization.

#### *4.5.5.3. Key Technology and Process Development*

Alkali-Melting Coupling with Steam Exploded Technology to Produce Oxalic Acid

As has mentioned that Alkali-melting coupling with steam exploded technology has the shortcomings like low conversion (less than 35%), high alkali consumption and serious environmental pollution. In order to solve these problems, our group successfully developed a KOH-melting recycling cleaner production system of oxalic acid on the basis of cleaner production technology. The alkali-melting oxidation of steam exploded straw is more adequately than untreated straw, which can reach to 70%. This technology system not only be able to achieve the production of oxalic acid, and the resulting waste can used to produce organic fertilizers after treated by compounding and drying.

*1) Principle of KOH-Melting Recycling System*

Using ionized O2- as high temperature solvent, the steam exploded straw generates oxalic acid, formic acid, acetic acid, propionic acid, phenol and aromatic hydroxy acid in the alkali-melting solution. By adding calcium oxide solution, the reaction of formic acid, acetic acid or propionic acid with calcium oxide does not generate precipitation, and the content of phenol and aromatic hydroxy acid are slight, only oxalic acid can react with alkali to generate precipitation of calcium oxalate. Then, the calcium oxalate precipitation reacts with sulfuric acid solution to form oxalic acid. The remaining excess alkali solution is concentrated and returned to recycling system for reutilization.

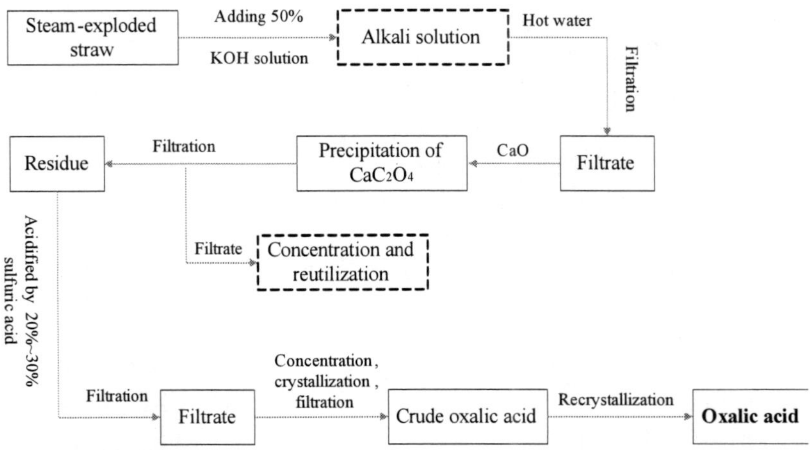


Figure 4.20. Preparation of oxalic acid by KOH-melting recycling system with steam explosion straw as raw material

*2) Technological Process of KOH-Melting Recycling System*

Technological process of oxalic acid production by KOH-melting recycling system is shown in figure 4.9. Main ingredients of steam exploded straw are cellulose, hemicellulose and lignin. In theory, in the alkali solution, 2,3-dialdehyde group on cellulose occurred rupture reaction, semi-cellulose occurred peeling reaction, acidic phenolic hydroxyl of lignin has reaction in alkali solution, all these reactions made straw alkali-degraded and further generates crude oxalic acid. Therefore, production of steam-exploded straw in alkali-melting oxidation system is theoretically feasible.

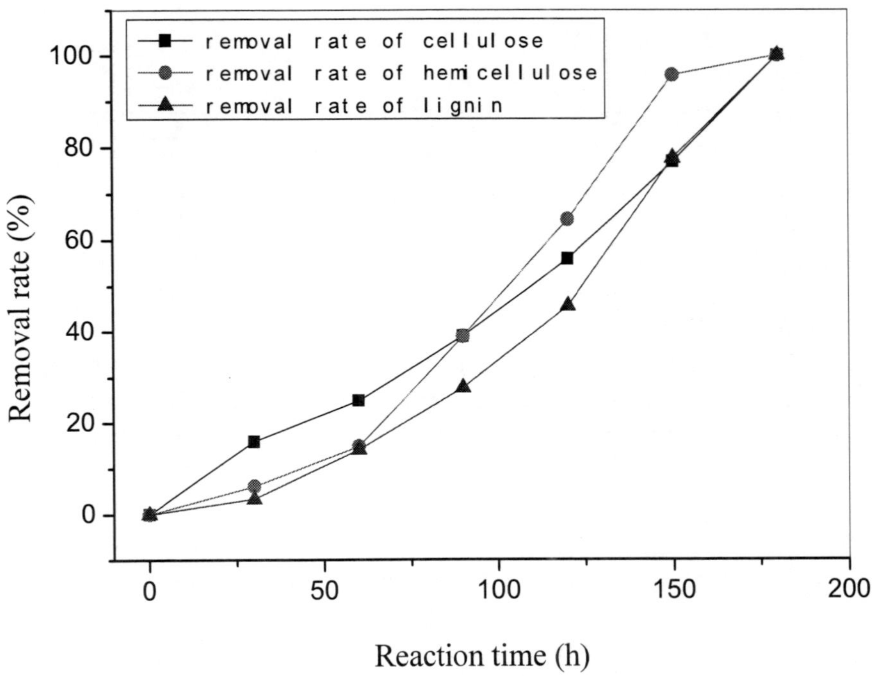


Figure 4.21. Removal rate of main integrants of raw material

Figure 4.21 indicated that the removal rate of cellulose, hemicellulose and lignin increased with the extension of reaction time. The comparison of different antioxidants on KOH alkali-melting recycling systems showed that in the same concentration of oxidants the yield of oxalic acid using potassium permanganate as oxidant is the highest and the best concentration is 30% (Figure 4.22). The oxalic acid yield is 0.75 g oxalic acid / g straw by using 20% K2MnO4 as oxidant, and 0.82 g oxalic acid / g straw by using 30% H2O2 as oxidant. With steam-exploded straw as raw material the oxalic acid yield is 1.26 times of that without pretreatment. Oxalic acid yield produced by KOH-melting recycling system is 3.01 times of that produced by NaOH-melting recycling system.

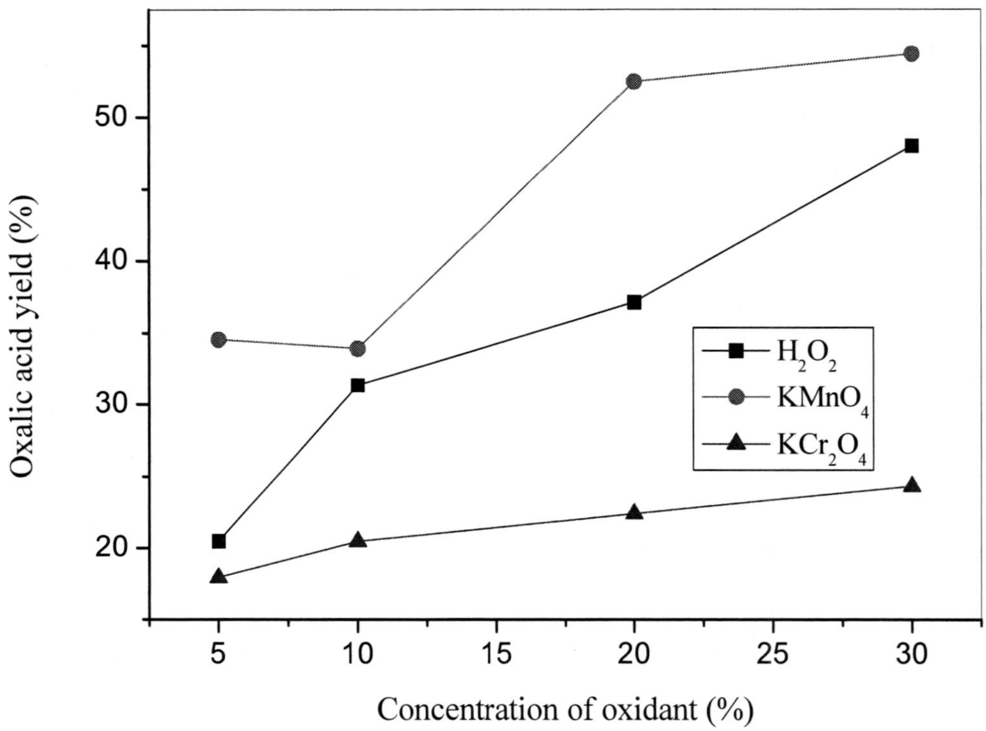


Figure 4.22. Oxalic acid yield using different oxidant

*5) Material Balance*

Figure 4.23 shows the material balance of oxalic acid production by KOH-melting recycling system, which indicates that raw material consumption of production each 1 t oxalic acid is 0.37 t KOH, 0.45 t CaO, 0.43 m$^3$ $H_2SO_4$, 17.5 t $H_2O$ and 0.20 t $KMnO_4$. 1 t straw can produce 0.6 t oxalic acid, that is, 0.84 t oxalic acid crystals.

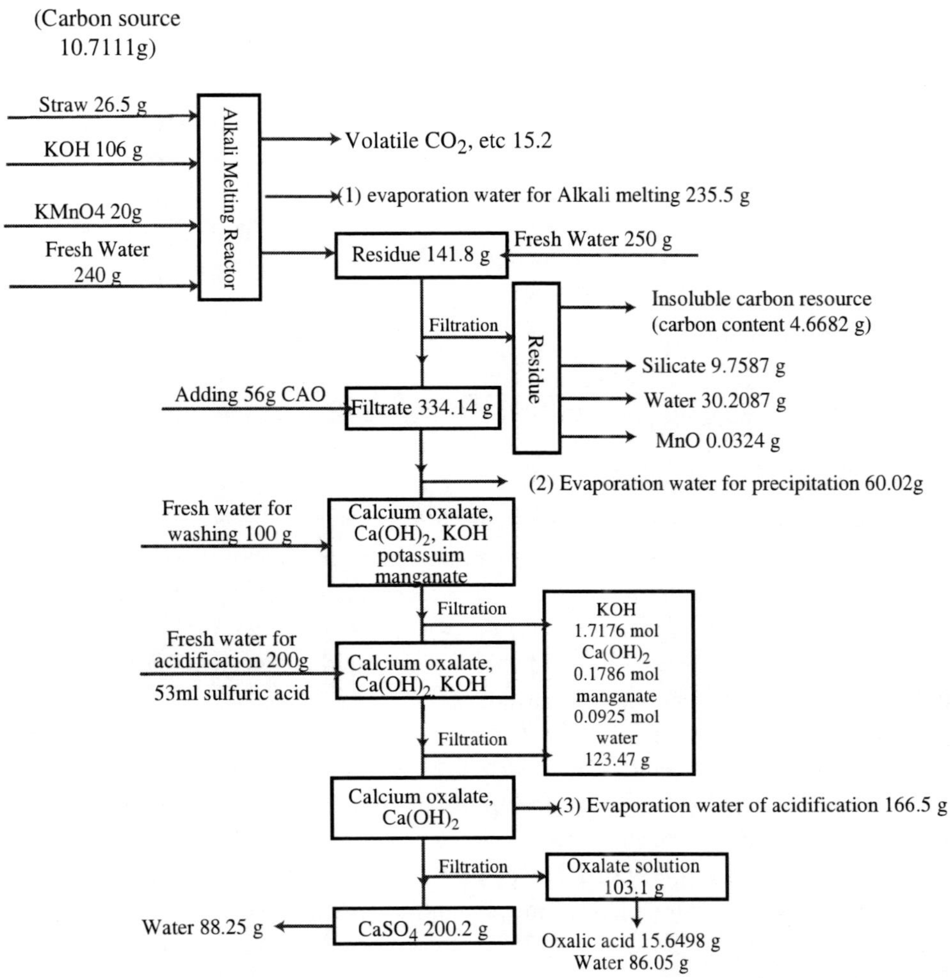


Figure 4.23. Material balance of oxalic acid production by KOH-Melting recycling system

*4) Alkali Recovery System*

Alkali recovery system of oxalic acid production by KOH-melting recycling technology is mainly used continuous centrifugation and counter-current rinsing process. The alkali recovery system can be present as figure 4.24. Alkali solution containing calcium oxalate is directly into centrifuge device (1) which divided into high concentration and low concentration alkali solution by continuous centrifugation and counter-current rinsing process. The high concentration alkali solution sends off directly into alkali solution tank (2) and adjusted concentration for further reutilization. The low concentration alkali solution sends into dew evaporating enrichment device (3), which enriched by continuously recycling evaporation and then sends into alkali storage pool (2) after reaching to setting concentration.

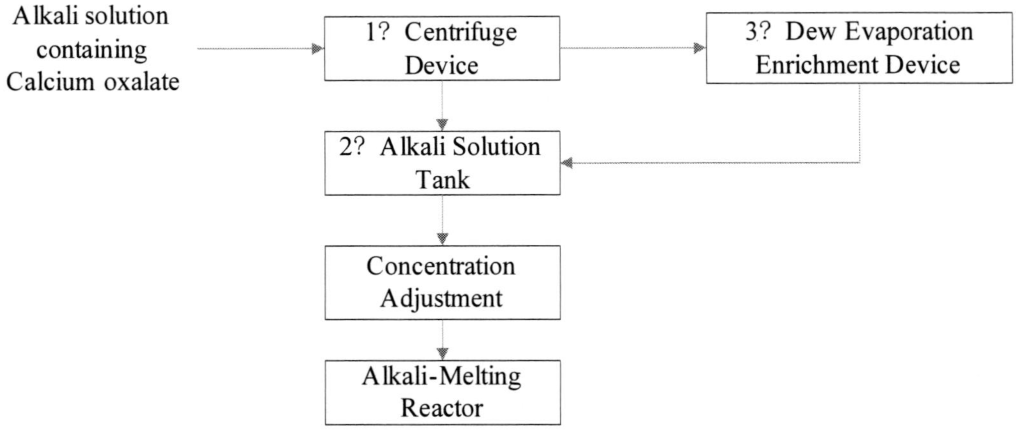


Figure 4.24. Technological process schematic diagram of alkali recovery system

Microbial Fermentation of Oxalic Acid

Due to increasing demand in hydrometallurgy and for wider applications in other processes, production of oxalic acid from a cheap carbon source by fermentation appears more attractive as an eco-friendly, non-hazardous microbial process yielding oxalic acid at lower cost. Accordingly, fermentative production of this acid has been studied with renewed interest in the recent years. Among the variety of fungi producing oxalic acid, there mainly include *Sclerotinia sclerotiorum* [85], *Coriolus versicolor*, *Heterobasidion annosum*, *Pleurotus florida*, *Phanerochaete chrysosporium* [86], and *Aspergillus niger* [87, 88], etc. *Aspergillus niger* is much more efficient than other microbes and has been used most widely. Mandal and Banerjee immobilized mycelia (bioparticles) of *A. niger* in polyurethane foam, the yield of oxalic acid is 150 g/l [89]. However, the microbial fermentation of oxalic acid is still in its infancy stage and remains to be further study.

## 4.5.6. Lactic Acid

Lactic acid (2-hydroxy prop ionic acid) is the most widely occurring carboxylic acid in nature, which is expected to become an alternative mass-produced chemical instead of chemical products. Lactic acid is chiral and has two optical isomers. One is known as L-(+)-lactic acid or (S)-lactic acid and the other, its mirror image, is D-(-)-lactic acid or (R)-lactic acid. L-(+)-Lactic acid is the biologically important isomer since that human only has L-lactate dehydrogenase. Only L-lactic acid can be metabolized fully by human, and do not have any toxic side effects. The most promising application of lactic acid may yet be in the manufacture of biodegradable and biocompatible polylactate polymers, an environment-friendly alternative to conventional non-biodegradable plastics derived from petrochemicals.

As a result, lactic acid is generating more and more interest and the market [90]. Currently, the worldwide consumption of lactic acid is estimated to be 130 000-150 000 (metric) tones per year. There are several major manufacturers of fermentative lactic acid, including Purac (Netherlands), Galactic (Belgium), Cargill (USA), Sterling (USA) and several Chinese companies. L-lactic acid production capacity of ADM Company (USA) is

9,000 to 18,000 tons / year, 9000 tons / year for Ecochem Corporation (joint venture of DuPont and Conagra Inc.), and 7300 tons / year for Sterling Chemicals Inc.

#### *4.5.6.1. General Situation of Lactic Acid Industry Progress*

*Production Method*

Production methods of lactic acid include chemical synthesis, fermentation and enzymatic methods. As the chemical synthesis of lactic acid synthesis using toxic acetaldehyde and hydrocyanic acid as raw materials, therefore about 90% of lactic acid was obtained by microbial fermentation. According to Wee et al. (2006) [91], raw materials for lactic acid production should have the following characteristics: low-cost, high production rate, high yield, little or no byproduct formation, ability to be fermented with little or no pretreatment and year-round availability. Moreover, further expansion of the lactic acid industry as a competitor in the production of petroleum-based plastics will depend in part on the availability of low-cost substrates, such as hexose and pentose sugars derived from lignocellulosic residues [92, 93]. Fermentative production of L-lactic acid includes three main methods: direct fermentation method, separate hydrolysis and fermentation (SHF), simultaneous sacchrification and fermentation (SSF) [93].

1) Direct fermentation means that lignocellulose directly produce lactic acid without pretreatment of acid or enzymatic hydrolysis. The advantages of this method is sample equipments and low production cost, while the lactic acid yield is low and easily to produce byproducts. Recently, the development of co-fermentation provides a new direction for the direct fermentation method.

2) SHF is the most popular method for lactic acid production, in which the sacchrification and fermentation is processed independently. Lignocellulose firstly degraded into glucose by chemical or enzymatic hydrolysis and then glucose is used as carbon source to produce lactic acid by bacteria or fungi. The advantages of this method is hydrolyzed completely by control the optimal conditions of sacchrification and fermentation separately. However, the fermentation is inhibited by many factors such as end-product, low cell concentration, substrate inhibition, etc.

3) SSF can combine the hydrolysis of cellulose and fermentation of lactic acid, which conducive to reducing mixed fungi pollution, accelerating hydrolysis rate, shortening fermentation period, and saving recovery of enzyme. The optimum fermentation temperature, pH and microanaerobic or aerobic fermentative characteristics of many lactic acid bacteria have the consistent effect of cellulase. In the process of SSF, the continuous consumption of hydrolysate makes the sugar concentration in a low level, which can reduce the feedback suppression of end-product efficiently and increase the hydrolysis and extraction rate of cellulose. Because of no production of $CO_2$ the theoretical conversion rate is 100%. Therefore, SSF has many advantages on the production of lactic acid.

*Fermentation Mechanism*

L-lactic acid fermentation mechanism can be classified into three groups: homofermentative, heterofermentative, mixedfermentative [93-95].

1) Homofermentative mechanism is that lactic acid bacteria convent glucose into pyruvate through Embden-Meyerhof pathway (EMP pathway), and pyruvate almost exclusively reduced to lactic acid by the catalysis of lactate dehydragenase. Since the end-

product of glucose metabolism is only lactic acid, two lactic acid molecules are produced from each molecule of glucose with a yield of more than 0.90 g/g [91].

2) Heterofermentative mechanism is that lactic acid bacteria catalyzed glucose into ethanol and $CO_2$ as well as lactic acid through Bifidus and PK pathway. In this type of fermentation, one molecular of glucose can produce one molecular acetic acid, one molecular $CO_2$ and one molecular lactic acid, so the conversion rate of glucose to lactic acid is only 50%.

3) Mixed-lactin fermentation is that happened under special circumstances, such as restrictions on the concentration of glucose, increase of pH or decrease of the temperature. It has the same EMP pathway of fermentation, but the metabolism pathway of pyruvate is changed, which in addition to lactic acid generation but also generate by-products such as formic acid.

*Fermenting Strains*

Microorganisms that can produce lactic acid mainly include bacteria and fungi [96, 97].

Lactic acid bacteria can be divided into three genuses: 1) *Bacillus*: *Lactobacillus acsei, Lactobacterium thermophilum, Lactobacterius salivarius, Lactobacterius sake, Lactobacterius acidophilus, Lactobacterius delbrueckii, Laetobaeillus amylophilus, Laetobacillus Plantarum, etc.* 2) *Lactobacillus*: *Streptoccoccus, Streptoccoccus thermophilus, Streptoccoccus cremoris, Streptoccoccus salivarius*. 3) *Streptoccoccus*: *Bacillus coagulans*. Lactic acid fungi mainly include *Rhizopus nigricans, Rh. jaranicus, Rh. frifiici, Rh. chinensis, Rh. batata, Rh. oryzae, Rh. nodosus, Rh. japonicus, Rh. fonkinensis, Rh. thermosus, Rh. arrhizus*, *Rh. elegens*, and so on. *R. oryzae* is the ideal strains for lactic acid fermentation with some advantages in that it requires only a simple medium and produces L (+)-lactic acid, but it also requires vigorous aeration because *R. oryzae* is an obligate aerobe.

Recent advances in metabolic engineering of microorganisms may provide more opportunities for selective and efficient production of optically pure lactic acid through the improvement of future strains. Lactic acid production and sugar conversion rate of recombinant *Rhizopus oryzae* is 80~140 g / L and 68%~85% respectively using glucose as carbon source and 90 g / L and 86.0% respectively using corn starch as carbon resource [97-99].

#### *4.5.6.2. Problems*

*1) High Production Cost*

With the production capacity expansion of companies (such as ADM, Gargill and Purac) and the improvement of production technology, the cost of lactic acid production reduced greatly. However, there are still several issues that need to be addressed in order to produce lactic acid biotechnologically within the targeted cost, such as the development of high-performance lactic acid-producing microorganisms and the lowering of the costs of raw materials and fermentation processes. The market prospect of lactic acid is very optimistic and it is reported that over the next 10 years, lactic acid products will jump to be commodity chemicals [91].

*2) Low Productivity*

The accumulation of lactic acid in the fermentation process has an obvious product inhibition effect. If the fermentative pH below 5.0 the yield of lactic acid is only about 1.6%. With the continuous generation of lactic acid, the broth pH value will continue to decrease, so that the growth of lactic acid bacteria production of lactic acid will be inhibited.

*3) Lag of Fermentative Technology*

Lactic acid fermentation technology mainly is batch fermentation, continuous and semi-continuous (fed-batch) fermentation, which hinders large-scale industrialization of lactic acid. Presently, most manufacturers produce lactic acid from calcium carbonate by using batch fermentation technology, of which the process is simple but lactic acid extracting rate is only about 45% and company with a large amount of waste residue. Therefore, many countries are exploring a new fermentation process. Continuous fermentation simplifies the streamline operations and improves the utilization of equipment, which enhances the productivity, automation degree and product quality. However, the yield and conversion rate of continuous fermentation is low than intermittent fermentation.

#### *4.5.6.3. Key Technology and Process Development*

Recently, in order to improve the productivity of lactic acid and improve the separation process, many new technologies are developed according to each unit operation of lactic acid production, such as cell immobilization technology, in situ product removal technology, etc. The in situ product removal technology includes solvent extraction fermentation, adsorption, membrane fermentation, and so on. These technologies turn intermittent fermentation of lactic acid into continuous fermentation, so as to enhance the productivity of lactic acid.

*Cell Immobilization Technology*

*R. oryzae* can get high purity L-lactic acid and have amylolytic enzyme activity, converting starch directly to L (+)-lactic acid. However, due to *R. oryzae* is a kind of filamentous fungi with complex growth state. The growth state of *R. oryzae* is obviously affected by environmental factors, such as the characteristics of bacteria itself, nutrition supply, medium pH, stirring and ventilation, etc. As the mycelium of R. oryzae easily agglomerated and entangled in the mixing tank, which increase the mass transfer resistance and power consumption. Immobilization can increase the cell concentration and lactic acid production rate. In the immobilized cell or biological particles, cells are restricted in a specific carrier and retain the high biological activity [98].

Commonly used methods include sodium alginate-immobilization, calcium alginate-immobilization, and polyurethane absorption-immobilization and bio-rotating absorption-immobilization. Cells embedded by alginate-immobilization can grow normally and have very good operating stability. However, whereas the shortcomings of alginate-immobilization such as the complexity of operation, low pollution resistant ability and easy adhesion of embedded, some soft high molecular weight materials like sponge, cotton cloth, carrageenan, gelatin and so on are used in the immobilization of *R. oryzae* by using the natural absorption of mucellium or spore.

Co-immobilization system can exert the synergistic effect of enzyme and cell, has sustained and stable production capacity of lactic acid. Xia Liming et al co-immobilized spores of Aspergillus niger, enriched with cellobiose, and cells of Lactobacillus bacillus in calcium alginate gel beads to establish a new series bioreactor with the combination of co-immobilization system and enzymatic hydrolysis [100]. The produced lactic acid concentration and the yield of lactic acid from cellulose were 55.7 g • L-1 and 91.5% respectively. Under the fed-batch process, the final concentration of lactic acid and productivity increased to 106.7 g • L-1 amd 1.270 g • L-1• h-1, respectively, which the dosage of cellulase per gram substrate reduced by 25%. The coimmobilized cells were quite stable and efficient, and could be repeatedly used for at least 12 batches [99].

*In Situ Product Removal Technology*

Lactic acid production processes traditionally suffer from end-product inhibition. The inhibition mechanism of lactic acid is probably related to the solubility of the undissociated lactic acid within the cytoplasmic membrane, which eventually influences the transmembrane pH gradient and decreases the amount of energy available for cell growth. Therefore, to alleviate the inhibitory effect of lactic acid during the fermentation, it must be removed selectively in situ from the fermentation broth. In situ product removal (ISPR) refers to the method that rapidly removes metabolism products. Productivity of ISPR is higher than batch fermentation, which realizes the automatization, and continuous fermentation.

1) Membrane Recycling Fermentation

Membrane recycling bioreactor maintains a relatively high concentration of cells by combining separation and fermentation process. The productivity can significantly be improved by the recycling of cells and removing lactic acid continuously. Types of cell recycling membrane include membrane-dialysis (relying on the proliferation exclusion), electrodialysis (relying on ion exclusion), microfiltration and ultrafiltration (relying on the molecular exclusion). Membrane reactor provide a green platform for the lactic acid fermentation which without introducing other chemicals and avoiding the effect of chemical composition on the product quality, so that reduce the cost and difficulty of follow-up and purification.

2) Extractive Fermentation

Extractive fermentation refers to the technology that reduces the end product suppression of fermentation by removing product with organic solvents. The advantages of extractive fermentation are low energy consumption, good selectivity, no sundry bacteria contamination. However, the shortcoming of Extractive fermentation is the toxicity of organic solvent on microorganisms. Commonly used methods to reduce the toxicity of solvents include separating solvent and cell with membrane, immobilization, embedding plantoil in the immobilized carrier. Chen and Lee studied the lactic acid production of *Lactobacillus delbrueckii* in a fermenter-extractor by employing a microporous hollow fiber membrane (MHF) and using Alamine 336 as extractant for lactic acid. They proposed that by extractive fermentation pH of fermentation process changed from an initial value of 5.0 down to 4.3, which significantly improved the overall performance of the simultaneous saccharification and extractive fermentation [101]. Recently, the extractive fermentation method has been developed to aqueous two-phase [102].

3) Absorptive Fermentation

Absorptive fermentation refers to the fermentation process using absorbent such as activated carbon, ion-exchange resins, and so on, for the removal of fermentation product to remove the product inhibition. Due to the disadvantages of activated carbon adsorbent such as small absorption capacity, poor selectivity and reproducibility, usually adopts ion exchange resins as absorbent.

4) Electrodialysis Fermentation

Electrodialysis fermentation with ion exchange membranes was often used for in situ removal of lactic acid [103, 104]. Min-Tian et al [105] had previously developed a continuous electrodialysis fermentation system for the production of lactic acid. In their study, the system of electrodialysis fermentation with a level meter was the most efficient system and a higher yield could be obtained if the glucose concentration in the broth could be controlled to remain at a lower level.

## 4.5.7. Bio-Ethylene

As the largest chemical products, the industrial scale, output and technical level of ethylene have become an important symbol to measure the chemical industry level of country. Ethylene is mainly derived from hydrocarbon cracking, gas oil recovery, ethanol dehydration, coal or oil direct cracking, heavy oil catalytic cracking, etc. However, in recent years, the irreversible trend of world oil resources shortage results in high production costs and obvious decline of commercial competitiveness of ethylene. On the other hand, the demand of ethylene derivative industries such as chemicals, energy, industrial materials is keep growing, which result in an increasingly striking contradiction between supply and demand of ethylene. Therefore, it is necessary to seek a sustainable development pathway to solve the negative effects results from oil shortage, which provide biological ethylene wide development opportunities.

### *4.5.7.1 General Situation of Lactic Acid Industry Progress*

*1) Research Status of Bioethanol*

Bio-ethylene is produced by the catalytic dehydration of ethanol which fermented by microorganisms using renewable biomass as raw materials. So, the industrial development of bio-ethylene must base on the high-performance and low-cost industrialization of bio-ethanol. At present, bio-ethanol is attracting attention of the whole world. Many countries have put the biomass ethanol as a long-term development plan to research and popularization, which provide protection for the development of bio-ethylene.

Bioethanol refers to the ethanol obtained through the microbial fermentation by using biomass as raw material. Biomass materials include crops contain sugar and starch such as maize, sorghum, wheat, barley, sugar cane, sugar beet and potatoes, in addition lignocellulose such as municipal solid waste, bagasse, tree trunks, wood chips can also be used to produce bio-ethanol fuel. At present, 60% of the world ethanol use molasses or grain as raw material. However, the production of ethanol by grain has many problems. Production of bio-ethanol using lignocellulose as raw material will gradually become a hot topic of the high-tech research and the most intense competition field of industrialization.

However, the most controversial problem of the development of cellulosic ethanol is still the production cost, compared with ethanol generated from gasoline and grain, still remains too high. In recent years, substantial amount of research has been focused on the production technology of cellulosic ethanol such as pretreatment of raw materials, low-cost and high efficient production of cellulase, coupled enzymatic fermentation technology, etc, which make the production cost of cellulosic ethanol reduced greatly. But so far, production cost of cellulose eth**anol** is still very high. On the other hand, the concentration of ethanol from cellulose is very low, usually only 10% (mass) or so. Due to many impurities in the fermentation broth such as methanol, acetic acid, glycerol, and so on, the distillation and purification of the ethanol consume a lot of energies. Therefore, it is an important development aspect that catalyze the dehydration of bioethanol to generate bioethylene by one step without strict separation and purification of low concentration bio-ethanol [106].

*2) Research Status of Bioethylene Dehydrated by Bioethanol*

Production ethylene from biomass refers to a process that biomass hydrolyzed into sugar, sugar produces ethanol by fermentation and ethanol generates ethylene by dehydration. Prior to the petrochemical development ethylene was mainly produced by the dehydration of ethanol, which demanded the volume fraction of ethanol generally higher than 95%. Compared to oil-ethylene technology, this technology has many shortcomings such as small handling capacity, low production capacity of equipment, high-energy consumption, etc. As a result, the vigorous development of the petrochemical industry makes ethylene production of ethanol dehydration is gradually phased out. However, the increasing depletion and rising price of oil resources caused the rising cost of oil ethylene. Recent years, ethylene preparation technology by bio-ethanol dehydration gradually be taken seriously and made significant development. At present, the international ethanol-ethylene plant mainly in Brazil, India, Pakistan, Peru, etc, for example the production capacity of India is 64,000 tons / year. Although the production scale of ethanol-ethylene obviously lower than the oil ethylene plant, its strong vitality should be paid special attention [105, 106].

*3) Technical and Economic Analysis*

Analyzing from the current oil price, ethylene production from bio-ethanol has a larger profit margin. The cost of preparation ethylene by ethanol dehydration is almost similar to oil ethylene, even slightly lower. If using low concentration of ethanol fermentation broth (about 10% mass fraction) as raw materials, the production costs can be further reduced. Furthermore, ethylene preparation by ethanol dehydration has many advantages such as simple process, simple composition of products, easy separation, which reduce the total investment, shorten the construction period, and accelerate the proceeds.

#### *4.5.7.2 Problems*

*1) Cost of Bioethanol*

Compared with ethanol generated from gasoline and grain, the production cost of ethanol produced from cellulose is still remains too high, especially in the aspect of raw material pretreatment, cellulase and ethanol distillation cost. For example the cellulase cost of ethanol using straw as raw material accounts for 30~50% which is 8% in grind-ethanol. Due to the low concentration of ethanol in products the ethanol distillation cost accounts for about 20% which is 14% in grind-ethanol.

*2) Production Equipment and Technology of Ethylene from Bio-Ethanol*

Ethanol dehydration equipment is built on the follow-up ethanol production line, of which the technology is relative antiquated and high energy consumption. Therefore, it is necessary to promote the technical innovation and equipment reformation according to the characteristics of ethylene. At present, ethanol dehydration equipment mostly use tubular fixed-bed reactor, which has the shortcomings such as complicated production technology, high cost, and inconvenient load and unload of catalyst. Compared with the tubular fixed-bed technology, fluidized bed technology conduce to higher single pass conversion and ethylene yield and lower postprocessing equipment cost.

*3) Low Catalytic Efficiency of Catalyst*

Problems of ethanol dehydration catalyst mainly are high reaction temperature, easy coking deactivation, short lifetime, higher concentration requirement of ethanol, etc.

#### *4.5.7.3. Key Technology and Process Development*

Key technologies of biological ethylene mainly include low-cost ethanol production technology, ethanol catalytic dehydration technology, as well as the coupling process of integration technology [107].

*Low Cost Bio-Ethanol Production Technology*

Basic technology of bio-ethanol includes raw materials collection, pretreatment, hydrolysis, fermentation and purification.

1) Cost and Energy Consumption of Raw Material Pretreating

Pretreating process is one of key technologies, which not only decide the bio-ethanol production cost but also one of the key steps of biomass ethanol industrialization. Effective pretreatment can reduce the used cellulase amount and the concentration of microbial inhibitor such as acid and formaldehyde, which result in high efficient ethanol microbial fermentation and low distillation cost of ethanol. In addition, acid hydrolytic technology of lignocellulosic raw materials is one of the earliest and most commonly used pretreatment technologies, which have relatively high pollution and high pollution controlling cost. In general, pretreating must meet several necessary conditions such as increasing enzymatic combination rate, avoiding the degradation and loss of carbohydrates, reducing hydrolysis and fermentation inhibitors; enhancing performance price ratio.

Facing multitudinous pretreating technology of lignocellulosic materials, how to assess and rationally use these methods becoming a hot research field due to each of them have their own strengths and weaknesses. Chen put forward a reasonable fractionation method for the pretreating of biomass, which is not just limited to the use of cellulose, but the maximal and high value-adding utilization of each component based on comprehensive utilization. Therefore, it is necessary to estimate the biomass pretreating technology from the technical, economic and environmental objectives. Try the best to make pretreating technology feasible in technique, energy-saving, consumption reduction and pollution-free in technology, reasonable in economy.

*High Efficient Ethanol Dehydration Technology*

Ethanol dehydration efficiency depends largely on catalyst and alumina is the most popular industrial catalyst for ethylene. However, due to the demanding conditions, high reaction temperature, high demanding ethanol volume fraction (90%) result in higher energy consumption of whole process. Therefore, it is crucial for bio-ethylene from bio-ethanol to develop long-life and high efficient catalyst, which has high selectivity and high conversion rate of low concentration ethanol under low temperature [108]. In addition, in order to reduce energy consumption researchers are still exploring new suitable catalyst converting ethanol into ethylene in liquid state. Because catalyst can react under room temperature and pressure and has many advantages like high specificity, high efficiency, moderate reaction condition and low equipment requirement, etc, many scholars take enzymatic catalyst as a breakthrough [109].

*Coupling Process of Integration Technology*

The existing ethylene producing system by ethanol dehydration technology is conducted after the production and distillation of ethanol, which has low ethylene yield and high-energy consumption. In addition, biomass ethanol contains a variety of impurities, in which took place various reactions. The traditional ethanol dehydration technology fails to meet the industrialization requirements of bio-ethylene. It is very important to integrate the process of ethanol production from biomass with ethylene production from ethanol dehydration, to couple the ethanol flow characteristics in the reactor with the catalytic reaction, to develop new type reactor, to increase the ethylene yield and catalytic selectivity.

## REFERENCES


[1] Yihong, S.; WangShuai; Chen, L.; LiuHaicha, Biomass-derived polyols as new bio-platform molecules for sustainable production of fuels and chemicals. *Progress in chemistry* 2007, 19, (2/3), 131-141.

[2] Enze, M., Developing biorefinery by utilizing agriculture and forestry biomass resources: striding forward the "Carbohydrate" era. *Process in chemistry* 2006, 18, (2), 131-141.

[3] Lu, L.; Beihai, H.; Runcang, S.; Ruofei, H., High-value chemicals from lignocellulosic biomass. *Progress in chemistry* 2007, 19, (7/8), 1206-1216.

[4] Czernik, S.; Bridgwater, A. V., Overview of applications of biomass fast pyrolysis oil. *Energy & Fuels* 2004, 18, (2), 590-598.

[5] Shufen, Z.; Jinzong, Y., Development chance of biomass fine chemicals. *Modern Chemical Industry* 2006, 26, (4), 1-5.

[6] Danner, H.; Braun, R., Biotechnology for the production of commodity chemicals from biomass. *Chemical Society Reviews* 1999, 28, (6), 395-405.

[7] Kamm, B.; Kamm, M., Biorefineries--multi product processes. *Adv Biochem Eng Biotechnol* 2007, 105, 175–204.

[8] Zhen, L.; Liang, J.; Shaoyi, J., Extract technique of Chinese Medicine efficacious ingredient and its application. *Chemical industry and engineering* 2005, 22, (6), 450-456.

[9] Perva-Uzunali, A.; kerget, M.; Knez; Weinreich, B.; Otto, F.; Grüner, S., Extraction of active ingredients from green tea (Camellia sinensis): Extraction efficiency of major catechins and caffeine. *Food Chemistry* 2006, 96, (4), 597-605.

[10] Feiyue, L.; Yinsong, X., Analysis on products of photocatalytic oxidation of Lignin on nano-semiconductor and Ii's mechanism. *Journal of Cellulose Science and Technology* 2006, 14, (3), 24-31.

[11] Wei-dong, Y.; Ming, D.; Jinping, Z.; Jingwen, W., Prmiary study on Low-molecular-mass products from electro-oxidation of lignosulfona. *Chemistry and Industry of Forest Products* 2006, 26, (1), 121-123.

[12] Bridgwater, A. V., Biomass fast pyrolysis. *Thermal Science* 2004, 8, (2), 21-49.

[13] Tripathi, A. K.; Iyer, P. V. R.; Kandpal, T. C., A financial evaluation of biomass-gasifier-based power generation in India. *Bioresource Technology* 1997, 61, (1), 53-59.

[14] Dasappa, S.; Paul, P. J.; Mukunda, H. S.; Rajan, N. K. S.; Sridhar, G.; Sridhar, H. V., Biomass gasification technology-a route to meet energy needs. *Current Science* 2004, 87, (7), 908-916.

[15] M. T. Klein; Virk, P. S., Modeling of lignin thermolysis. *Energy Fuels* 2008, 22, (4), 2175-2182.

[16] Hongzhang Chen; Jian Xu; Jianzhong Yao; Li., Z., Studies on multi-level conversion of liquid fuel with straw. *Chinese Journal of Bioprocess Engineering* 2005, 3, (1), 9-13.

[17] Ayhan Demirbas; Arin, G., An Overview of Biomass Pyrolysis *Energy Sources, Part A: Recovery, Utilization, and Environmental Effects* 2002, 24, (5), 471-482.

[18] Demirba, A., Biomass resource facilities and biomass conversion processing for fuels and chemicals. *Energy Conversion and Management* 2001, 42, (11), 1357-1378.

[19] Zhang, Y.; Ikeda, A.; Hori, N.; Takemura, A.; Ono, H.; Yamada, T., Characterization of liquefied product from cellulose with phenol in the presence of sulfuric acid. *Bioresource Technology* 2006, 97, (2), 313-321.

[20] Alma, M. H.; Yoshioka, M.; Yao, Y.; Shiraishi, N., Some Characterizations of Hydrochloric Acid Catalyzed Phenolated Wood-Based Materials. *Journal of the Japan Wood Research Society* 1995, 41, (8), 741-748.

[21] Zhang Q H; J., Z. G., Liquefaction of wood by using phenol or polyhydric alcohols. *Journal Of Beijing Forestry University* 2003, 25, (6), 71-76.

[22] Pan, H.; Shupe, T. F.; Hse, C. Y., Characterization of liquefied wood residues from different liquefaction conditions. *Journal of Applied Polymer Science* 2007, 105, (6), 3740-3746.

[23] Kobayashi, M.; Asano, T.; Kajiyama, M.; Tomita, B., Analysis on residue formation during wood liquefaction with polyhydric alcohol. *Journal of Wood Science* 2004, 50, (5), 407-414.

[24] Yamada, T.; Ono, H., Characterization of the products resulting from ethylene glycol liquefaction of cellulose. *Journal of Wood Science* 2001, 47, (6), 458-464.

[25] Nagasawa, T.; Yamada, H., Microbial production of commodity chemicals. *Pure Appl. Chem* 1995, 67, 1241-1256.

[26] Straathof, A. J. J.; Panke, S.; Schmid, A., The production of fine chemicals by biotransformations. *Current Opinion in Biotechnology* 2002, 13, (6), 548-556.

[27] Settings, M., Conversion of furfuryl alcohol to levulinic acid or an alkyl levulinate. *Focus on Catalysts* 2007, 2007, (10), 8-8.

[28] Bozell, J. J.; Moens, L.; Elliott, D. C.; Wang, Y.; Neuenscwander, G. G.; Fitzpatrick, S. W.; Bilski, R. J.; Jarnefeld, J. L., Production of levulinic acid and use as a platform chemical for derived products. *Resources, conservation and recycling* 2000, 28, (3-4), 227-239.

[29] Girisuta, B.; Janssen, L.; Heeres, H., Green chemicals: a Kinetic study on the conversion of glucose to levulinic acid. *Chemical Engineering Research and Design* 2006, 84, (A(5)), 339-349.

[30] Benthin, B.; Danz, H.; Hamburger, M., Pressurized liquid extraction of medicinal plants. *Journal of Chromatography A* 1999, 837, (1-2), 211-219.

[31] Congbing, F.; Xiaochun, W.; Jiang, S. E.; Heqin, C. u. L., The Immobilized Cell Cultures of Pueraria lobata and the production of Isoflavonoids. *Food and Fermentation Industries* 2006, 32, (5), 23-27.

[32] Hurlbut, J. A.; Carr, J. R.; Singleton, E. R.; Faul, K. C.; Madson, M. R.; Storey, J. M.; Thomas, T. L., Solid-phase extraction cleanup and liquid chromatography with ultraviolet detection of ephedrine alkaloids in herbal products. *Journal of AOAC International* 1998, 81, (6), 1121-1127.

[33] Yi, X. P.; Chen, D. Y.; Liu, J. P.; Liu, Q. W., Determination of twelve elements in ephedrine extraction by microwave digestion-AAS. *Guang Pu Xue Yu Guang Pu Fen Xi* 2003, 23, (1), 81-3.

[34] Choi, Y. H.; Kim, J.; Kim, J.; Yoo, K. P., Selective Extraction of Ephedrine from Ephedra sinica Using Mixtures of C02, Diethylamine, and Methanol. *Chromatographia* 1999, 50, (11/12).

[35] Hongzhang, C.; Zuoha, L., Studies on the steam explosion of wheat straw Ⅱ. mechanisms of steam explosion of wheat straw. *Journal of Cellulose Science and Technology* 1999, 7, (4), 14-22.

[36] Hongzhang, C.; Yitao, Y., Application of Steam Explosion in Ephedrine Extraction. *Journal of China Pharmaceutical University* 2005, 36, (5), 414-416.

[37] Jiangtao, J.; Sheng, W.; Chunsheng, L.; Chunhua, Y., Research and Application Progress in Countercurrent Solvent Extraction. *Journal of Rare Earths (English version)* 2004, 22, (5).

[38] Kefan, Y.; Pengxiang, Y.; Weiyun, W., The application of dynamic countercurrent extraction equipment in extracting from natural plant. *Packaging and Food Machinery* 2005, 4, 25-28.

[39] Kechuan, Y.; Pengxiang, Y.; Weyun, W., Design and Development of New Extracting Equipment to Effective Composition from Natural Plant. *Light Industry Machinery* 2006, 24, (2), 138-140.

[40] Jinming, G., *Plant chemical.* Science Press: Beijing, 2003.

[41] Xin, C.; Xiaoyong, X.; Chaohong, H.; Lizeng, B.; Chunhui, J., Recycling leaching of Alkaloids from plant *Tripterygium Wilfordii. Journal of Chemical Engineering of Chinese University* 2007, 21, (4), 581-585.

[42] Chun, C.; Xiaojian, M.; Pei, C., Experimental study on production of levulinic acid new platform chemical from biomass. *Acta Energiae Solaris Sinica* 2007, 28, (4), 380-384.

[43] Singh, A.; Das, K.; Sharma, D. K., Integrated process for production of xylose, furfural, and glucose from bagasse by two-step acid hydrolysis. *Industrial & Engineering Chemistry Product Research and Development* 1984, 23, (2), 257-262.

[44] Dias, A. S.; Pillinger, M.; Valente, A. A., Dehydration of xylose into furfural over micro-mesoporous sulfonic acid catalysts. *Journal of Catalysis* 2005, 229, (2), 414-423.

[45] Mansilla, H. D.; Baeza, J.; Urzua, S.; Maturana, G.; Villasenor, J.; Duran, N., Acid-catalysed hydrolysis of rice hull: Evaluation of furfural production. *Bioresource Technology* 1998, 66, (3), 189-193.

[46] Chughtai, F. A.; Hameed, A.; Nosheen, S., Catalytic acid hydrolysis of rice straw to produce furfural and active carbon. *The Journal of Animal and Plant Sciences (Pakistan)* 2003, 13, (2), 87-89.

[47] Win, D. T., Furfural–Gold from Garbage. *AU J.T.* 8, (4), 185-190.

[48] Yanping, L.; Shuo, L.; Wenhai, L., Study on improvement of chemical furfural wastewater treatment process. *Modern Chemical Industry* 2008, 28, (1), 65-68.

[49] Hongjun, R., The way of saving energy in production of the furfural. *Chemical Technology Market* 2006, 29, (7), 41-45.

[50] Xiaohui, Z.; Jiaqi, L.; Yanlei, J.; Yaping, C., Reclaim process of acetic acid in waste water of furfural production. *Chemical industry and engineering* 2006, 23, (2), 142-146.

[51] Dongxiang, Z.; Lei, S.; Changhao, L., On the Combined Model of Furfural Production and Thermoelectric Disposal of Furfural Resid. *Fly ash comprehensive utilization* 2006, 2, 35-37.

[52] Bingyong, C., Furfural closed-cycle waste water handling technology. *China Water and Wastewater* 2007, 23, (10), 82.

[53] Chang, C.; Cen, P.; Ma, X., Levulinic acid production from wheat straw. *Bioresource Technology* 2007, 98, (7), 1448-1453.

[54] Chun, C.; Xiaojian, M.; Shuqi, F.; Hongliang, L.; Peilin, C., Advances in production of new type platform chemical levulinic acid by renewable resources. *New Chemical Materials* 2005, 33, (8), 69-70.

[55] Cha, J. Y.; Hanna, M. A., Levulinic acid production based on extrusion and pressurized batch reaction. *Industrial Crops & Products* 2002, 16, (2), 109-118.

[56] Ghorpade, V. M.; Hanna, M. A. Method and apparatus for production of levulinic acid via reactive extrusion. 5,859,263., 1999.

[57] lei, C. Study on effieient and environmentally friendly methods for preparation of levulinic acid. ZhejiangUniversity, 2004.

[58] Xinghong, D. Study on the Xylitol Fermentation from Hemicellulosic Hydrolysate of Corncob. Zhejiang University, Hangzhou, 2006.

[59] Karhumaa, K.; Sanchez, R. G.; Hahn-H?gerdal, B.; Gorwa-Grauslund, M. F., Comparison of the xylose reductase-xylitol dehydrogenase and the xylose isomerase pathways for xylose fermentation by recombinant Saccharomyces cerevisiae. *Microbial Cell Factories* 2007, 6, (1), 5.

[60] Mervi H. Toivari; Laura Ruohonen; Andrei N. Miasnikov; Peter Richard; Penttilä, M., Metabolic Engineering of Saccharomyces cerevisiae for Conversion of d-Glucose to Xylitol and Other Five-Carbon Sugars and Sugar Alcohols. *Appl Environ Microbiol* 2007, 73, (17), 5417-5476.

[61] Nigam, P.; Singh, D., Processes for Fermentative Production of Xylitol--a Sugar Substitute. *Process Biochemistry* 1995, 30, (2), 117-124.
[62] Dong Lihui; Zhou Xiaoyun; Zhaohui, Z., Research Progress on Bioconversion of Xylitol. *Bulletin of science and technology* 2004, 20, (1), 37-41.
[63] Carvalho, W.; Silva, S. S.; Santos, J. C.; Converti, A., Xylitol production by Ca-alginate entrapped cells: comparison of different fermentation systems. *Enzyme and Microbial Technology* 2003, 32, (5), 553-559.
[64] Carvalho, W.; Santos, J. C.; Canilha, L.; Silva, S. S.; Perego, P.; Converti, A., Xylitol production from sugarcane bagasse hydrolysate Metabolic behaviour of Candida guilliermondii cells entrapped in Ca-alginate. *Biochemical Engineering Journal* 2005, 25, (1), 25-31.
[65] Fouad, S. B.; Tresinari, D. S. D.; Silverio, S. S., Biotechnological production of xylitol in a three-phase fluidized bed bioreactor with immobilized yeast cells in Ca-alginate beads.
[66] Ramakrishna, S. V.; Prakasham, R. S., Microbial fermentations with immobilized cells. *Current Science* 1999, 77, (1), 87.
[67] Santos, D. T.; Sarrouh, B. F.; Rivaldi, J. D.; Converti, A.; Silva, S. S., Use of sugarcane bagasse as biomaterial for cell immobilization for xylitol production. *Journal of Food Engineering* 2008, 86, (4), 542-548.
[68] Oh, D. K.; Kim, S. Y., Increase of xylitol yield by feeding xylose and glucose in Candida tropicalis. *Applied Microbiology and Biotechnology* 1998, 50, (4), 419-425.
[69] Kwon, S. G.; Park, S. W.; Oh, D. K., Increase of xylitol productivity by cell-recycle fermentation of Candida tropicalis using submerged membrane bioreactor. *Journal of Bioscience and Bioengineering* 2006, 101, (1), 13-18.
[70] Yang Chunyu; Wang Xia; Su Haijun; Jiang Boying; Ping, X., Recent advances in xanthan gum biosynthesis. *Modern Chemical Industry* 2005, 25, (2), 21-24.
[71] García-Ochoa, F.; Santos, V. E.; Casas, J. A.; Gómez, E., Xanthan gum: production, recovery, and properties. *Biotechnology Advances* 2000, 18, (7), 549-579.
[72] Shuqi, F.; Xiaojia, M., Technologies of saving energy and reducing consumption in xanthan gum production. *Chemical Engineering (China)* 2004, 32, (4), 71-74.
[73] Krishna Leela, J.; Sharma, G., Studies on xanthan production from Xanthomonas campestris. *Bioprocess and Biosystems Engineering* 2000, 23, (6), 687-689.
[74] Lo, Y. M.; Yang, S. T.; Min, D. B., Effects of yeast extract and glucose on xanthan production and cell growth in batch culture of Xanthomonas campestris. *Applied Microbiology and Biotechnology* 1997, 47, (6), 689-694.
[75] Papagianni, M.; Psomas, S. K.; Batsilas, L.; Paras, S. V.; Kyriakidis, D. A.; Liakopoulou-Kyriakides, M., Xanthan production by Xanthomonas campestris in batch cultures *Process Biochemistry* 2001, 37, 73-80.
[76] Cacik, F.; Dondo, R. G.; Marqués, D., Optimal control of a batch bioreactor for the production of xanthan gum. *Computers and Chemical Engineering* 2001, 25, (2-3), 409-418.
[77] Smith, I.; pace, G., Recovery of microbial polysaccharides. *Journal of Chemical Technology and Biotechnology* 1982, 32, 119-129.
[78] Gonzales, R.; Johns, M. R.; Greenfield, P. F.; Pace, G. W., Xanthan Gum precipitation using ethanol. *Process biochemistry* 1989, 24, (6), 200-203.

[79] Galindo, E.; Albiter, V., High-yield recovery of xanthan by precipitation with isopropyl alcohol in a stirred tank. *Biotechnology Progress* 1996, 12, (4), 540-547.

[80] Amallah A; Tuttiett B; AW, N., Agitator speed and dissolved oxygen e□ects in xanthan fermentations. *Biotechnology and Bioengineering* 1998, 57, 198-210.

[81] Becker, A.; Katzen, F.; Pühler, A.; Ielpi, L., Xanthan gum biosynthesis and application: a biochemical?/genetic perspective. *Applied Microbiology and Biotechnology* 1998, 50, (2), 145-152.

[82] Haifeng, S.; XuXilin; Weijun, C., Study on the production of xanthangum by two-staged fermentation. *Journal of South China University of Technology (Natural Science)* 1999, 27, (6), 114-117.

[83] Chun, C.; Xiaojian, M.; Guizhuan, X.; Shuqi, F.; Hongliang, L., Study on fermentation of xanthan gum in airlift fermentor. *Food Science* 2004, 25, (2), 111-114.

[84] Yang, S. T.; Lo, Y. M.; Chattopadhyay, D., Production of cell-free xanthan fermentation broth by cell adsorption on fibers. *Biotechnology Progress* 1998, 14, (2), 259-264.

[85] Guimaraes, R. L.; Stotz, H. U., Oxalate Production by Sclerotinia sclerotiorum Deregulates Guard Cells during Infection 1 [w]. *Plant Physiology* 2004, 136, (3), 3703-3711.

[86] Dutton, M. V.; Evans, C. S.; Atkey, P. T.; Wood, D. A., Oxalate production by Basidiomycetes, including the white-rot species Coriolus versicolor and Phanerochaete chrysosporium. *Applied Microbiology and Biotechnology* 1993, 39, (1), 5-10.

[87] Schrickx, J. M.; Raedts, M. J. H.; Stouthamer, A. H.; Vanverseveld, H. W., Organic Acid Production by Aspergillus niger in Recycling Culture Analyzed by Capillary Electrophoresis. *Analytical Biochemistry* 1995, 231, (1), 175-181.

[88] Santoro, R.; Cameselle, C.; Rodríguez-Couto, S.; Sanromán, á., Influence of milk whey, nitrogen and phosphorus concentration on oxalic acid production by Aspergillus niger. *Bioprocess and Biosystems Engineering* 1999, 20, (1), 1-5.

[89] Mandal, S. K.; Banerjee, P. C., Submerged production of oxalic acid from glucose by immobilized Aspergillus niger. *Process Biochemistry* 2005, 40, (5), 1605-1610.

[90] Bouchoux, A.; Roux-de Balmann, H.; Lutin, F., Investigation of nanofiltration as a purification step for lactic acid production processes based on conventional and bipolar electrodialysis operations. *Separation and Purification Technology* 2006, 52, (2), 266-273.

[91] Wee, Y.; Kim, J.; Ryu, H., Biotechnological Production of Lactic Acid and Its Recent Applications. *Food Technology and Biotechnology* 2006, 44, (2), 163.

[92] Zhou, S.; Causey, T. B.; Hasona, A.; Shanmugam, K. T.; Ingram, L. O., Production of optically pure D-lactic acid in mineral salts medium by metabolically engineered Escherichia coli W3110. *Appl. Environ. Microbiol.* 2003, 69, (1), 399-407.

[93] Yanmei, H.; Liwen, J.; Junping, J.; Haijun, L., Progress of Producing L-lactic acid for Using Lignocellulose Material as Raw Material. *Academic Periodical of Farm Products Processing* 2007, 11, 27-30.

[94] Anan, H.; Canming, L.; Dehua, L., Microbial Production of L-lactic acid. *Chemical engineering and Equipment* 2008, 2, 59-63.

[95] Rong, W.; Yuan-liang, W.; Guoping, C.; Yanfeng, L., Advances in Fermentation of L(+)-lactic Acid by Rhizopus Oryzae. *Journal of Chongqing University (Natural Science Edition)* 2004, 27, (12), 95-98.

[96] Dien, B. S.; Nichols, N. N.; Bothast, R. J., Recombinant Escherichia coli engineered for production of L-lactic acid from hexose and pentose sugars. *Journal of Industrial Microbiology and Biotechnology* 2001, 27, (4), 259-264.

[97] Dien, B. S.; Nichols, N. N.; Bothast, R. J., Fermentation of sugar mixtures using Escherichia coli catabolite repression mutants engineered for production of L-lactic acid. *Journal of Industrial Microbiology and Biotechnology* 2002, 29, (5), 221-227.

[98] Senthuran, A.; Senthuran, V.; Mattisson, B.; Kaul, R., Lactic acid fermentation in recycle batch reactor using immobilized L casei cells. *Biotechnol Bioeng* 1997, 55, 841-853.

[99] Xueliang, C.; Liming, X., Synergetic saccharification and lactic acid fermentation by coimmobilized cells using cellulosic materials. *Journal of Chemical Industry and Engineering* 2008, 59, (1), 167-172.

[100] Xueling, S.; Liming, X., Lactic acid production by coimmobilized cells using cellulosic hydrolysate *Journal of Zhejiang University (Enzineering Science)* 2007, 41, (5), 823-826.

[101] Chen, R.; Lee, Y. Y., Membrane-mediated extractive fermentation for lactic acid production from cellulosic biomass. *Applied Biochemistry and Biotechnology* 1997, 63, (1), 435-448.

[102] Kwon, Y. J.; Kaul, R.; Mattiasson, B., Extractive Lactic-Acid Fermentation in Poly (Ethyleneimine)-Based Aqueous 2-Phase System. *Biotechnology and Bioengineering* 1996, 50, (3), 280-290.

[103] Kim, Y. H.; Moon, S. H., Lactic acid recovery from fermentation broth using one-stage electrodialysis. *Journal of Chemical Technology & Biotechnology* 2001, 76, (2), 169-178.

[104] Gyo Lee, E.; Moon, S. H.; Keun Chang, Y.; Yoo, I. K.; Nam Chang, H., Lactic acid recovery using two-stage electrodialysis and its modelling. *Journal of Membrane Science* 1998, 145, (1), 53-66.

[105] Min-tian, G.; Koide, M.; Gotou, R.; Takanashi, H.; Hirata, M.; Hano, T., Development of a continuous electrodialysis fermentation system for production of lactic acid by Lactobacillus rhamnosus. *Process Biochemistry* 2005, 40, (3-4), 1033-1036.

[106] Pan Feng; Wu Yulong; Zhang Jianan; Junmao, T., Development status of ethylene production by catalytic dehydration of fermented alcohol. *Modern Chemical Industry* 2006, 26, (2), 27-30.

[107] Tiegang, H.; Keke, C.; Jianan, Z.; Yonggang, W., Research progress of catalytic dehydration of bio-ethanol into ethylene. *Modern Chemical Industry* 2007, 27, (2), 96-99.

[108] Huang Yingming; LI Heng; Huang Xinjiang; Hu Yaochi; Yi, H., Advances of bio-ethylene. *Chinese Journal of Bioprocess Engineering* 2008, 6, (1), 1-6.

[109] Qing-hua, et al, Development of Preparation of Ethylene from Biological Ethanol Catalytic Dehydration. *Guangzhou Chemistry* 2007, 32, (4), 60-64.

*Chapter 5*

# PROCESS ENGINEERING AND APPLICATIONS OF PLANT-BASED MATERIALS

## 5.1. OVERVIEW

### 5.1.1. Brief Introduction of Plant-Based Materials

Plant-based materials belong to a kind of new materials which are produced from renewable bio-resources including crops, trees, other plants through biological, chemical or physical means. It has been a research hot point in recent years. In fact, as early as in the 1970s, when the oil crisis broke out, countries such as USA and Brazil have begun the researches of producing chemicals from renewable resources in order to reduce the amount of fossil fuels use. However, along with the decrease of oil price, much research work on the substitution of petroleum industry has been suspended. Today, the whole world faces the oil crisis again which causes human put their eyes on the renewable bio-resource once more. Meanwhile, the utilization of renewable bio-resources which are mainly plants can realize the close cycle of carbon element to the greatest extent and reduce the $CO_2$ emission. Therefore, there is significant ecological importance to develop renewable plant-based chemical industrial technologies.

The research frontiers of plant-based materials fields include: the synthesis of macromolecules from C2, C3, C4 and etc. platform chemicals and their derivatives, the design, characterization, and combination of functional biomass macromolecules, the preparation and capability evaluation of green catalysts, the application of bio-materials, the developments of substitutions for oil-based chemicals such as natural epoxy resin, phenolic resin, waterborn epoxy resin, bio-plastic and so on, the intermediate verification of the middle and downstream process of bio-platform chemicals and bio-materials productions.

### 5.1.2. Category of Plant-Based Materials

The concept of bio-refinery was first brought out in 1982 [1]. Bio-refinery is a new industrial model which produces energy products and chemicals by utilizing renewable bio-resources. The innovative developments of chemical, biologic and mechanic technologies would greatly improve the present level of utilization of renewable raw materials. Bio-refinery will play an important role in the development of environmental sustainable chemical industry and energy industry. Essentially speaking, modern bio-refinery almost equates to petroleum refinery. Bio-refinery cell factories are natural product lines which can turn the biomass resources instead of raw petroleum into all kinds of chemicals like fuels, materials, or platform chemicals through a series of biochemical paths (similar to the unit operations of cracking, reforming and so on in the petroleum refinery ) [2, 3].

Through different treatments, such as microbial fermentation, acid hydrolysis, physical chemical methods, plant biomass like starch, cellulose, and other carbohydrates can be degraded into small molecule compounds such as glucose and xylose. Then these small molecule compounds can be effectively converted to essential platform chemicals of the petroleum chemical by means of microbial fermentation or chemical hydrolysis (platform chemicals from C1 to C4 are methane, ethanol, lactic acid and crylic acid, fumaric acid, succinic acid and butanol respectively, aromatic platform chemicals are benzene and phenol). These platform chemicals can further be converted to plant-based products through chemical unit operations. Another way to prepare plant-based products is to utilize the natural macromolecules extracted and purified from plant biomass. Presently, plant-based products cover products from many areas: bio-plastic, plant-based printing ink, dye, pigment, oil, paint, sorbent, detergent, industrial bond, bio-polymers and glue, compound materials, starch and cellulose derivatives, bio-medical materials and etc.

### 5.1.3. Applications of Plant-Based Materials

With the increasing sense of both the importance of environment protection and pressure of fossil energy crisis, human are eager to utilize renewable, environmentally friendly plant-based materials. In the textile area, plant fibers are important raw materials which are extracted from natural or cultivated plants. There are many kinds of plant fibers. Cotton and bast are commonly used in the textile area. Because of their superior capabilities of pliability, permeability, durability, water absorption, pure and fine hand feeling, natural plant fibers are popularly utilized in textile industry and it has the trend to substitute chemical fibers in the near future. In the area of sorbent material, plant-based carbon material and carbon fiber are applied more and more widely due to their outstanding porous absorption and durability. As to degradable materials, using plant as raw material to prepare plastic resin is a trend of gradual interest. Plant-based material is superior to the traditional plastic in its degradable, renewable and environmentally friendly characteristics, which entitle it the perfect material to prepare one-off dishware and give an end to the worldwide "white pollution" problem. As plant-based bonds, products from Chinese lacquer, rosin, cellulose derivatives, starch-based materials etc. are applied in more and more areas.

## 5.2. IDEAS AND METHODS IN THE PRODUCTION PROCESS OF PLANT-BASED MATERIALS

Plant-based raw materials are similar to petroleum since that both of them have complicated components. Consequently, the developing process of plant-based material is similar to that of petroleum-based materials. The raw material should be divided into several suitable components, and then these components are processed into different products under different treatments which form the whole product system. The basic principle in the petrochemical is as follows: to produce pure product which has clear chemical components and can be easily operated from the hydrocarbons in the refining process. A highly effective production line can be constructed based on the product system, in which the basic chemicals, intermediate products and complex products can be produced. This principle can apply either to petroleum refine or to bio-refinery industry. However, biomass, comparing with petroleum, has the capability of automatic synthesis, and there have different composition proportion in the C:H:O:N. That is to say, only if the biotransformation and chemical transformation are combined together can biomass make great contribution to the bio-refinery industry.

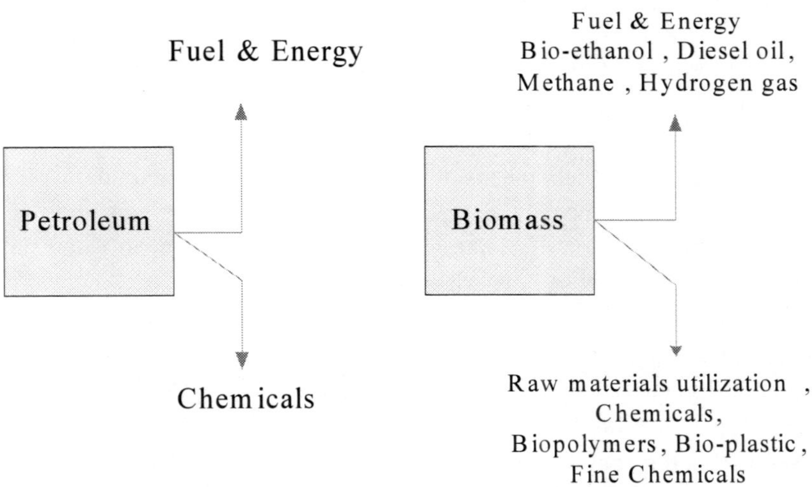


Figure 5.1. Research and development thoughts of plant-based materials

### 5.2.1. Analysis of the Common Key Problems in the Preparation Process of Plant-Based Materials

Comparing to the relatively mature developed process of petroleum-based products, the preparation process of plant-based materials is far more complicated than that of the petroleum-based materials. According to the great developments of petroleum industry in the past decades, each unit operation technology has reached high levels of sequential operation, in other words, reactions can be effectively controlled and the target products can be gained from molecule design, which facilitate the processing of petroleum-based materials. The development level of plant-based materials processing is relatively lower. At present, developments of plant-based materials mainly concentrate on the following three aspects.

One is to process the whole plant biomass to board materials under physical treatments. Another is to obtain extra-cellular products with special functions by microbial fermentation using elements like carbon, nitrogen enriched in plant biomass as substrates. The last one is to make use of some certain components of plant after been chemically or biologically treated or modified. There have many problems among these aspects of utilizations, such as lacking the cognition of the plant components complexities, concentrating on single utilization of one component and single technology, and in short of systematization of process.

Plant biomass have complex and inhomogenous structures, and different components are interlaced and restricted with each other, both of which make it impossible to use plant biomass in a whole because it is impossible to guarantee the homogeneity of raw materials. Since there is successful example of graded utilization of petroleum resource in the development process of petroleum-based products, the similar concepts of graded utilization and comprehensive utilization can be applied into the development of plant-based materials preparation using plant biomass resource. The essence of graded utilization refers to the differential utilization of each component of plant according to its inherent characteristics. For example, cellulose is the main substrate for fermentation, however, the existence of hemicellulose and lignin not only influence the accessibility of enzyme and microorganism to the substrate and finally influence both the quantity and quality of product, but also cause huge waste of resources (hemi-cellulose can be the raw material for preparing xylo-oligosaccharides, xylitol, furfural, etc, and lignin can be modified into functionalized macromolecule material or used as power fuel). Therefore, based on the capability requests of target materials, adopting the concept of graded utilization of raw material according to their composition characteristics, which means that selecting suitable method to extract target components and make developments and utilizations of other components, are the necessary steps to realize the total utilization of biomass effectively and economically and further carve out a brand new phase of plant-based materials industry.

In a word, the most urgent problems that should be solved now in the development of plant-based industry are the mode of thought and cognition. Original mode of thought which is characterized as simple, opposite and competitive should be upgraded into comprehensive and harmonious. Developments of biomass industry call for comprehensive cognitions from different angles such as the nature, the development principles of human society, the situation of China and the future of the world. Conversion of plant biomass resource is a complex and systematic project. New thoughts and technical course of the further development of biomass transformation technologies needs the systematic integration of existed single technologies on the basis of deep understanding of plant biomass characters, as well as of the analysis and summarization of the development status of biomass transformation technologies.

### 5.2.2 Principles of Plant-Based Material Preparation

The basic principle is to choose suitable process route aiming at target product. Commonly there are two ways: (1) To degrade the natural macromolecules of the plant raw material into small molecules platform chemicals through chemical or biological transformation, then design reaction types aiming at target product, and gain the plant-based products either through polymerization or molecule condensation; (2) To modify the plant biomass raw material by biological or chemical means, which may cause changes to the surface hydrophilic/hydrophobic characteristics and other surface characteristics to produce functionalized materials.

## 5.2.3 Production Process of Plant-Based Materials

Plant biomass raw materials can be degraded, modified, or polymerized to plant-based products. Referring to the chemical materials processing, the production process of plant-based materials can be classified into three main routes.

### *5.2.3.1 Biotransformation Process*

In this section, we will introduce the refining process of plant biomass raw materials by taking lignocellulose as an example. Lignocellulose raw materials can be separated into four main parts of cellulose, hemicellulose, lignin and ash after fractionation process. Firstly, cellulose is degraded into glucoses by acid hydrolysis or enzymatic hydrolysis, then glucoses can be fermented to produce platform chemicals like ethanol, acetone, butanol etc, all of which are useful reactants to prepare different kinds of plant-based products through cracking, condensation or polymerization processing. Utilization route of hemi-cellulose is similar to that of cellulose, and lignin can be modified directly to prepare dispersants, sorbent, bonds and other chemical materials.

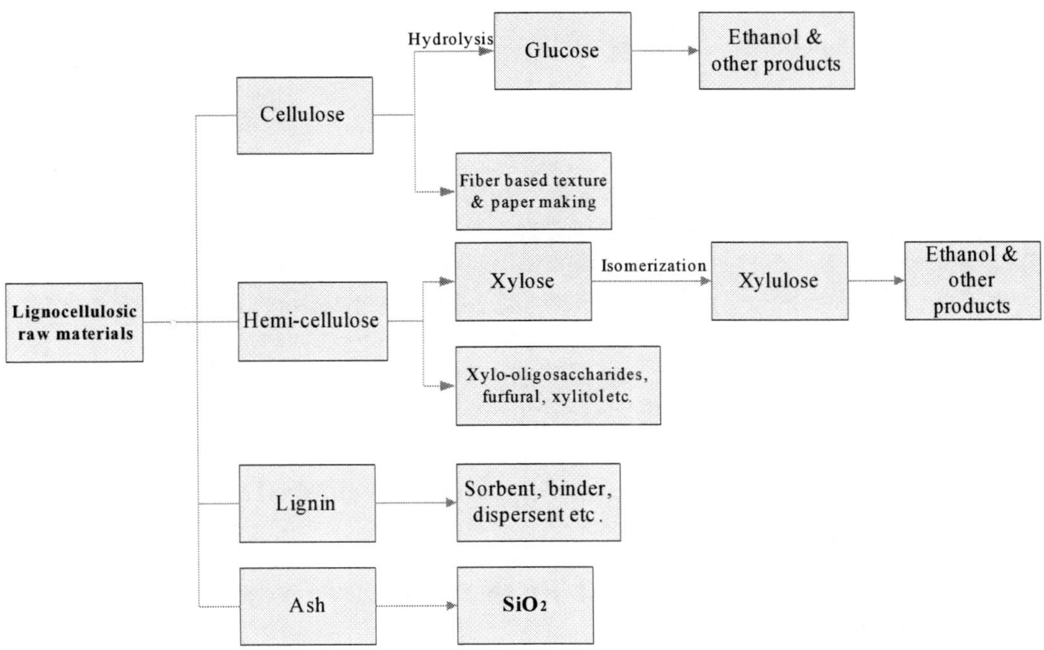


Figure 5.2. Biotransformation processes of lignocellulosic raw materials

### *5.2.3.2. Thermochemical Transformation Process*

By treatments of pyrolysis, liquefaction, gasification and etc, plant biomass raw materials can produce coke, bio-oil, gas fuel and etc, which can be further synthesized into intermediate products like diesel oil, methanol, ammonia and etc, and then prepare plant-based products through polymerization or cracking.

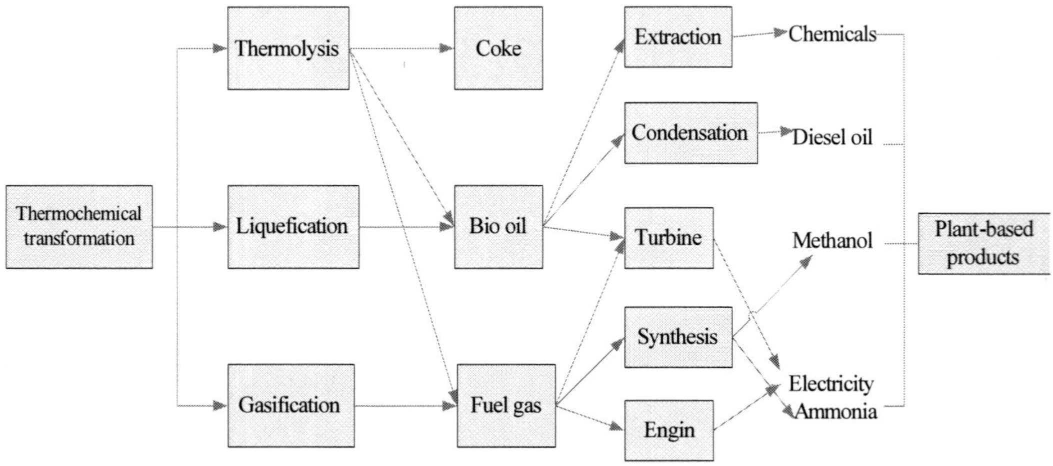


Figure 5.3. Thermochemical transformation processes of plant-based materials

#### *5.2.3.3. Direct Modification of Plant-Based Biomass Raw Materials*

After activation, the plant-based biomass raw materials can etherify or esterify with suitable reactants to gain raw products with different degree of substitution, then we can gain the pure functional products through separation and refining of the raw products.

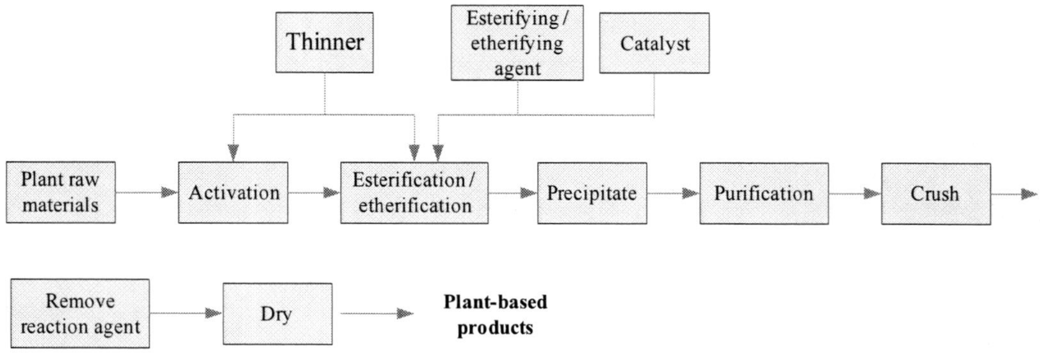


Figure 5.4. Modification and transformation processes of plant raw materials

## 5.3. Production of Plant-Based Materials

China is rich in biomass resource, and according to the chemical compositions and structural characteristics of different plants, a variety of materials can be produced through physical, chemical, biological or a combination of these methods. In this chapter, we will focus on the following four categories to introduce the current research and application of a wide range of plant-based materials and their production techniques.

## 5.3.1. Modified Lignocellulosic Materials

### *5.3.1.1. Cellulosic Materials*

For a long time, our over-reliance on fossil energy leads to increasing annual consumption, together results in increasingly serious energy shortage. On the other hand, due to the difficulty of degradation of most functional polymer materials prepared by petrochemical synthesis, particularly with the rapid development of the plastics industry, the "white pollution" caused by plastics brought about huge harm to the manufacture and living environment of human beings. To address the shortage of raw materials and environmental issues, people gradually shift attentions to renewable resources [4].

Natural cellulosic materials are the most abundant renewable resources in the world and each year the total amount of plant-based materials produced through photosynthesis is up to about 10 billion tons. However, the vast majority have not been reasonably developed and utilized [5]. Cellulose is a multi-hydroxy-glucose polymer, which possesses different characteristics after specific physical or chemical modification. The modified cellulose can present as different forms of powder, sheet, film, fiber, solution and so on.

As the cellulose is insoluble in common solvents, in the cellulose industry the cellulose is previously transformed into derivatives and then be made use of. For instance, in the industries of regenerated cellulose fibers, plastics, film, painting, and so on, products are produced from cellulose acetate or ether derivatives by dissolving them in the solvent and then be machine-shaped. At present, the cellulose-based products are including leather finishes, films, cigarette filter tow, surfactant, paint thickeners, oil-displacing agent, suspension polymerization stabilizer, tackifier and a series of functional materials [6].

*Cellulose Acetate*

Cellulose acetate is the first commercial cellulose derivative among all the cellulose derivatives, as well as a continuously developing organic cellulose ester. In the field of cellulose fibers, cellulose acetate is the second largest species only following the viscose fiber. Cellulose triacetate is widely used as film base, X-ray film base [7], plastics 8], textile fibers, and so on. Cellulose diacetate is used as membrane material [9, 10], such as desalination membrane [11], water purification membrane filter [12, 13], ultrafiltration membrane [13, 14], nanofiltration [15], reverse osmosis membrane [13, 16-18], and so on. Meanwhile, after been spun cellulose diacetate can be widely used as cigarettes filter tips [19-22] and textile materials.

(1) Cellulose Acetylation Mechanism

Cellulose acetate is the reactive product of wood pulp or cotton with acetic anhydride, in which the hydroxyl was substituted by ester base. It is a general name for all the cellulose acetate having the degree of esterification ranged from 1 to 3. If all the three hydroxyls in each glucose residue are esterified, the degree of esterification is 3, and the final product named as cellulose triacetate. When the degree of esterfication is down to 2.7 ~2.0, it is called cellulose diacetate [23]. Acetylated chemical reaction equation is as follows:

$$\left[C_6H_7O_2\begin{matrix}\diagup OH\\ -OH\\ \diagdown OH\end{matrix}\right]_n + 3n(CH_3CO)_2O \longrightarrow \left[C_6H_7O_2\begin{matrix}\diagup OCOCH_3\\ -OCOCH_3\\ \diagdown OCOCH_3\end{matrix}\right]_n + 3n\,CH_3COOH$$

Hydrolytic reaction equation is as follows:

$$\left[C_6H_7O_2(OCOCH_3)_3\right]_n + (3\text{-}x)n\,H_2O \longrightarrow \left[C_6H_7O_2(OCOCH_3)x(OH)_3\text{-}x\right]_n + (3\text{-}x)n\,CH_3COOH$$

Where, x is the degree of esterification.

The course of chemical reactions is as follows: the reaction is two-molecule nucleophilic substitution reaction (SN2), firstly the $H^+$ attacks the oxygen atoms in the carbonyl group, making carbonyl carbon have more positive charge, then nucleophile Rcell—OH attacks carbonyl carbon atom and nucleophilic to form a middle transition state, which is the speed-control reaction, finally the transition state complex removes an acetate and transforms to cellulose acetate [24].

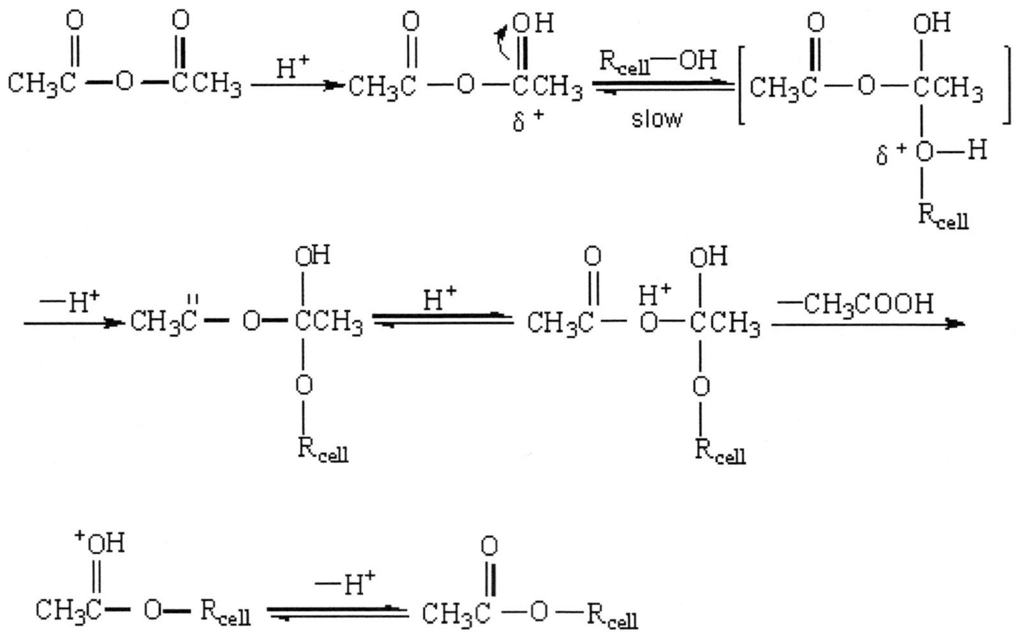

Figure 5.5. Course of cellulose acetylation reaction

(2) Technology Overview

At present, industrial cellulose acetate is generated mostly from high-level wood pulp and cotton fiber in which the α-cellulose content is more than 95% by two main methods, one is using glacial acetic acid as the solvent, the other is using methylene chloride as the solvent. Liquid-solid two-phase synthesis is the traditional method of producing cellulose acetate as described in figure 5.6, including pretreatment of cellulose, acetylation, hydrolysis of the cellulose triacetate, precipitation and extraction of cellulose acetate, and the recovery of glacial acetic and acid anhydride.

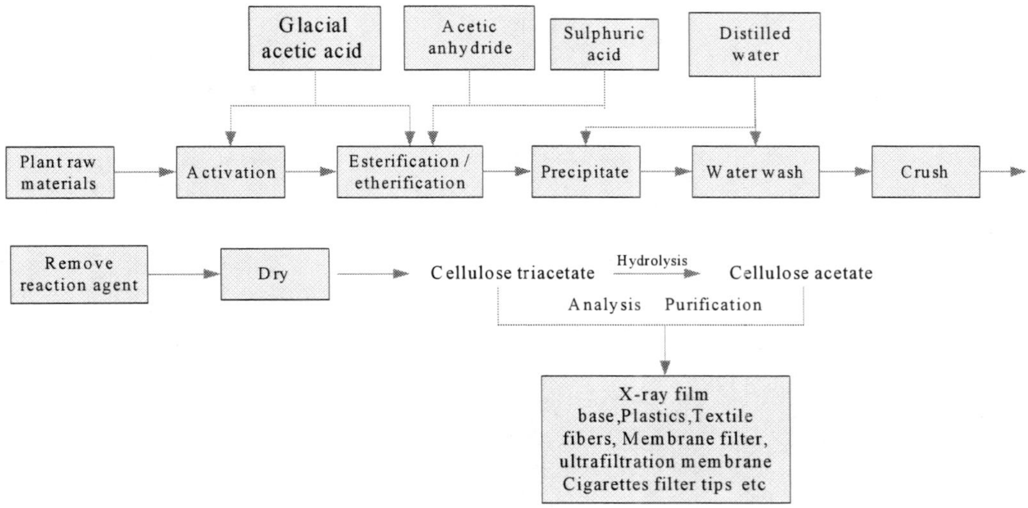


Figure 5.6. Production technology and brief flow chart of cellulose acetate

However, this process is not only high cost and complex [25], but also caused serious environmental pollution [26, 27]. On the other hand, the complicated pretreatment of cellulose will make the consequent processes much more complex. Current studies mainly focus on the preparation processes (gas-solid-state acetylation and homogeneous acetylation) and the choice of raw materials. Gas-solid acetylation is originated from timber modification researches, and gas phase reaction in high-temperature will improve the wood surface hydrophobicity. Chinese patent 94113778.3 [28] introduced a method of modifying lignocellulosic materials in gas phase and designed the process line. In 2001 Meijie adopted a method of preparing cellulose acetate in gas-solid phase [29]: the so-called gas-phase method refers that the acetic anhydride and catalyst was confined in the container and gasified under high temperature, and then the gasified products infiltrate into cellulose and carry out acetylation. The character of the gas-phase method is that it is unnecessary to add glacial acetic acid as reaction medium and the whole reaction is taking place in the reactor with the pressure of lower than 0.06 MPa using sulphuric acid as catalyst. Due to no need of large amount of glacial acetic acid, the cost of acetic acid and its recovery is reduced a lot, and the dosage of catalyst is also cut down to 40~50% of its original amount, which can effectively reduce the residue sulphuric content in the products and enhance thermonstability of product, however, the degree of esterification homogeneity is not high.

Homogeneous acetylation of cellulose is to dissolve the cellulose in the solvent, and then to carry out the acetylation reaction in uniform phase. There have been reported a lot of cellulose solvent including both of the water solvent system and non-water solvent system [30]. Beatriz et al [31] dissolved cotton, sisal fiber in DMAc/LiCl [32, 33] (DMAc-N, N-Dimethylacetamide) to conduct homogeneous acetylation. Wujin et al [33] dissolved cellulose in the ionic liquid 1-allyl-3-methylimidazolium chloride (AMIMCl), and then conducted acetylation of cellulose in the homogeneous conditions. Yan Lifeng et al [34] studied the homogeneous acetylation of cellulose acetate in the presence of solid acid catalyst, which was more environmental friendly comparing to the above mentioned solvents. To dissolve cellulose in the ionic liquids firstly and then to carry out acetylation is a good way to prepare high-quality cellulose acetate. It is also environmental friendly, in addition, the reaction media can recycle and reuse and the final product uniformity is better than that of the liquid-solid two-phase synthesis. And it is also an important trend of green chemistry development. However, how to prepare cheap ionic liquids with better dissolution capability is still worthy further study.

How to reduce the cost of raw materials has always been the focus of the studies. As early as 1946, American scholar Stamm studied on the acetylation of wood. There also had many domestic reports about wood acetylation [35, 36]. After several decades of development, the researches are mainly concentrated on the timber modification. The modified timbers have improved hydrophobility, texture, and stability. Chinese patent 93106682.4 [37] introduced a method of preparing fine cellulose acetate with good spinnability by using wood pulp having α-cellulose content of 85% to 93% as raw materials. Then the acetylated lignin, hemicellulose and sugars could be separated according to their differences in solubility from cellulose acetate. The synthesis cost of this method is reduced and the yield rate can reach as much as 84%. As to the researches of direct acetylation of straw, the main concentration is limited to the surface modification, which can improve its hydrophobility [38].

There are little reports on the preparation of acetyl cellulose acetate based on pretreated straw until 2006. USP, 6984731 [39] introduced the acetylation of corn-cob after been pretreated with high temperature steam and high pressure. Atanu Biswas et al [40] explored the feasibilities of preparing cellulose acetate by using different raw materials such as corn fiber, rice husks and wheat straw pretreated with acid. As far as saving resources and protecting the environment be concerned, all the above mentioned studies had significant meanings. Firstly, the raw materials used in theses studies are cheap and available; secondly the transformation process could turn trash into cash, which add high value on the agricultural wastes. However, the main problems of these studies are that the process route is complicated and it is difficult to reach industrialization.

(3) Multi-Level Conversion of Acetylated Straw

To solve the problems on current process, we investigated the feasibility of using cheap agricultural waste straw as raw materials to carry out acetylation reaction. The raw straws firstly were activated by steam explosion pretreatment, and after acetylation reaction, different products can be separated according to their different physiochemical characteristics. The use of straw as raw materials for acetylation reaction to prepare high value-added cellulose acetate not only has the advantages like low cost and simple process, but also has great significance on saving resources [41], and most important, it is helpful to contribute to the establishment of eco-industrial model which employs cellulose acetate as leading products.

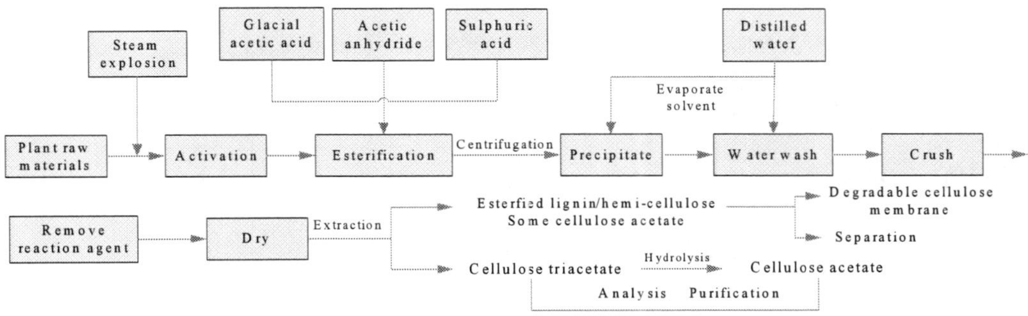


Figure 5.7. Acetylation processes of straw and its components fractionated flow

*Carboxymethyl Cellulose Sodium*

Carboxymethyl cellulose sodium (CMC) is the most widely used and the most representative ionic cellulose ether. CMC is a kind of cellulosic ether derivatives which is obtained from chemical modified cellulose. As a result of the poor water-solubility of its acid-form, it is generally made in the form of sodium, that is, carboxymethyl sodium. CMC was invented in 1918 by a German named E. Jansen who applied to invention patent [42]. With the continual in-depth understanding of CMC's functions, in worldwide, USA, Germany and Japan have developed the industrialization of CMC production. China invested the first factory in Shanghai celluloid factory in 1958 to produce CMC. CMC solution has multiple functions such as solution thickening, bonding, film, maintaining water, emulsion and suspension, and etc. Therefore CMC is widely used in detergents, food, textile, printing and dyeing, paper, petroleum, mining, cosmetics and other industries [43]. Because of its wide applications [43], CMC is famous as "Industrial Monosodium Glutamate".

(1) Carboxymethylation Mechanism

Reaction of preparing CMC belongs to the type of nucleophilic substitution reaction, it is also known as the Williamson reaction. Reaction of alkali cellulose with halocarbons (such as halogenated methane) is a typical example of this kind of reaction. Such reaction is irreversible, and the degree of substitution and its distribution is controlled by the reaction speed.

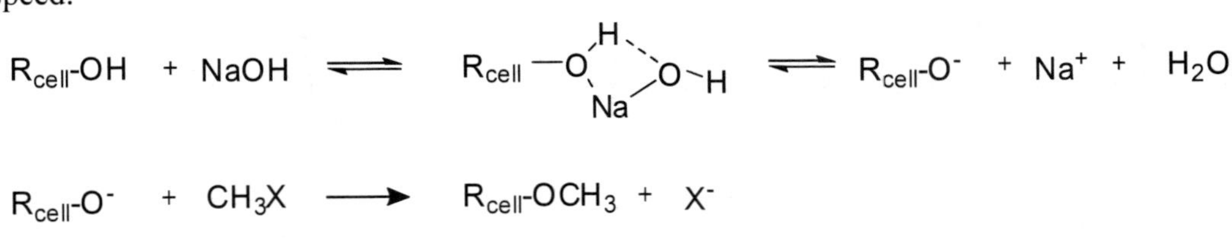


Figure 5.8. Williamson reaction diagram [43]

Where, Rcell-OH represents the cellulose molecules, $CH_3X$ represents Halogenated alkanes

The reactive principle of CMC can be described using the following chemical reaction equation:

1). cellulose is basified into alkali cellulose:

$$Cell\text{-}(OH)_3 + xN_aOH \longrightarrow Cell\text{-}(OH)_{3-x}(O^- N_a^+)_x + xH_2O$$

2). Chloroacetate is transformed into sodium chloroacetate:

$$ClCH_2COOH + NaOH \longrightarrow ClCH_2COONa + H_2O$$

3) Alkali cellulose reacts with sodium chloroacetate:

$$Cell\text{-}(OH)_{3-x}(O^- N_a^+)_x + nClCH_2COONa \longrightarrow Cell\text{-}(OH)_{3-x}(O\,N_a)_{x-n}(OCH_2COO^- Na^+)_n + nNaCl$$

4) Neutralization of acid and alkali:

$$Cell\text{-}(OH)_{3-x}(O\,N_a)_{x-n}(OCH_2COO^- Na^+)_n + (x\text{-}n)CH_3COOH \longrightarrow$$

$$Cell\text{-}(OH)_{3-n}(OCH_2COO^- Na^+)_n + (x\text{-}n)CH_3COONa$$

(2) Producing Process of Carboxymethyl Cellulose

There are mainly two methods for CMC production, that is, water solvent method and menstruum method.

1) Water solvent method

Water solvent method is the early common technology of producing CMC, and the reaction is taking place in the presence of free alkali and water. This method is relatively easy and the required equipments are simple, therefore it is commonly used to prepare low and middle class CMC, but also can prepare advanced CMC by improving the extraction process. There are two kinds of manufacturing techniques in this water solvent method, the intermittent process and the continuous process, and the basic production process is shown in Figure 5.9.

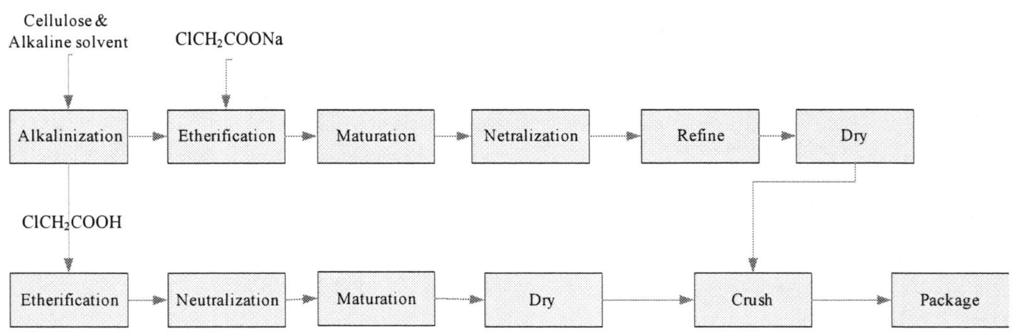


Figure 5.9. CMC production process in water solvent method [44]

2) Menstruum method

Menstruum method is also called organic solvent method, and the most prominent feature of the method is to carry out basification and etherification reaction in the organic solvent medium. According to the adding amount of organic diluents, menstruum method can be divided into low-bath-ratio kneading method and high-bath-ratio slurry method. The basic process of organic solvent method is similar with that of water solvent method except that the etherification and basification are carrying out in organic solvents. Using organic solvent as reaction medium can decrease the side effects and improve the uniformity of reaction, and gain fine CMC products with high degree of substitution and better uniform substitution.

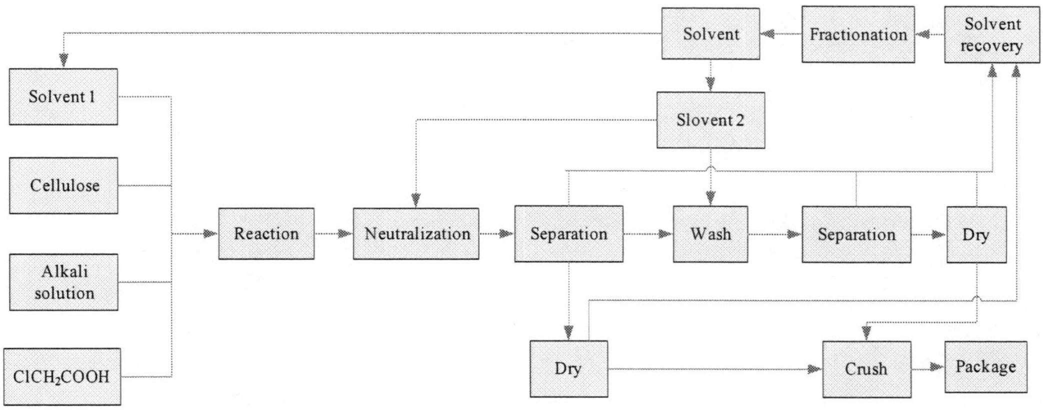


Figure 5.10. CMC production process in organic solvent method [43]

3) Factors of carboxymethylation reaction

Factors of fiber quality, basification process, acidification process, water content, dosage of ether, the reaction temperature and reaction time, as well as the way of adding mode of alkali and etherifying agent can affect the carboxymethylation reaction. And cellulose fiber has complex shape and supramolecular structure which result in various degrees of heterogeneity of reaction between cellulose and chemical reagents. The substitution and distribution of cellulose ether directly affect the final performance of CMC.

### (3) Multi-Level Conversion of CMC Based on Fractionated Straw

At present, raw materials for producing industrial CMC requires high α-cellulose content and high polymerization degree, which means to choose wood pulp and cotton or the raw materials pretreated with strong acid or alkali. The first choice limits the resource of raw material, while the second choice has the problems such as high cost, high pollution, low integration of process, etc. If there is a simple pretreatment to activate straws and then using the activated straw to conduct carboxymethylation reaction, finally to separate the different compositions through their distinct physiochemical properties, it will be a promising process for total utilization of biomass since that it not only prepares enzymatic hydrolyzed CMC which conduce to the consequent fermentation application, but also the lignin and hemi-cellulose can be modified to useful industrial chemicals.

Based on the above thoughts, we brought out the idea of oriented transformation of fractionated straw, which is used as the basic principle on establishing eco-industrial mode of straw with CMC as its main product (Figure 5.11). According to this eco-industrial mode network, it not only decreases the cost of target products due to the sustainability of raw material, but also obtains high value-added products due to the total utilization of biomass.

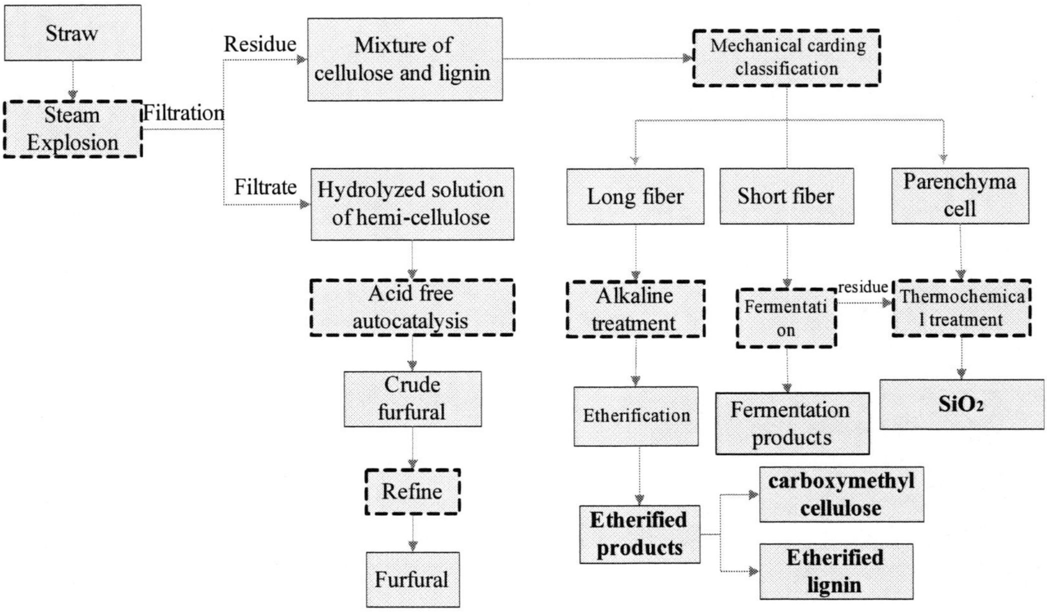


Figure 5.11. Total utilization of straw with CMC as main product

*Prospects of Industrial Development of Cellulose Materials*

With the rapid development of science and technology, researches and applications of cellulose and its derivative materials have made tremendous progress. The significance of cellulose as a synthetic organic polymer raw material is super-highlighted. More and more people are fully aware that the cellulose resource has obvious superiority than fossil energies such as coal, petroleum, natural gas, etc The renewable cellulose resource is not only just a renewable resource, but also can be used as the raw material for various functional products. Meanwhile, with the integrated development of agricultural wastes, the sources of cellulosic materials will be much wider, which will boost the development of cellulose industry.

### *5.3.1.2. Plant-Based Materials for Meliorating Desertified Land*

Desertification is the most serious ecological problem in China. Although in recent years China has taken actions on management and improvement of desertification, now we are still in a stalemate of desertification and management. Since 1990s, China began to manage the desertified lands with straw, and the early practice was similar to that of foreign countries, which were mainly using straws as the sand barriers by burying them into sands or just covering them on the sand. But it can not fundamentally solve the problem for it is just a support measure to control the desertification.

Drought, vegetation damage and poor soil are the main driving forces of desertification. Therefore scholars all over the world consent that the fundamental solution to the desertification is to improve the soil, enhance the water conservation capacity and rebuild vegetation. They have done great amount of researches on this point. Among them, the chemical materials were once the hot research point. However, the cost of chemical materials is too high, and absorptions of chemical are limited to a large number of plants, moreover, it has the potential to pollute the groundwater after a long time application and harden the soil which is directly harmful to the ecological environment. Under such circumstance, it is urgent to investigate the biodegradable materials which not only have highly bibulous capacity but also can increase the fertility of soil.

*Brief Introduction of Plant-Based Materials for Meliorating Desertified Land*

At present, when adding straw as a supplement to those materials which are used to meliorate desertified land, the straws are often only mechanical disintegrated. As a result of the simple pretreatment, it costs a long time to make compost and the quality of products is usually poor. Yin Xingyi [45] made some improvements on the soil fertility, however, the straw could not obviously increase the water reserves of soil and the energy consumption of process because of the complex product processing. Wang Zhenyu et al used straws to prepare bioactive water retention agent [46, 47], but the preparation process is including the treatment of acid and alkali which would cause environment pollution to a certain extent. Besides, the imbalance of straw nutrient content asks for the extra addition of some certain proportion of inorganic fertilizer, organic nitrogen source, humic acid and rare metal compounds [48], which does not fundamentally reduce the cost of the product.

Straw is the most abundant renewable resources of the world, according to statistics, every year around the world there are over 2.9 billion tons of straws, and only in China the crop straws each year are more than 700 million tons. The straw is rich in elements. Usually the contents of C, H, O, N are about 50%, 6%, 44% and 0.05~0.4 %, respectively. As a result, researches again gradually pay attention to straw for its potential anti-desertification ability. According to the current researches, the effect on anti-desertification of straw-based materials is not inferior to that of the chemical materials, besides, the straw-based materials have the advantages of availability of raw materials, lower cost, long activity duration, and multifunction in comparison with the chemical materials. In this section we will focus on the preparation process of straw as a raw material to produce functional materials to meliorate desertified land.

*Research on the Process Engineering of Plant-Based Materials for Meliorating Desertified Land*

Aiming at solving the current problems of producing plant-based materials to meliorate desertified land, such as the poor pretreatment process, heavy pollution, the high cost of adding other nutrition components and etc, we developed a new method with steam exploded straws. After the pretreatment of steam explosion and solid state fermentation, the quality and property of straws are improved, meanwhile the efficiency of straw-base bio-fertilizer is enhanced. Therefore decreasing the dosage of chemical fertilizer will greatly reduce the cost and the environment pollution.

The technical process is described as follows (Fig. 5.12).

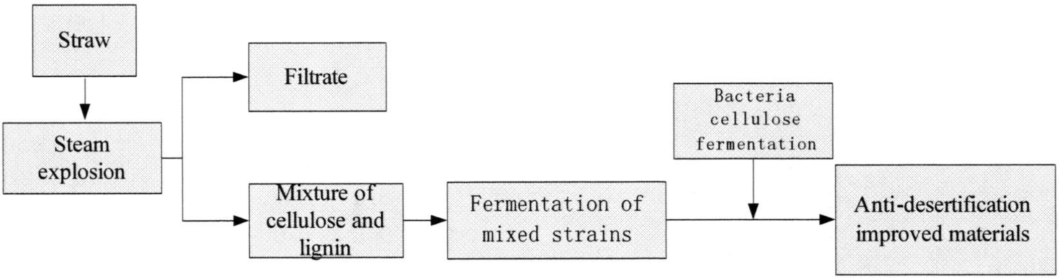


Figure 5.12. Technical processes for producing plant-based materials using straw to meliorate desertified land

(1) Preparation of steam exploded straw: cut the whole straw into 3-5 cm long pieces and then add water to infiltrate the raw materials, after infiltration exploded the straw in steam explosion reactor.

(2) Solid-state fermentation of composite strains

① Preparation of seed of strain.

② Solid-state fermentation of composite strains. Firstly smash the steam exploded straw and add a certain amount of wheat bran, stirring mixed evenly, and then conduct the solid-state fermentation of substrate. The substrate is firstly added with some water to reach the certain solid-liquid ratio, and then sterilized the substrate.

(3) Fermentation of bacterial cellulose: inoculate the *Acetobacter xylinum* on the substrate and ferment the bacterial cellulose.

(4) After fermentation, mix the fermented products and bacterial cellulose together and dry at room temperature to prepare the straw-based materials for meliorating desertified land.

Comparing the properties of plant-based materials for meliorating desertified land using steam exploded straws to crushed straws, the water absorption and water retention rate of the former are 32~ 40% and 80~100% higher than that of the latter, respectively. Besides, the fermented products have the bio-fertilizer function and the disease resistance capability. In the final product organic content is >60%, humic acid content is >15%, active bacteria number is > 0.3 billion/gram, and the contents of N,P,K are all > 5%.

*Prospect of Plant-Based Materials for Meliorating Desertified Land*

In China, the area of desertified land is 1,740,000 square kilometers which is accounting for 18% of the Chinese total land area, which affects nearly 4 million people's production and lives. Each year the direct economic loss of land desertification amounts to more than 500 billion RMB, which seriously restricts the sustainable development of economy, society and environment. Therefore, the development of new environmentally friendly materials to promote the manage process of desertification is not only a good way to retrieve the huge economic loss, but also can benefit the current generation and the future generations, in addition, it can also push the development of the related researches which is favorable for the development of biomass resources.

### *5.3.1.3 Lignocellulose-Based Board Technology*

*Brief Introduction of the Lignocellulose – Based Board Technology*

Board technology mainly uses crop straws as raw materials by adding adhesive to form boards or profiles with various shapes through crushing and restructuring, unit restructuring, whole straw reorganization of the plant [49, 50]. Lignocellulose–based board can be used as furniture, construction, packaging, and conductive materials to replace part of the wood. The light wall board prepared from the wheat (rice) straws possesses properties of light, sound insulation, heat preservation and so on can be widely used in quick housing, temporary construction and so on. The house constructed by this board has superior anti-seismic capability. Therefore, they are widely used in construction in many states and provinces in the North America, and there is a growing trend in use in Latin America and Europe.

Currently, the technologies of using straws to prepare eco-materials need not pretreat natural straws, so the addition of some thermosetting resins and plastic materials such as formaldehyde are necessary which account for a large proportion of the entire production costs (about 60%). The high cost of resins not only makes the product incompetent with other plastics materials and decorative materials economically, but also causes the second environment pollution. Such material can not be regarded as the real eco-materials. Thus, the development and utilization of synthetic eco-materials based on natural raw materials is not only the requirement of sustainable development, but also the trend of this area. In addition, the eco-materials should possess satisfied properties and harmonious with the environment, and the raw materials and production process should also meet the requirements of green chemistry.

*The Status Quo of Lignocellulose-Based Board Technology*

(1)Biomass-Based Light Board

The characteristics of biomass-based light board include: a) Raw materials have advantages like short regeneration period of renewable, sustainable availability, great amount, widely distribution, and low cost. b). The content of $SiO_2$ is high in plant stalks, which can improve water resistance and durability. Take wheat straws as example, the ash content of rice hull accounts for 19% of its weight, and $SiO_2$ content accounts for 98% of ash weight, therefore wheat straw and rice husk is a common used raw material of light boards. c). Straws are energy-saving raw materials and there are no pollution of formaldehyde, volatile organic compounds (VOC), ammonia and radionuclide. d). The density of straw is low and the packing density is usually in the range of 0.1-0.25 g / cm3. These advantages are the overwhelming properties that other light materials such as industrial ashes are beyond compare.

The use of agricultural residues to produce boards can be traced back to early 20 century. In 1905 German once prepared boards from the mixture of straws and adhesive. And then the Americans in 1930~40s carried studies on the feasibilities of using straws to prepare the insulation boards and latter in 1970s to produce structural boards. Bagasse is firstly used to produce man-made boards, and it was widely considered as the only agricultural residues that can prepare high-quality composite board. In 1921 the USA built the first bagasse board factory. After 1940s, preparation of artificial boards using stalks of economical crops such as waste of flax achieved different levels development. Stalks of cotton have similar fiber structure with hardwood, but higher content of extractions and polysaccharides in the chemical composition, which lead to the poor water resistance and anti-fungus capability. At present, the most widely studied and applied raw materials for man-made boards are wheat straws and rice straws. With the application of isocyanate resin in the field of straw-based artificial board, the adhesive problem is well solved, and the board performance also is greatly improved, besides, there is no release of formaldehyde. In the fiber separation areas, the researches are mainly concentrated in the following fields: the double helix extruder method, hyperbaric defibrator method, extruder combined with hyperbaric defibrator method, etc. The aim of all these methods is to destroy the surface wax and silicon compounds layer of straw through strong shear. However, there are still problems of raw materials pretreatment, adhesive adding system, transporting systems as well as the slab releasing system.

From 1970s, Chinese researchers have carried out the studies on the preparation of man-made boards using raw materials such as rice husk, rice straw, wheat straw, bagasse, flax, and other scraps. As for man-made boards production using bagasse, flax and cotton stalk, the production technologies have succeeded and been spread in large scale. In the late 1990s, Chinese drew a lesson from the successful foreign experience of wheat straw board production with isocyanate and began the large-scale industrialization of straw-based board. Meanwhile raw material of wood-based board industry in China has become increasingly tight. All of these situations spurred the research and development of man-made straw-based boards into the second fast development stage. In recent years, China has gained the most rapid increasing amount of man-made boards in the world. According to the information from 2007 Beijing International Bonding Technology Seminar, in 2007 total output of boards in China has reached 74,000,000 m3. However, the cheap urea-formaldehyde resin which has the potential risk of releasing free formaldehyde is still the mostly used synthetic resin adhesive, accounting for 60~70% of the wood adhesives, which limits the utilization of environmentally friendly adhesives. As for the developing trend of products, it is necessary to develop low-cost environmentally friendly adhesives or high-grade formaldehyde-free straw-based boards for the industrial development of biomass-based light boards.

(2)Biomass-Based Structural Boards

Progress of modern architectural technology is based on light and high-strength materials to reduce the self-weight of the building, thereby reducing the cost of materials and dealing with a corresponding increase in floors. Biomass-based man-made boards possess fine construction functions such as heat keeping, heat insulation, noise resistance, aseismatic and so on and can be used as the walls of nonweight-bearing frames, ceiling, closet wall, floor, construction moulding boards and so on [51]. In foreign countries, Hungary sets the cement flakeboard into the kind of materials which have 100 years of serviceable life, so it is no doubt the well weatherability.

At present, biomass-based structural boards are mainly divided into two categories according to cementing material used First, the organic cementing materials, mainly are synthetic resins such as urea-formaldehyde resin adhesive UF, phenolic resin glue PF, melamine plastic resin MF, modified melamine urea-formaldehyde resin, as well as unsaturated polyester plastic and so on. Second, inorganic cementing material, such as cement, gypsum, magnesium oxychloride cement, calcium silicate, calcium carbide and so on. According to the kinds of biomass and cementing material, as well as the rate of water content, the production methods of structural boards can be divided into: Hot-press method, the cold press method, casting method, extrusion method, high pressure vacuum extrusion method, flow slurry method and so on. Taking the inorganic cementing material for example, the production methods for partition materials are as follows.

1) Semi-dry extrusion molding method, which mixes the cementing materials (cement, magnesia materials, gypsum, etc.), active industrial solid waste (fly ash, slag, slag, etc.), lightweight stuff (such as ceramic particles, perlite, etc.) and water, and stir them into a uniform mixture of semi-dry (keep the moisture content between 17%-20%) mixture, and then mold them in the extruder. The extruding modes are more than one ways: multi-screws extrusion, single screw extrusion; push board extrusion; vibration mode and so on.

2) Vacuum press extrusion process, which extrude and form the linear boards with multi-sections in the driving forces of high-pressure and high shear through the process of stirring and kneading in a vacuum press extruder, using ordinary Portland cement or cement-Si-Mg as the cementing materials, mixing with industrial and agricultural wastes, such as fly ash, stone, slag, straw, synthetic fiber, paper fiber, and other filler as strengthen stuff, adding with suitable amount of plasticizer and water. The advantages of this method are obvious: a. products possess uniform texture, smooth surface, high-density, and high-intensity; b. it is easy to produce a variety of products with cross-sections by replacing the die modes, and the laminate void of boards can be square, rectangle, round, and the laminate void rate can reach to 50%~65%; c. the squeezed stuff possess plasticity, low moisture content, and small shrinkage, which can not only be used as inside partition but also the outside wall with excellent architectural features.

3) Semi-dry flat-press forming process: Semi-dry process means the mixture has only a small amount of water which can only meet the requirements of the hydration of cementing materials and the adsorbed water in the mixture composition, so that the mixture is semi-rigid material that can be solidified in mechanical process. Semi-dry process has the advantages such as energy saving , water saving , without waste water, and the corresponding products also with the virtue of solid, high intensity, low water absorption, good dimensional stability, etc, since that the board is consolidated under certain pressure. Therefore, the production of cement flakeboard is a typical process of this method.

4) Casting mould forming process: In this process the stuff used to casting is required to be fluid, and the liquid/solid ratio is usually between 0.41~0.51. The modes of casting include manual assembly, flat-mould forming, and vertical-mould forming.

(3)Biomass Fiber/Plastic Composites

Biomass fiber / plastic composite, also called Wood-Plastic Composite (WPC), is a kind of new environmentally friendly composite using different forms of plant fiber as a filler or reinforcement which is mixed with thermoplastic resin (PP, PVC, etc. ) or other materials [52]. Because of a combination of advantages of both the wood and plastic, the WPC began to be industrialized in the 1980s in the United States, and mainly applied in the areas of building products, infrastructure, transportation and industrial commodities. The key of producing WPC with high-strength and low-density is to maintain the physical and mechanical properties of materials [53] to meet the requirements of light polymers by reducing the density of materials, which has become the new issues of WPC. The WPC after foam not only reduces the density but also partially improves the physical and mechanical properties. Due to these unique advantages WPC has great potentiality in many fields, such as components with high-intensity and light-weight, sports equipments, insulation materials and thermal insulation materials for packaging and construction industries, etc. In recent years, researches on the microporous foaming technology have progressed from single polymer foaming to the microporous foaming of composites, however, the biomass fiber / plastic composite materials research foam started late. At present, the foaming of biomass fiber / plastic composites are mostly based on PVC, and foam extrusion technology is commonly used [12]. Foaming technology can be divided into two categories according to the basic methods of foam formation: physical foaming and chemical foaming. The development of biomass fiber / plastic composite materials results from the growing emphasis on the environmental problems. Because of the advantages such as wide range of raw materials, cheap production costs, increasingly improved performance and environmental friendliness, biomass-based fiber / plastic composite materials has become the first choice of wood and plastic substitutes and will be widely applied in more and more fields.

*Studies on the Process Engineering of Eco-Boards Based on Steam Exploded Straws*

Currently, the existing producing technologies of eco-materials usually use natural straw as raw materials without pretreatment. In that case, to produce products, such as the plant-based fiber board, plant-based fiber wrapper, disposable tableware box, etc, in order to satisfy the application requirement, it is necessary to add some thermosetting resins (such as formaldehyde) and other plastic materials to control the dimensional stabilities of materials. The use of a large number of chemical adhesives and additives not only increase the cost of production but also result in serious environment pollution. To produce boards from straws without adding adhesives is mainly reported as follows: the straw is simply heated, during the thermal curing process only hydrolyzed reducing sugar is used as adhesive, without making use of the adhesiveness of lignin and the performance of cellulose is not improved either. Therefore, the production process still does not realize industrialization due to high cost and unstable performance of products.

To solve the problems stated above, researcher Chen of Institute of Process Engineering, Chinese Academy of Sciences deeply studied the similarities and distinctions of straw and wood in the aspects of chemical constitution and structures. Then he carried out the studies of straw denaturalization and thermal curing process and brought about a new concept of straw denaturalization and an integrated process of straw denaturalization which can produce high quality eco-boards. The technological line is shown as follows in Figure 5.13:

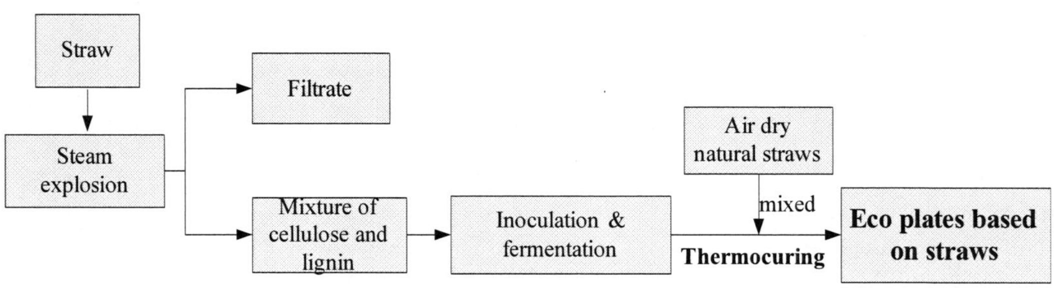


Figure 5.13. Technical course of using steam explosion straw to prepare eco-boards

Denaturalized straws are those straws with modified structures and properties through the physical, chemical or biological treatments. Cellulose, hemicellulose and lignin in the straw can partly degrade and form biological heat-curing adhesive resin, which entitle the straw fiber auto-adhesive capability and can meet the molding requirements of constructional boards. Straws after steam explosion treatment have distinct increase of active group in the lignin part which is similar to the polyphenol that can act as natural adhesives, and the hemicellulose can degrade into soluble sugar, dehydrated sugar, furfural and so on which can replace the formaldehyde for cross linking reaction with lignin, and the cellulose itself does not degrade but increase the degree of crystallization. The technological line includes two main types of molding techniques. One is heat-curing technology which refers to oriented control heat-curing process by adjusting the water content of the material and to maximize the formation of biological resin and be uniformly distributed in the straw fiber surface by rearranging the hydrogen bond of straw cellulose. The other is fiber re-forming technology, that is, under a certain temperature and pressure to cement the straw fiber with a biological adhesive resin which is produced through the heat-curing process, and to form constructional boards with certain mechanical strength which can meet the requirements of the construction. Due to no resin addition, using straws to prepare eco-materials is very attractive to the ecological environment and economy. It is obviously that the new technological line not only simplifies the production process but also decrease the production cost by removing the resin stirring procedure and relative equipment. The product can be completely degraded in nature and further used as either fertilizer or forage without causing any environment pollution. After effective denaturalization treatment, the strength of the eco-materials is greatly improved.

The use of straw denaturalization technology can transform the waste straw into useful eco-materials, such as green constructional materials and biodegradable materials. Moreover, the long fiber part of the steam exploded straws can be condensed and processed into environment friendly products such as corrugating, green wall materials, green fiberboard, green packaging materials, disposable tableware box. Experts predict that housing area of China can reach 1,500,000,000 $m^2$ per year, if per square meter of floor area needs 10-square-meter building board, every year the need of building board is 15,000,000,000 $m^2$. It's a very substantial market, and is expected to be the new economic growth point in the new century.

*Outlook of the Lignocellulose-Based Eco-Boards*

National project of "Natural Forest Protection Project" does not allow cutting a large number of natural forests, but it also provides a great opportunity for the application of the straw-based eco-boards. To produce eco-boards and constructional materials using biomass resources as raw materials is an effective way to realize biomass high-value transformation as well as maintain the ecological balance. The current problems such as the chemical adhesives pollution, asymmetric adhesive distribution, and unstable performance should be solved fundamentally. To advocate the development of green adhesives and green straw-based boards (e.g. steam exploded denaturalization technology) is called from the industry. Though there are some problems in the technology, with the joint efforts of government, research institutes and manufacturing enterprises the gratifying situation of biomass integrated application will be in the near future.

#### *5.3.1.4. Liquefaction Modification Technology of Lignocellulose*

At present, plant resources have not been reasonably used and straw as one of the major crops is only at a low level of utilization stage. Liquefaction modification technology is an effective transformation technology up to now, if we are taking straw resources as renewable and valuable mineral deposits and we are trying all possible methods to explore useful resources. According to the heterogeneous characteristics of different components in the lignocellulose, our research group chose straws as raw materials to investigate the selective liquefaction and the further utilization of liquid and solid products.

*Overview of Liquefaction Modification Technology*

Liquefaction modification of lignocellulose refers that transforming lignocellulose in organic reagent under certain temperatures to generate small molecule liquefaction products which can be used as the raw materials to synthesize polymer materials [54]. At present, the commonly used organic solvents include phenol, polyols, and so on. The liquefied products using phenol as solvent can be used to synthesize resin materials, and the liquefied product using polyol as solvent can be used for synthesis of biodegradable polyurethane foam [55]. As a commonly used organic solvent in lignocellulose liquefaction, phenol often plays a major role as follows [56, 57]: (1) Nucleophiles. In liquefaction process it can cause fracture of the ether bond and the open-loop of epoxy compound; (2) Reaction reagent. Phenol can react with some active small molecule and functionalize them with phenolic characters. (3) Solvent. Small molecules generated in the liquefaction process can be dissolved in the phenol solution and be significantly delayed to the re-condensation reaction.

Phenol plays different roles in the liquefaction of cellulose, lignin and hemicellulose, thus they have different liquefaction mechanisms. Cellulose is constituted of glucosidic bonds and glucoses, hemicellulose is mainly constituted of pentoses and glucosidic bonds, and lignin, possessing three-dimensional structures, is constituted by the guaiacol-and-clove, hydroxyphenyl groups combined with each other through certain chemical bonds (α-O-4, β-O-4). The different monomers and distinct structures of lignin determine different liquefaction process.

*Liquefaction Mechanism of Lignocellulose*

Liquefaction Mechanism of Cellulose, Hemicellulose in Phenol

As shown in Figure 5.14, when there is no catalyst, phenol acts as nucleophiles to destroy the glucosidic bonds of the cellulose firstly, which resulted in the degradation of cellulose into glucose, then glucose is further transformed into 5-hydroxymethyl furfural (HMF) through open-loop and dehydration reaction, and the generated HMF can further react with phenol to synthesize resins.

When a catalyst such as sulfuric acid exists, liquefied products of cellulose become more complicated, but one thing all the liquefied products in common is that they are combined with certain amount of phenolic groups. Lin, etc [58] studied the liquefaction mechanism of cellulose in the phenol using the cellobiose as model, which indicated that cellobiose is converted into four main categories of materials in the liquefaction, as shown in Figure 5.15. Matters of Group I are likely the original liquefied product of cellobiose and both the glucosidic bond and the reducing ends have been replaced by phenol, making cellobiose degraded, and the liquefied products all combined with one phenol molecule. Matters of Group II are the derivatives of pyranoid glucose-phenol, and the protonated O in the pyranoid ring is replaced by the phenol and transformed into the matters of Group II which could be further generated into matters of Group III and IV. Besides, they predicted the liquefaction mechanism of cellulose that glycosidic bond was broken in phenol, which may also result in the open-loop and degradation of pyranoid glucose (Fig 5.16). It can be seen that the acid catalytic liquefaction of cellulose was the degradation and open-loop process of cellulose under the act of phenol and acid, and the generated liquefied products were all combined with one or more molecules of phenol, thus, the activities of liquefied products were greatly enhanced.

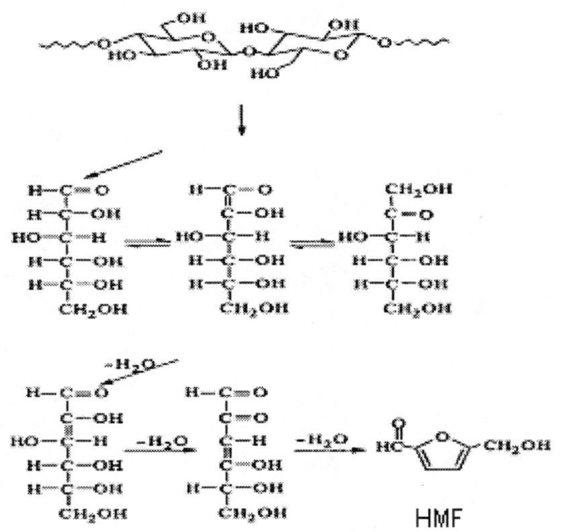


Figure 5.14. Liquefaction mechanism of cellulose by phenol without catalyst [60]

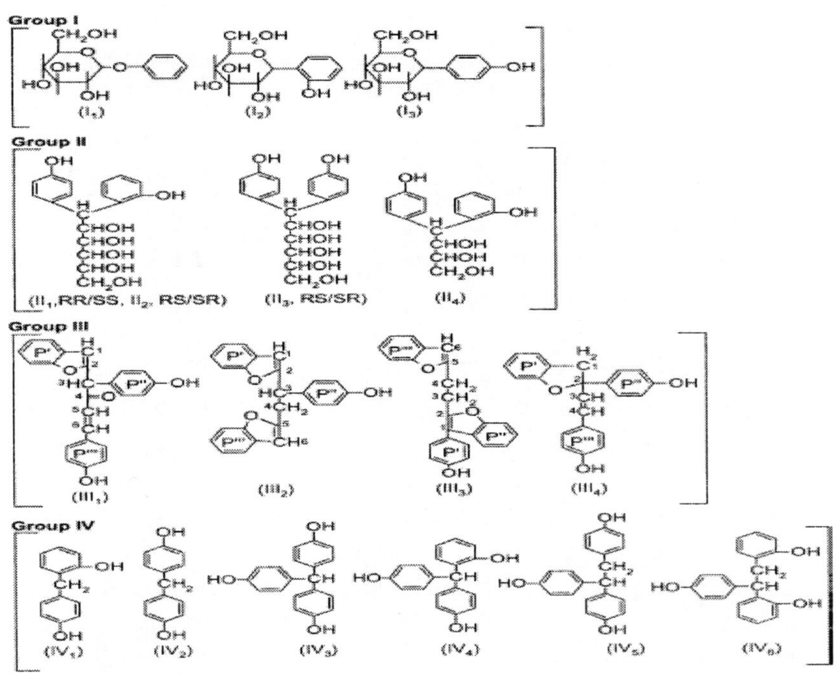


Figure 5.15. Structures of major components obtained in the liquefaction of cellobiose by phenol with $H_2SO_4$ [58]

In addition, in the latter part of cellulose liquefaction, polymers may also be regenerated, enhancing the liquid residue rate [59]. When phenol is insufficient in the latter part, existing liquefied products many further cross link with each other and generate new polymers and become part of the liquid residue.

The liquefaction mechanism of hemicellulose is similar to that of cellulose.

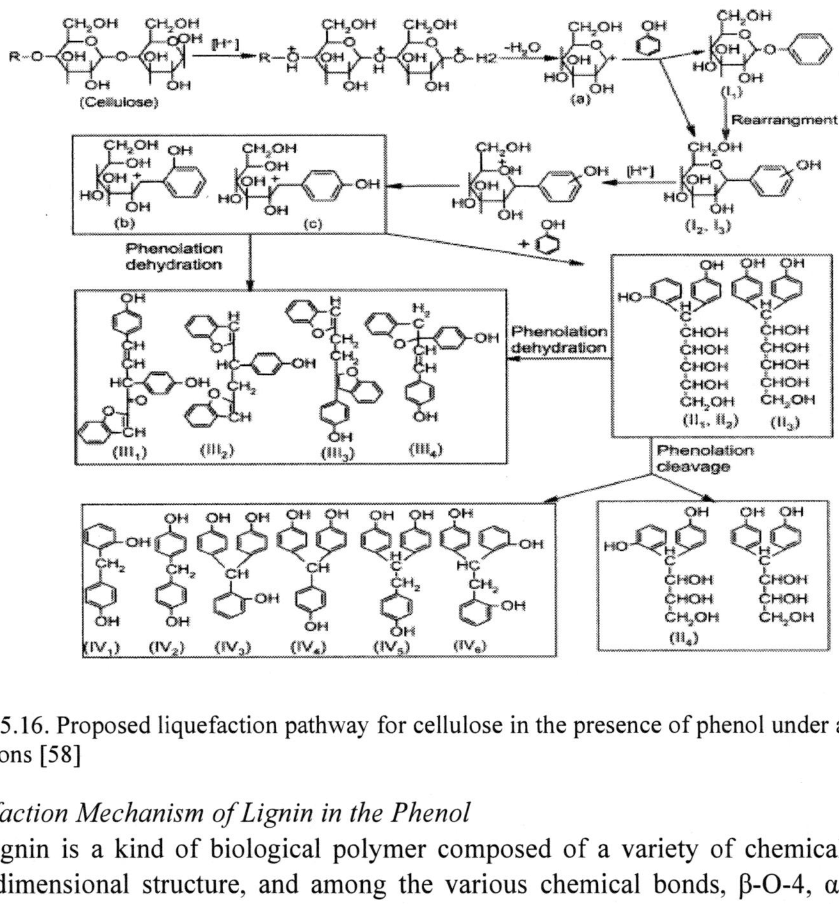

Figure 5.16. Proposed liquefaction pathway for cellulose in the presence of phenol under acid-catalyst conditions [58]

*Liquefaction Mechanism of Lignin in the Phenol*

Lignin is a kind of biological polymer composed of a variety of chemical bonds with three-dimensional structure, and among the various chemical bonds, β-O-4, α-O-4 are the most common chemical bonds. Therefore, researchers often use the compounds with β-O-4, α-O-4 chemical bonds as the model compounds to study the liquefaction mechanism of lignin.

Lin, etc. [51, 61-64] investigated the liquefaction mechanism of lignin using model compound Guaiacylglycerol-beta-guaiacyl ether (GG). Under the situation without catalyst, with the increase of temperature, β-O-4 bonds of GG are broken and generate low-molecular-weight intermediates of coniferyl alcohol and guaiacol free radicals. Coniferyl alcohol with conjugated double bonds in branched-chain can react with phenol free radicals and generate a large amount of phenolic products. And the guaiacol free radicals are mainly reduced to guaiacol. When liquefied with catalyst, the main liquefied products include guaiacylglycerol-α-phenyl-β-guaiacyl ethers, guaiacol, triphenylethanes, diphenylmethanes, benzocyclobutanes and phenylcoumarans. And the types of products have nothing to do with the type of acid catalyst. The generated liquefied products were all combined with one or more molecules of phenol, thus, the activities of liquefied products were greatly enhanced.

*Selective Liquefaction Process of Lignocellulose*

Necessity of Selective Liquefaction Process of Lignocellulose

Liquefied lignocellulosic products have strong reactivity due to the replacement of high active functional groups, they can be used to synthesize resin adhesives and coatings (such as phenolic resin, epoxy resin) and can also be use to generate biodegradable plastics such as polyurethane [65]. The great progress of using liquefied lignocellulosic products to prepare polymer materials has promoted the utilization of lignocellulose. However, there are still many problems in lignocellulose liquefaction. In particular, in order to achieve the full use of raw materials, the lignocellulose is liquefied to extreme extent that all the compositions in the lignocellulose are liquefied, that is, complete liquefaction. Due to the drastic reactive conditions (strong acid, high-temperature) of complete liquefaction, treating lignocellulose is treated as homogeneous material to reach its total utilization. However, the chemical compositions and structures of the three main compositions (cellulose, hemi-cellulose and lignin) are distinct which directly result in the asynchronous liquefaction of lignocellulose. Compared with the lignin and hemi-cellulose, cellulose is more difficult to liquefy [66]. Therefore we can take advantage of this characteristic to liquefy the hemi-cellulose and lignin as completely as possible while selectively retain cellulose. The liquefied products can be used to synthesize polymer materials and the cellulose can be hydrolyzed to prepare other chemicals, which will increase the value of lignocellulose. We called this kind of liquefaction as selective liquefaction.

Selective Liquefaction Process of Lignocellulose

We designed specific research route shown in figure 5.17 to explore the feasibility of realizing the high value-added utilization of lignocellulose through selective liquefaction. Firstly, the straw is selectively liquefied into liquid products and insoluble solid residues. Secondly, according to the different components and characteristics of liquefied products, high-value added process is developed. The liquid products (mainly hemicellulose and lignin) are used to synthesize polymers like resin adhesives, and the insoluble solid residues (mainly cellulose) can further be hydrolyzed by cellulase. The performances of the synthesized resin adhesive entirely meet the standards of the GB/T14732-2006 (industrial adhesives in wood industry: urea-formaldehyde, phenol-formaldehyde, melamine formaldehyde resin) except some improvements on the appearance of the product. It is proved that by further improvement of technology the high value-added utilization of liquefied products can be realized.

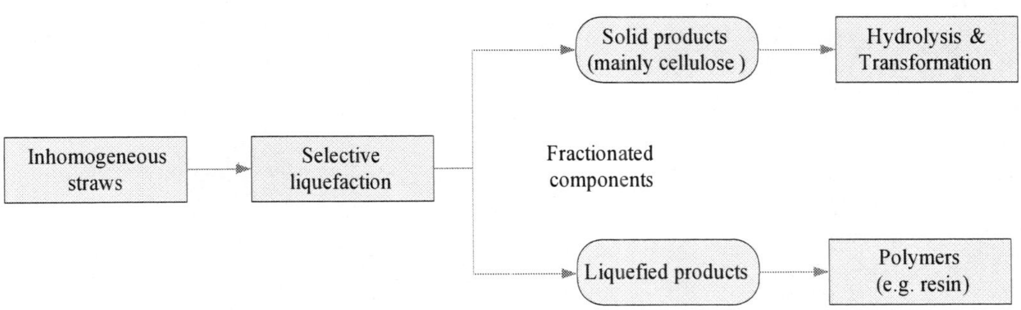


Figure 5.17. Technical route for selective liquefaction with phenol

### *5.3.1.5. Carbon Fibre Material*

*Brief Introduction of Carbon Fibre*

Carbon fibre (CF) is one kind of inorganic macromolecule fibre in which carbon content reaches as high as above 90%. It is produced from organic fibre such as PAN fiber, asphalt fibers, viscose etc. or low molecule weight hydrocarbon gas by means of oxidation and carbonization treatments. Especially, those having 99% content of carbon or more are called graphite fibres. As early as in 1879, Edison had invented the use of CF as filament. However, the CF that possesses certain mechanic capabilities did not come out until 1950. Ever since then, CF entered into a new phase of development as the enhanced materials for the complex materials. At present, more than 92% of CF productions are used in complex materials.

CF is a high strength and high modulus without any change in high temperature for a long time and its specific heat value and conductivity is between the metal and non-metal materials. It even has a low coefficient of thermal expansion, excellent corrosion resistance and good transmission for X-ray. As a fiber type carbon material, CF has the general properties of fiber materials including resistance to high temperature and abrasion, as well as good conductor for heat and electrical transfer. Besides, CF has many characteristics distinguished from ordinary carbon materials such as the anisotropy in its shape, tenderness, high strength along the fiber axis, etc. The high specific strength results from its small specific weight. The graphite structure is the reason that CF has so many good physical properties.

The chemical capabilities of CF are similar to that of carbon. It can not be oxidized by common acid or alkali except the super oxidant. In air, when the temperature is higher than 400 °C the CF can be easily oxidized to generate CO or $CO_2$. When the atmosphere is without air or oxygen,, compared with other materials, CF has outstanding thermo-resistance that begins to decline when the temperature is higher than 1500°C, at which the $Al_2O_3$ crystals capability of other materials has declined greatly. In addition, CF can also endure low temperature, for example, it can not be embrittled even in liquid nitrogen temperature. It can also endure oil, resist radiation, absorb toxic gas and decelerate neutron etc. Just like every coin has two sides, it has some disadvantages either. For example, it can not bear strong concussion and is easy to be destroyed, it will be oxidized in strong acid, and when combined with metal always occurs in metal carbonization, cementite and electrochemical corrosion, etc. Therefore, it is necessary to perform surface treatment before use.

*Classification of Carbon Fibre Materials*

There are many kinds of commercial CF which could be categorized by their precursor fibre type, properties and applications [67]. According to the different properties, CF can be labeled as (1) High quality CF, which includes high strength CF, high modulus and middle modulus CF; (2) Low modulus fiber, which contains non-flammable, carbon and graphite fiber. Based on the type of precursor fiber, CF could be classified into (1) Polyacrylonitrile-based CF; (2) Viscose CF; (3) Asphalt CF; (4) Lignocellulose CF; (5) Phenolic-aldehyde and other organic CF. Also, CF could be sorted by its functions, such as (1) CF resistance to strong force, (2) CF used in flame resisting, (3) Active CF, (4) Conductive CF, (5) Lubricating fiber, and (6) Wear resistance CF.

*Synthetic Process of Carbon Fiber*

The only way to produce CF is to use the organic fiber with carbon such as nylon silk, nitrilon silk as raw material since the elementary carbon can neither melt at high temperature, nor dissolved in any solvent. After blending the raw material with plastic resin, the mixture was carbonized under rare gases at high pressure and temperature. Nowadays, CF is mostly used as enhanced plastic composite material.

Most of the commercially available CF is produced from the carbonizing of the polyacrylonitrile. The procedure could be divided into three parts. The first section is called pre-oxidization, in which the polyacrylonitrile is heated in air for ten to many hundred minutes under 200-300°C aiming to transfer the liner structure of polyacrylonitrile into trapeziform structure with thermo-resistance. In the second step which is entitled as carbonization, the product obtained above is treated in inert atmosphere for tens of minutes with 1200-1600°C in order to produce the CF. The utilized inert gases are usually high pure nitrogen, argon and helium. Finally, the fiber would be graphited for tens of second at 2000-3000°C in order to form the CF, also named as graphite fiber.

*Production Process for Plant-Based Carbon Fiber*

At present, the main raw material used for producing CF is polyacrylonitrile, asphaltum or other fossil fuel materials. As the shortage and soaring prices of fossil fuel resources, the danger to human health during the process of the synthesis due to the carcinogen contained in the petroleum based materials, to decrease the output cost while improve the CF application in civil industry and realize the transformation of the CF, it has aroused more and more emphasis on the transformation of raw material from fossil fuel resources to biomass resources.

Plant-based material is a good stuff for producing CF as it contains abundant cellulose and lignin, making it not only reduce the manufacturing cost, but also hold a significant strategic importance as a reproducible resource both in the social economy and environment. The raw material of plant based CF mainly based on cellulose, lignose, resin, and starchy. In this part, we will introduce the development in research and application respectively below.

1) Cellulose Based Carbon Fiber

Presently, the general way for extracting cellulose from plant based raw material is chemical separation method. Then, the cellulose was melt or dissolved for spinning and the cellulosic thread is carbonized to get CF. According to the fiber type after spinning, cellulosic CF could be categorized into pectic CF and Lyocell type CF.

Pectic CF is produced by the carbonization of pre-synthesized pectic fiber come from cellulose [68, 69]. The detail technical process is illustrated in figure 5.18.

Lyocell carbon type fiber is a new kind of fiber developed with the Lyocell fiber, which is a dissolvable fiber. Lyocell fiber, called as the green fiber in the 21st century, is a kind of reproducible cellulosic fiber, which can be spun with wet method, dissolving in N-Morpholine methyl-N-oxides (NMMO) solution. The castoff can be degraded naturally and the reclaim rate of amine oxide is 99.5%. The technology was first patented by the Germen Corporation Akzo-Nobel in 1980 and named officially by the international committee of artificial fiber and synthetic fiber in 1989. Now, nearly 120 thousand tons Lyocell CF could be manufactured around the world distributing in USA, England, Austria, Germany, India, South Korea and Chinese Taipei. Pure Lyocell woof possess a pearl like gloss and the intrinsic innervation makes it seems filmy and good pendency. In addition, Lyocell fiber is a real eco-fiber with excellent structure property, lower pollution and energy compared with the traditional pectic fiber.

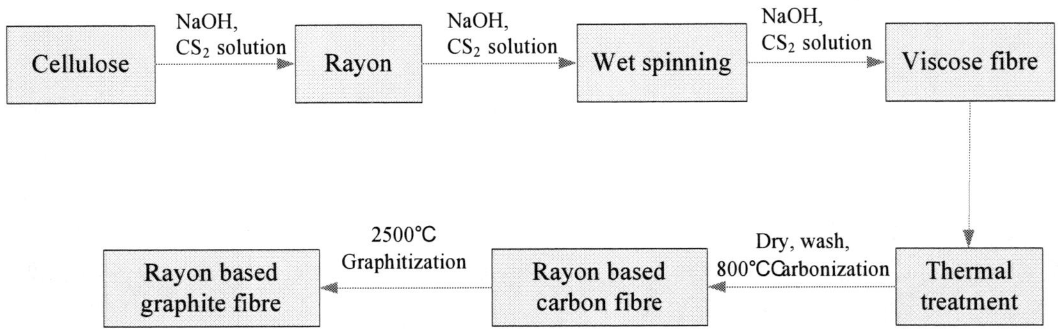


Figure 5.18. Transformation of pectic carbon fiber

2) Lignin-Based Carbon Fiber

Lignin is the second larger component except cellulose in plant based raw material. The pulp obtained from the paper mill and the abandoned lumber, dissolved in alkali solution after been whetted and desalted, is mixed with polythene and then the dried mixture is carbonized under inert atmosphere at high temperature to get lignin-based C. Shimizu et al [70] obtained the lignin-based CF by spinning the melt lignin at high temperature firstly and then hydrocracking it under vacuum. Kadla [71] produced the lignin-based CF utilizing the sulfate lignose as the raw materials. The extender generally used included polyethylene, polypropylene, oxide polyethylene and so on [72].

3) Wood Resin-Based Carbon Fiber

Wood resin-based CF emerged recently along with the development of the technology of lumber liquefaction. The liquefaction products of wood under acid condition were used to synthesize phenol-aldehyde resin, which was further spun into cellulosic silk under melt state. The CF was attained by carbonization of the silk which was pre-solidified [72].

## 5.3.2 Microbial Fermentation Materials

### *5.3.2.1 1,3.-Propanediol*

1, 3-propanediol (PDO) is an important organic chemical. It can substitute glycol and butanediol to synthesize polyester, poly ammonia ester, emulsion, antiseptic, wash and etc, and can be also used as intermediate in the organic synthesis. Among all of these applications, the most prominent one is the synthesis of polyester PTT [Poly (trimethylene terephthalate)]. PTT is a new kind of polyesters which possesses excellent capabilities and is the condensation polymer of PTA and PDO. PTT fiber possesses almost all of the most prominent capabilities of other fibers like the elasticity of nylon, the fluffiness of acrylic fibre, the anti-dirt of terylene, the wearability and dyeing in normal temperature. All of these advantages entitle it wide application in the areas of texture and engineering plastics, therefore it is predicted to be one of the most important new fibers in the 21st century.

However, presently, the high cost of PTT limits its wide utilization. As the key monomer raw material for PTT fiber products, the industrialized level of PDO directly affects the industrial development of PPT. Therefore, researches on the developments of low cost synthesis processing of PDO have been the hot research since 1990s.

As early as in 1942, Hatch etc of Shell Company in America had applied the patent about using acrolein to synthesize PDO [73]. During the 1980s and 1990s, Degussa Company in German developed another process of synthesizing 1, 3-PDO from acrolein and realized its industrialization in Europe. Presently, there are three main methods to synthesize PDO: the ethylene oxide carbonylation method which was industrialized in 1996 by Shell Company in America [74, 75], Acrolein hydration hydrogenation method developed by Degussa Company in German [76] and microbial fermentation method.

*Technical Level of PDO*

1) Ethylene Oxide Carbonylation Method

PDO can be synthesized directly, or by two-step method, by ethylene oxide and syngas through hydroformylation Two-step method is firstly to generate 3 -hydroxy-propionaldehyde (HPA) through hydroformylation reaction, and then prepare PDO by catalytic hydrogenation through a fixed bed. Reaction formulas are as follows:

(1) Ethylene oxide reacts with the $H_2$ and CO to prepare HPA in the presence of catalyst

$$\underset{\diagdown\,O\,\diagup}{H_2C-CH_2} + CO + 2H_2 \longrightarrow HOCH_2CH_2CHO \qquad (3\text{-HPA})$$

(2) Separated HPA is turned into PDO through hydrogenation

$$HOCH_2CH_2CHO + H_2 \longrightarrow HOCH_2CH_2CH_2OH \qquad (1{,}3\text{-PDO})$$

To develop a high-active, high selective, easy-recovery and recycling catalyst is the key of ethylene oxide carbonylation process. Presently, the two main catalyst categories are cobalt and rhodium.

The latest patent of Shell Company proposed a so-called ethylene oxide two-step method which indicated that by using improved cobalt-double phosphine ligands catalyst and catalyst system to prepare HPA through carbonylation, and then producing PDO through the traditional catalytic hydrogenation method is an efficient method with high yield of PDO.

Modified phosphine catalysts call for higher cost and in the production process some of the phosphine ligand may be dissociated which results in the difficulty of catalyst recovery. Shell turned to non-phosphorus ligands of carbonyl cobalt catalyst. One way to prepare non-phosphorus ligands is to add ruthenium catalyst to carbonyl cobalt catalyst. Slaugh et al [77] took phosphine-free ligand of carbonyl cobalt as catalyst and carbonyl ruthenium as assist catalyst, EO conversion rate and selectivity of HPA can reach 76% and 86.1% respectively. Knifton[78] adopted a new ligand to replace the phosphine ligand and also achieved a better catalytic effect, selectivity and yield of HPA was 74% and 66%, respectively. Lanzhou Institute of Chemical Physics, Chinese Academy of Sciences and the Shanghai Petrochemical Company Institute also has studied on cobalt catalyst system, but has not yet reported for industrial applications. Rhodium-based catalysts mainly include complex metal catalyst, phosphine-rhodium catalyst and ion catalyst. Murphy [74, 79, 80], Briggs [81, 82] et al had carried out relevant researches on these three types of rhodium-based catalysts, which indicated that the Rhodium catalyst has higher recycling rate, but has many disadvantages like expensive, harsh reation condition and high cost due to the use of noble metal.

2) Acrolein Hydration Hydrogenation Method

Degussa Company in Germany has developed an industrial production course of PDO using acrolein as raw material that included two major steps:

(1) Acrolein is hydrated into 3-hydroxy-propionaldehyde (HPA):

$$H_2C{=}CHCHO + H_2O \longrightarrow HOCH_2CH_2CHO$$

(2) HPA is hydrogenated into PDO in the presence of catalyst:

$$HOCH_2CH_2CHO + H_2 \longrightarrow HOCH_2CH_2CH_2OH$$

Acrolein can be hydrated to HPA in the presence of solid acid catalyst, especially chelating resin catalyst, and HPA can be hydrogenated directly or in two step into 1,3-PDO in the presence of Ni, Pt or Ru as catalyst by controlling the reaction temperature. Acrolein hydration and HPA hydrogenation reaction is the key to the process, which directly relates to the yield and quality of products, and the key technology is the catalyst [83]. There are two types of acrolein hydration process, homogeneous catalysis and heterogeneous catalysis. In the homogeneous catalytic system, the earlier used catalyst for the hydration of acrolein into HPA included $H_2SO_4$、$H_3PO_4$ and other inorganic acid and acid salt etc [73], which occurred low yield rate, poor selectivity and vulnerable to side effects. In recent years, the homogeneous catalytic system of acrolein hydration have made some new progress, for example, by using the buffer solution with acetic acid, propionate or acrylic acid, and trimethylamine, TEA, or pyridine propylamine [84] as catalyst, or the buffer solution with iminodiacetic acid, imino dimethyl phosphoric acid, N-phosphono-methyl-glycine and its salt used as catalyst allows the conversion of acrolein reach 60.7% and 93%, HPA selectivity can reach 80.3% and 62.0% respectively [85]. Compared with the homogeneous catalytic system, heterogeneous catalytic system gained more researches and more significant results because of its simpler consequent process. In the hydration process of HPA, there has good results in the acrolein convertion by using weak acid cation exchange resin [75, 86, 87], chelating ion-exchange resins [88, 89], polyamide / poly-resin acid [90] as catalysts. As for modified-type of phosphoric acid chelating resin, the selectivity of main product can reach as high as 91% [91, 92]. However, due to the high cost, poor thermal stability, liable to inactivation, and difficult to regeneration of ion-exchange resins, there are also researches on the inorganic catalysts vectors $TiO_2$ of active site [93], molecular sieve catalyst [94, 95]. Inorganic carrier catalytic system is easy to prepare and very stable, active in high temperature as well as renewable, which greatly reduce production costs.

A variety of side effects was companied with the hydrogenation of HPA, so catalyst with high selectivity to the reactant is required. The use of improved Raney nickel [96], as well as loaded precious metal catalyst [97-99] for two-stage hydrogenation reaction showed that the catalyst can maintain catalytic activity after many times of hydrogenation and obtain high single conversion rate of HPA and high selectivity of 1,3-PDO.

Heilongjiang Petrochemical Research Institute was charged with the Sinopec's project of 50 t/a PDO production by acrolein hydration hydrogenation method and the pilot test has received the acceptance of Sinopec's experts by the end of March 2006. It showed that China has mastered the key technologies of PDO production by acrolein hydration hydrogenation method.

3) Production of PDO via Microbial Fermentation Method

Chemical synthesis of 1,3-PDO is based on crude oil, and the cost is increasing with the decreasing of fossil energy resource amount and the increasing of oil price consequently. Meanwhile, there are always harmful and poisonous intermediate chemicals generated during the process which can easily pollute the environment. PDO produced by microbial fermentation is using renewable resource as raw material, such as starch (sugar). It is more easily to operate and gain the target product while the reaction conditions are relatively mild and has less pollution to the environment. Therefore, researchers all over the world are interested in developing this method to produce PDO.

There are abundant amounts of plant biomass resources, in which the cellulose and starch can be degraded into glucoses and provide rich carbon source for the fermentation process. However, there are no natural microorganisms that can transform glucose into PDO directly [100]. Natural microorganisms can only degrade cellulose into glucose, glucose into glycerol, and then glycerol into PDO. Since 1881 Freund found that *Clostridium Pasteur* can ferment glycerol to PDO, the reported PDO producing bacteria included the intestinal bacteria like *Klebsiella pneumoniae, Citrobacter freundii* and *enterobacter agglomerans*, the fusobacteria like *Clostridium butyricum* and *Clostridium Pasteur*, and so on. Because of the relative high yield and PDO productivity, *Klebsiella pneumoniae* and *Clostridium butyricum* gain more attention from researchers [101].

Microbial fermentation of PDO can be classified into two categories according to different substrates. One is that intestinal bacteria transform glycerol into PDO and the other is transform glucose into PDO. The latter can be further classified into two kinds. One is to produce PDO in two step fermentation and the other is to construct genetically engineered bacteria to produce PDO.

Aiming at making full use of the superfluous glycerol, EU countries (such as Denmark, Germany etc.) have achieved good progresses on using intestinal bacteria to transform glycerol into PDO and 2, 3-butanediol. Glycerol can be transformed into PDO through bacteria in anaeroabic condition. References [102] reported that intestinal bacteria can transform above 60% glycerol in batch continuous fermentation and the final PDO mass concentration can reach above 50 g/L. In theory, using glycerol as the only substrate to ferment 1,3-PDO can obtain 0.72 mol 1,3-PDO per mol glycerol at most, so the highest transformation rate is 72% and the final 1,3-PDO mass concentration is 50~60 g/L, which is still not competitive comparing with the chemical 1,3-PDO in cost [103] .

To reduce cost, some cheaper substitute like glucose was used to ferment PDO. In general, there are two ways to use glucose as substrates, one is to use mixed bacteria and the other is to ferment in two steps [104, 105], that is to say, using two kinds of bacteria to transform glucose to glycerol, glycerol to PDO in order. Hartlep found that recombinant *E.coli* was more suitable to glycerol fermentation than *Pichia* [106]. Tsinghua University,

Dalian University of Technology, and Jilin Petrochemical Corporation carried on deep researches on production of PDO via microbial fermentation method and probed out a process of using aerobic and anaerobic organisms to transform glucoses [107] or crude starch [108] to PDO in two segment fermentation. Among of them, Tsinghua University developed a successful process of using double strains to transform glucoses and crude starch (e.g tapioca) to PDO. This method at first time introduced dialysis desalination [109] and flocculants pretreatment of fermentation broth into the extraction process [106, 110]. This process has been proved work well in pilot scale experiment in 5000 L fermentor and now they are preparing thousand tons of PDO product project with Henan Tianguan Corporation and Heilongjiang Chenneng bioengineering Co. LTD. DuPont Company, cooperated with Tate&Lyle Company (English) and Genencer Company (America), has invested a pilot scale equipment to produce PDO using corn as substrates. Using corn as raw material to produce PDO via two step fermentation method can reduce one quarter cost comparing with the cost of petrochemical method.

The most promising way for PDO production is the recombinant cell which possess both of the glycerol and PDO producing genes and then it can transform glucose to PDO in one-step fermentation. There are mainly three ways to recombine cells: (1) to clone gene dhaB and dhaT that can transform glycerol to PDO into glycerol producing strain. Prokaryote and eukaryote have different gene structures and coding/decoding systems which call for different gene construction to the two gene fragments. Laffend etc [111] and Cameron etc [112] cloned PDO producing gene of *Klebsiella pneumoniae* in *Saccharomyces cerevisiae* respectively, but the expressed PDO synthetases activity is very low and the PDO yield is only in trace amount. Raozhiming [113, 114] in Jiangnan University constructed recombinent *Saccharomyces cerevisiae* in which the gene fragments of yqhD and dhaB that were from *E. col.* and Pneumonia bacteria respectively were cloned in series. The PDO yield of this recombinant yeast for 70 h using glucose as substrate was 1.5g/L fermentation broth. The success of glucose fermentation recombinant yeast points a direction for the future constructions of gene engineering strains.

(2) To clone gene GPPF2 that can transform sugar to glycerol into PDO producing strains. Glycerolphosphatidase is key enzyme which can catalyze 3- glycerolphosphate into PDO in the glucose metabolic process. If this gene can be successfully expressed in the PDO producing strains, it is entirely possible to connect the two routes of glucose metabolism and PDO production together and realize the direct transformation from glucose to PDO. Norbeck [115], Cameron [112] and Cowriters have separated the coding gene of Glycerolphosphatidase, and the gene reconstruction work is on the way.

(3) To clone gene dhaB, dhaT, GPPV2 into other microbial cells which can grow in media with glucose as the sole organic constituent, such as *E coli* that the most common and clear studied gene engineering strain. The most prominent advantage of *E coli.* as gene engineering strain is that it is more easy to construct because there are no gene regulation barriers in the glycerol and PDO metabolic process to overcome since the *E coli.* can produce neither glycerol nor PDO. DuPont Company and Genencor Company carried on deep researches and walk futher in this area. They successfully construct high yields of PDO gene engineering strain by using *E. coli.* as host to express the two key enzymes [116, 117].

*Prospect of PDO*

According to the forecast of Shell Company, the PTT demand can exceed 1000 KT by the year of 2010. As the key monomer of PTT, the mass production of PDO has a commercioganic future. However, the market share speed of PDO has much to do with its price.

As petroleum-based chemicals, either ethylene oxide method or acroleinor method call for great investment, and the catalytic system is complicated and the chemicals used are usually toxic which are not only harmful to the operation workers but also have great potential to pollute the environment. Though the cost of PDO via microbial fermentation method is not competitive comparing with that of chemical synthesis method, it points out a valuable progress aspect especially the two-step fermentation method and the construction of recombine gene engineering strain. The microbial fermentation method not only accord with the requests of green industry, but also provide a satisfied way to make full use of plant biomass resource easily and solve the shortage of fossil energy. We have confidence in the biotransformed PDO because it has so many advantages mentioned above and it will be more competitive than the petroleum-based PDO with the development of related technologies.

### *5.3.2.2. Poly- β-Hydroxybutyrate (Phb)*

*Foreword*

Plastic has been widely applied in lives and industry of people and agriculture for its light weight, strong toughness, corrosion resistance and low price. Meanwhile, the disposal of waste plastic has been a big problem. There are as many as 25 million tons of plastic waste piled up in nature every year which heavily endanger the living place of human beings, moreover, a plenty of potential pollutions to the groundwater or air may be caused during the filling up and burning processes of plastic. With the increasing environment protection sense of human and the further decreasing trash filling space, many countries attach great emphasis on the research and development of environment friendly degradable plastics [118, 119]. At present, researchers from world wide countries are concentrating on the development of five main categories of bio-polymers: 1) strengthened fiber compound; 2) starch-based materials; 3) plant-based polymers; 4) microorganism-based polymers; 5) bio-resin, dope and bond. Among them, Polyhydroxyalkanoate (PHA) gains most attentions, which is a polymer origin from plants and microorganisms [120]. PHB is a widely studied PHA, which is produced by bacteria as the energy storage particle when there are surplus carbon sources [121, 122]. PHB-based plastics possess similar capabilities as the fossil fuel-based plastics and are named as "bio-plastics" because they can be totally degraded by many kinds of microorganisms.

Since Lemoigne firstly extracted poly-3-hydroxybutyrate from *Bacittus megatherium* cells, there has more than 90 kinds of bacteria were found to possess PHAs, including Gram positive bacteria and Gram negative bacteria. In recent 10 years, there are rapid increase of researches on PHAs , such as the biosynthesis related microbiology, biochemistry, molecule biology and physical properties of PHAs. The wide and deep researches on PHB provide solid theoretical foundations for the development and utilization of bio-polyester materials [119].

*Synthesis and Degradation of PHB*

Fermentation is still the main means for production of PHB at present. PHB can be used as the carbon source and energy source by many bacteria. Some bacteria such as *Alicaligenes eutroplus, Protomonas extorquens* can produce PHB in extreme conditions that with abundant carbon source but lack of certain nutrition like N, P, K, Mg, O or S. However, there are also some other bacteria such as *A. latus* that can produce PHB without demanding nutrition. There are two routes for PHB synthesis: the three steps synthesis route and the five steps synthesis route [123]. The three steps synthesis routes are as follows: 1) β-ketone-S-lyase (catalyze acetyl-CoA to generate acetoacetyl-CoA; 2) By the action of NADPH-dependent acetoacetyl-CoA reductase, the acetoacetyl-CoA is reduced to D-(-) -3 -Hydroxybutyrate-CoA; 3) Monomer D-(-) -3-hydroxybutyrate-CoA is polymerized to PHB by the action of PHB synthase. The five-step pathway can be described as: Firstly, β-ketone-S-lyase catalyzes the generation of acetoacetyl-CoA from acetyl-CoA. Secondly, the NADPH-dependent acetoacetyl CoA reductase catalyzes the formation of L-(+)-3-hydroxybutyrate-CoA; L-(+)-3-hydroxybutyrate CoA is transformed into D-(-) -3 -Hydroxybutyrate-CoA after the catalysis of two dimensional specific acyl-CoA hydratase, and finally they are polymerized into PHB.

PHB can be degraded by many microorganisms in soil and seawater. The degradation velocity order is usually the fastest in anaerobic sewage and the lowest in seawater. Enzymolysis mechanism can be divided into two steps, the first step is the increase of -OH and –COOH amount on the PHB surface, and the second step is deploymerization conducted by bacteria. The high polymers are depolymerized into monomers by enzyme in suitable pH conditions and the lower the pH, the slower the depolymerization is. Depolymerase of *Pseudomonas* can deploymerize PHB into monomers, dimer, trimer and then they can pass the cell wall and take part in the metabolism. Once the PHB dimer enters human cells, they can induce the pseudomonas to produce and release more PHB depolymerase. Chromatogram analysis indicates that the decomposed products include butanoic, β- hydroxybutyrate, and some acetic acid. In aerobic atmosphere, except the extremely few hydroxybutyrate, most are oxidated into carbon dioxide and water [119]. Since that PHB can be completely degraded by natural microorganism, it gradually becomes the wonderful raw materials to synthesize bio-plastic.

*Technical Level of PHB Production Process*

At present, there are mainly two ways of PHB production, one is microbial fermentation and the other is plant transgenic technology. We will make a brief introduction of them.

*Microbial Fermentation*

Microbial fermentation is the main method for producing bio-plastic. ICI Company in England firstly utilized *Alcaligenes faecalis* to produce PHB in small batch in 1990. Now they have enlarged manufacture equipments with the capacity of 5~10 kt/a, which is the biggest poly-β-hydroxybutyrate manufacturer in the world. Linz Chemical Group in Austria occupies the second place and the output is 1 t/week. Besides, countries such as England, America, Japan, South Korea are carrying on the related researches.

In China, researches on the poly-β-hydroxybutyrate were carried on since 1980s, and now there are some progresses. Some institutes and universities have conducted small experiments, for instance, the Institute of Microbiology in Chinese Academy of sciences, biology department and chemical engineering department in Tsinghua University, Jiangnan University. However, there is no industrial application of extraction and purifyication of PHB [124]. Production process of poly-β-hydroxybutyrate includes two steps, fermentation and purification.

PHB fermentation process:

Post-Processing Of PHB:

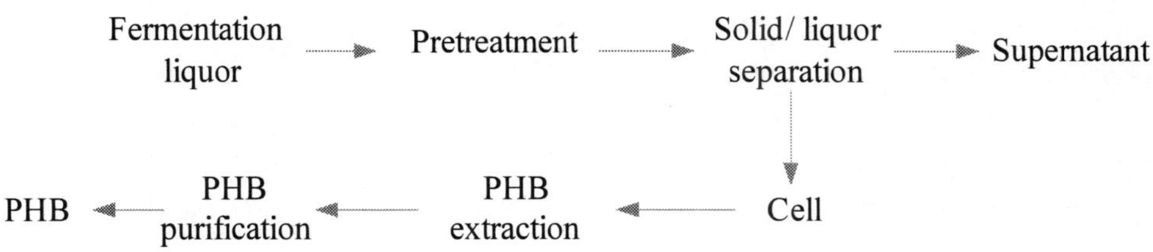


Strains used for fermentation are mainly *Alcaligenes faecalis*、*Azotobacter* and *Methylotroph*. Among them, though *Azotobacter* and *Methylotroph* can be cultured on honey and methanol respectively, the yield of *Azotobacter* PHB is low and the molecule weight of *Methylotroph* PHB is small, therefore, both of them are disused now. *Alcaligenes faecalis* can be cultured on glucose, and the yield of PHB can reach as high as 79.4% of cell dry weight after 52 h fermentation and the production intensity can reach 1.08 g/L•h in fed-batch fermentation. Yangming etc. [125] found that the metabolic process of strains can be divided into two processes: the prophase is strain growth phase, and there is great increase of strain biomass, the anaphase is PHB accumulation phase. In the prophase, if all nutriments including dissolved nitrogen are supplied sufficiently, strains can grow well and fast and the fermentation period can be shortened, which make sound foundation for the PHB accumulation. Because PHB is the outcome of bacteria grown under unbalanced conditions, in the latter phase of fermentation, if there is absent of one or more than one element in the substrate, bacteria will accumulate plenty of PHB, which can be achieved by either limiting oxygen or limiting phosphor in the anaphase.

Yupin etc [126] optimized the cultivation parameters of PHB fermentation and found that the optimum cultivation conditions are: 4.0% glucose, 0.3% ammonium sulphate, pH 7.2, optimum liquid volume ratio is 80 mL/250 mL, and optimum inoculums is 10%. The PHB concentration and dry cell weight can reach as high as 0.825 g/L and 1.734 g/L respectively after cultivating *alcaligenes faecalis* for 30 h under the optimum conditions. In addition, Z5-G is also a promising bacterium for the PHB production. Prof. Huang [127, 128] established the dynamic model of the PHB production based on Z5-G fermentation and improved the processing by using outside circulation air-lift reactor.

There are several technologies for the extraction of PHB from fermentation broth: organic solvent extraction, chloroform-sodium hypochlorous and enzymatic extraction. Organic solvent extraction consumes large amount of raw materials and energy while the yield rate is low, heavy pollution and difficult to operate. Treating PHB with chloroform-sodium hypochlorous can decrease the molecule weight of PHB. The purity of enzymatically extracted PHB is low. Besides the technologies mentioned above, in industry, in order to avoid the use of organic solvent, PHB was extracted using surfactant-sodium hypochlorous. In details, the microbe were firstly treated with surfactant and then the precipitates were washed with sodium hypochlorous. The purity of PHB can be above 97%. There are also other reports such as to break the cell wall mechanically to extract PHB.

*Transgenic Plant Method*

Transgenic plant method is not only a powerful tool to study the theoretical foundation of life sciences, but also provides a new way to achieve efficient, high-quality, high-yield agriculture. The use of genetically modified crops to improve the contents of protein, starch, fat, enhance the nutritional value of plants, enhance the anti-adversity endurance of crops and antiseptic capacity of fruits, and cultivate engineering plants antiseptic capacity and anti-herbicide capacity, which all have shown huge economic benefits.

In recent years, researches on PHB have gradually biased in favor of using genetically modified plants to synthesize PHB. In 1992 USA scientists for the first time tried to use plant to produce PHB. They inserted the genes of both PHB synthase PHBb and PHBc, produced from *Alcaligenes eutrophus* fermentation, into mouseearcress. However, it turned out not only low yield of PHB but also affected plant growth and development [129]. Nawrath, Somerville et al [130] in 1994 changed the strategy and located PHB in plastid. As the plasmid contains a large number of acetyl CoA which are the raw materials to synthesize PHB, and plasmid can also absorb a great deal of starch, it is suggested that plasmid can also store a large number of PHB. The study results showed that the cumulative amount of PHB could reach as high as 14% of dry weight and the transgenic plants grew normally. However, the study found that when PHB concentration was more than 300 ~ 400 ug/g in young leaves of plants, in the latter phase of the growth (50-60 days later) leaves might show a slight yellow phenomena. Too many accumulated PHB particles may be negative for the normal function of chloroplasts.

Presently, there are technologies to transfer recombined DNA of PHB synthase into plasmid of multiple plants, such as *Arabidopsis* [130, 131], tobacco [132] , cotton [133], alfalfa [134], flax [135]. It was reported that Bohmert et al [131] got the highest yield of biopolymer - 40% of dry leave weight. Traditional method is to transfer the three types of gene segments into plants separately, and then functionalized by combining together. The limitation of this method is obvious that it is unable to adjust the optimum ratio of the three types of gene segments and realize the best effect. Bohmert et al jointed the three gene segments in advance, and then transferred the combination into plasmid, and they monitored the final product by GC-MS. However, the PHB yield in single plant is still low which might because the concentration of biopolymer is very low in the plant, or the overall mass of biopolymer is relative low despite the relative high concentration [136, 137]. Matthew et al [138] did some researches on the PHB concentration distribution in the transgenic cane sugar plants and the outcomes showed that the content was lower in young leaves than that in the old leaves, and in each leave, the PHB content was lower in blade root than that in the blade tip, and the sequence was not changed with time.

*Prospect of PHB*

At present, microbial fermentation method is the main method of PHB production, but the cost is incompetent to that of chemical synthetic polymer and thus limits its large-scale commercial application. Therefore, the key to the development of fermentation is to raise production and decrease cost, which can be considered as follows: 1) Cultivate fine strains which can make use of low-cost carbon substrate and produce high-yield PHB [139-141]. For example, to find new natural strains through traditional methods of breeding, and to improve the ability of PHB synthesis by modifing the existing strains, also to build genetically engineered bacteria through the recombinant DNA technology; 2) Study microorganisms of different fermentation process and dynamics, to increase the density of cell culture, to shorten the fermentation cycle, to establish mathematics model for the fermentation process and achieve optimum control and automation of industrial fermentation production which can lay the foundation for the industrial design, scale-up, and optimal control; 3) Reform the PHB extraction process to reduce the extraction costs to a minimum and increase the purity of the target polymer; 4 ) Develop sludge in wastewater treatment as well as renewable carbon source in industrial and agricultural waste as fermentation substrate [142, 143]. Straw as an abundant carbon resource has great advantages on decreasing the cost of PHB fermentation process. The success of using transgenic plants to produce PHB shows that PHB price can drop to the level of petroleum products like polyethylene. Despite there are many obstacles in transgenic technology such as the difficulties of expression, function of bacteria gene and substrate transport in the plants, but the botanical system can correctly translate and process the eukaryotic protein and form activated molecule. Meanwhile, plant is rich in resources which can not only provide abundant substrates but also cut down the costs. Using Acetyl-CoA as substrate for the production of PHB, in theory, the yield can reach the production levels that of using starch, fat as substrate. Therefore, the use of transgenic plants to produce PHB, the biodegradable plastics to replace the chemical synthetic plastic and manage "white pollution" are promising.

By now, the urgent problems that should be solved in the transgenic technology are mainly as follows: 1) Further increase the content of biopolymer in the plants; 2) Ensure the normal growth of transgenic plants while reach the first aim. In recent years, along with in-depth study of these issues, these problems have gradually been recognized and understood that it is expected to be able to produce bio-degradable plastic in large-scale by transgenic plants very soon, and thus ease the energy and environmental crisis of the world.

#### *5.3.2.3 2,3. Butanediol*

2,3-butanediol (BD) is a colorless and tasteless chiral compounds, it has three forms of stereoisomers: dextro-type, levo-type and meso-type of isomers. The molecular weight of BD is 90.12 kDa, and the boiling point is between 180 and 184 °C, the freezing point is relatively low of - 60 °C.

The combustion value of BD is 27198 J g-1, which is competitive to methanol (22081 J g-1) and ethanol (29005 J g-1), therefore it is a valuable liquid fuel. L-forms of BD can be used as antifreeze agent due to its low freezing point. In addition, through condensation, polymerization reactions, L-type BD can produce some compounds such as styrene, octane BD diacetate and so on. As additives, it can be widely used in printing ink, cosmetics, lotion, antifreeze, fumigants, softeners, plasticizers, explosives and chiral vector of drugs, and so on.

In the areas of textile and materials manufacturing, it is the important raw material for the production of polybutene ester resin, γ-butyrolactone, spandex, and so on.

As a chemical intermediate, BD can be used to prepare an important industrial organic solvent, methyl ethyl ketone, which transformed into butanedione after dehydration. Butanedione possesses a high combustion value which has a wide range of uses in the fuel additive, also in the production of 2 - butene and 1,3 - butadiene which are widely used rubber monomer in the rubber synthesis industry. Esterified compound from BD is a precursor of polyethylene imine which can be used in drugs, cosmetics, lotion, etc. Diacetyl form of the BD gained through the catalytic dehydrogenation can be used as high-value food additives.

Due to such a widely utilization, BD is highly demanded in the international market, in particular liquid fuel additive.

*Technological Progress*

At present, in chemical industry, BD is obtained mainly through hydrolyzation of C4 carbohydrates which generate from oil cracking in high-temperature and high pressure. Compared with fermentation, chemical method still does not realize the large-scale industrialization and complete development not only due to the high cost, but also the cumbersome process and difficult operations. To produce BD through fermentation method is not only consistent with the requirements of green chemical engineering, also can overcome many difficulties in chemical production. Meanwhile, it is extremely important in realizing the great transition from oil refining to bio-refinery and gradually reducing the dependence on oil resources which is lasting exhausting. In more than 20 years, with the increasing price of oil, researches and applications on fermentation production of BD and its derivatives gradually attracted attention.

(1). Strains

In nature there are some strains of bacteria possessing the producing capacity of BD, including *Klebsiella, Bacillus, Enterobacter*, and so on. Different strains have different productivity of BD, among of them, *Klebsiella* and *Polymyxin bacillus* have greater capacity and potential, therefore, researches are mainly concentrated on them. Because of its wide range of substrate that can utilize glucose and xylose simultaneously and adaptation at different environments, *Klebsiella pneumoniae* has been studied much more deeply [144].

(2). Substrate

As to the fermentation substrates, many carbohydrates can be used for (bacteria, in particular) fermentation of BD, and the most commonly used substrates are glucose and C5 sugar such as D-xylose, L-xylose, D-ribose, D-arabinose, etc.

The studies on BD fermentation using glucose, xylose, fructose and lactose as substrates have been developed to a level close to the study of ethanol, which can meet the needs of industrialization. If using these sugars as substrates, a direct result is that the costs of BD production are too high to afford large-scale production. Thus, to find low-cost carbohydrates as a matrix to ferment BD is key to the low cost of production, which has become a hot research point.

Lignocellulose is the most abundant organic substance in the natural world. If it can be used as substrate to cultivate bacteria which can turn the carbohydrates of the lignocellulose resources into BD, it is expected to further reduce the cost of BD production [145]. As a result, many researchers that concentrated on using lignocellulosic hydrolysates to fermentate BD have been carried out. Bacteria which are able to use lignocellulosic degradation products as substrates to realize biological transformation possess a number of important properties, such as wider range of the substrates, higher yield, less by-products, strong anti-suppression and so on [146]. According to reports, *Klebsiella pneumoniae* is in line with the requirements of the strains described above that can use all major carbohydrates (hexose, pentose and some kinds of disaccharides) and alduronic acid which are from cellulose and hemicellulose degradation. In addition, carbohydrates from hydrolyzed hemicellulose in woods by acid hydrolysis or enzymatic hydrolysis can be effectively utilized [147, 148]. On this basis, Yu, etc. [149] also took cellulose and hemicellulose as substrates to culture both of *Trichoderma harzianum* E58 and *Klebsiella pneumoniae* in continuous fermentation. The principle is first to use *Trichoderma harzianum* E58 to degrade cellulose and hemicellulose to obtain glucose and xylose, and afterwards *Klebsiella pneumoniae* can transform the carbonhydrates into BD. According to Cao, etc. [150], when using pre-crushed corncob to carry out the pretreatment, hydrolysis and fermentation of BD studies, results showed that in the pretreatment, lignin and alkaline extracts can be separated from cellulose and hemicellulose through dissolution and hemicellulose partly hydrolyzed into xylose, finally *K. oxytoca* ATCC 8724 can grow well on the YMP medium which is using pretreated corncob as carbon source and realize the simultaneous saccharification and fermentation of BD.

*Key Technologies in BD Fermentation and Related Process Engineering Research*

From the research development, pretreatment of lignocellulose raw materials, as well as screening of lignocellulosic materials fermentable strains are the key to solve the high cost problem. Chen Hongzhang and Sun Fubao [151] successfully screened a BD producing strain, *Paenibacillus polymyxa* S-07, which possesses high capacity of wood fiber degradation. In addition, they developed a new pretreatment method, that is, to autocatalytic pretreatment of wheat straw by glycerin in ordinary pressure. After pretreatment, wheat straw can be removed more than 90% of the hemicellulose and more than 70% of the lignin, while retain about 95% of the cellulose. Enzymatic hydrolysis rate of wheat straw at 24 h and 48 h treatment is 90% and 92 % respectively and almost all of cellulose and hemicellulose can be hydrolyzed into reducing sugar [152, 153]. Using the pretreated wheat straw as substrates to conduct fermentation with the screening bacterium after 120 h, conversion rate of BD can reach up to 7.73%, and in the fermentation process, the strains degraded 56.7% of cellulose and 32.2% of hemicellulose in the substrate [151], and the lignin is basically still the same as before. In the guide of the "Integration Process" concept, an innovative process of BD production by a self-screened strain was developed using glycerin pretreated wheat straw as substrates. This process integrates cellulase preparation and enzymatic saccharification with BD fermentation, which is promising to simplify the BD production processes and reduce production costs.

The whole technique route is as follows:

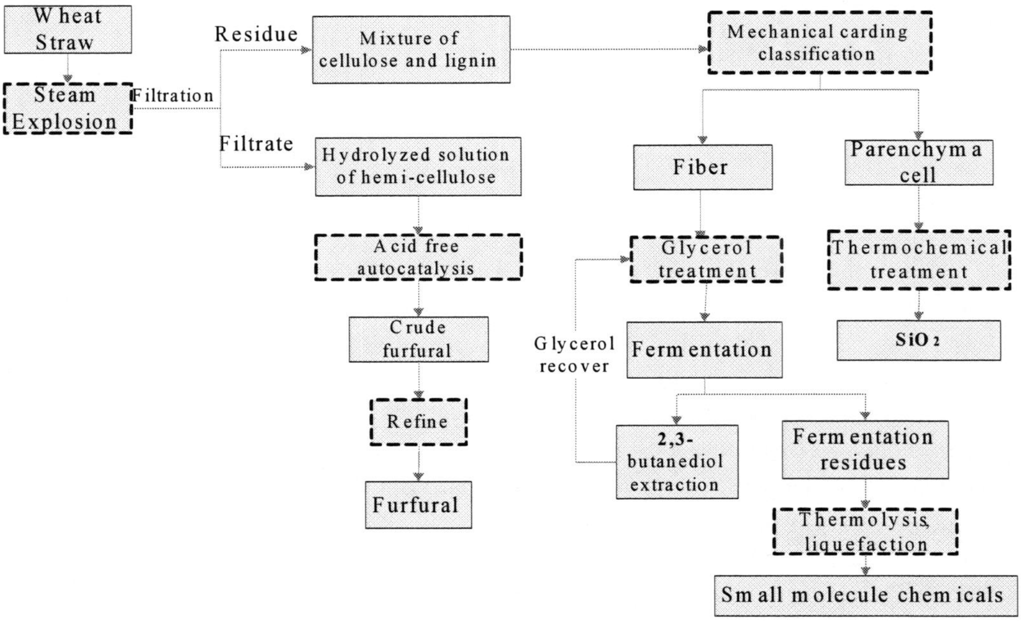


Figure 5.19. Multi-production of 2,3-butanediol using Glycerin pretreated straw as raw material through multi-level conversion

## 5.3.3. Modified Starch Material

### *5.3.3.1. Introduction of the Starch Material*

Environmental pollution brought by plastics was one of the most urgent problems of human beings confronted in today's world. To solve this worldwide problem and explore the source of non-fossil fuel based material, degradable material emerged as the time required. According to the way that material degraded, degradable material could be categorized as biodegradable, photodegradable, and chemical degradable materials. It was classified as nature degradable and synthetic degradable materials based on their different origin. The main nature degradable materials are starch, cellulose, chitin, and deacetylated polysaccharide.

The first explored and mass produced nature degradable plastic is starch plastic. Starch exists widely in seeds, leaves, tubers, roofs, fruits and dusts. The mostly commercial used starch is corn and waxiness starch, potato starch, wheat and cassava starch [154]. Starch, a kind of well bio-degradable macromolecule compound, could be used to produce fill-type starch-general plastic resin degradable plastic, double degradable plastic and complete starch thermoplastic. Starch plastics were also called starch-based plastics which generally refer to the materials that contained starch or its derivatives. Plastics that using nature starch as stuffing and nature starch or its derivatives as blended system were all belonged to starch plastics [155].

Starch plastics were initiated by British scientist G. Griffun, who proposed the standpoint in 1973 that the plastic could be degraded by stuffing starch into the petrochemical polymer. And he applied the first patent of the world on starch-filled polythene plastics.

#### *5.3.3.2 Preparation Process for Starch Material*

*Starch Filled Degradable Materials*

From the first patent gained by G. Griffin [156-158] for starch filled plastics, biodegradable materials draw the attentions all over the world and were further developed and improved in 1980s. Filled type materials was produced by mixing the filler, starch biodegradable materials, with general plastics which has good mechanical properties such as PE, PS, PVC [159] at a certain proportion. Since the starch is the polar macromolecule with many hydroxyl groups, so it can not dissolve in non-polar plastics. For that reason, the starch was needed to be modified to blend with high polymer. The modification method [156] could be categorized as physical and chemical methods. Physical modification includes minimize the grain of starch, destroy the structure by extrusion or modify the hydrophobicity by adding coupling agents, plasticizers or structure breach-agents. The essential of this kind of modification is to wrap a layer of coupling agent on every particle of starch uniformly result in covering the hydroxide group on the surface. Hence, the compatibility between starch and polymer could be greatly improved. Chemical modification technique is realized by introducing hydrophobic groups to starch molecule while these groups would enhance the solubility of starch in tree-type molecules. The modification could either be applied to the function group as etherification, esterification or to the free radical monomer.

The detailed method and principle of chemical modification are not described redundantly since that these all could be found in *Modification of Starch* written by Zhang Litian [160].

*Complete Starch Based Materials*

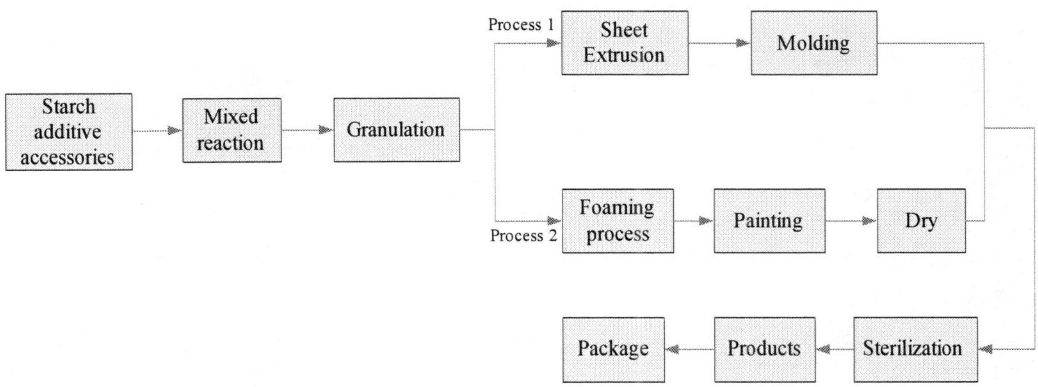


Figure 5.20. Production process of total starch based materials

Complete starch based materials possess a high content of starch, generally more than 85~95%, with the advantage of light, clean, innocuity, insipidity and environmental friendly which could contact with food directly and even be eaten. The castoff of these materials is used to produce feed. The manufacturing process could be divided into two sections. Firstly, the starch was denaturalized and then reacted with the additive, plasticizer, to produce master-stuff or to mold directly [161]. Presently, the technique utilized in China could be categorized as [156] (1) The altogether extrusion of starch and plasticizer; (2) The combination of active starch and degradable polymer; (3) Preparation of thermoplastic starch by grafting technology; (4) Modified etherization starch. The process is showed in figure 5.20.

#### *5.3.3.3. Outlook of the Development of Starch Materials Industry*

To solve the white pollution problem, the investigation of degradable plastic has attracted the attentions all over the world. Therefore the starch degradable material has a large potential use in (1) Commodities included shopping bag, litter bag, cosmetic container, toll and buffer material, (2) Gardening materials such as seedbed, cord, mesh, (3) Agricultural materials such as cover film, water keeping material, (4) Fertilizer, agricultural sustained-release materials and once-off medical materials..

Along with the reduction of the cost of starch based degradable plastic due to the progress of production technique, improvement of the product's properties and the scalable manufacturing, we believe that it would play an important role in the field of application of plastics, which would make a great contribution on the diminishing the environmental pollution caused by the plastics and the economic development of the rural areas.

### 5.3.4. Other Plant-Based Materials

#### *5.3.4.1. Rosin*

Rosin is the remnants of resin oil which is extracted from the Pinaceae plants or plants with tree trunks after distillation removing the volatile oil. Because of rosin's super characteristics of anti-corrosion, insulation, bonding and etc., since the beginning of last century, rosin and its modified products are widely used in coatings, adhesives, ink, paper, rubber, food additives and bio-products and many other fields. With the gradual depletion of fossil resources, how to use the natural renewable resource of rosin to replace some of the fossil resources has become an increasingly important research topic. At present, the utilization of rosin has transferred from direct production of primary products of rosin to the deep development of rosin derivatives. Deep process of rosin not only can improve the application performance of production, but also increase the added value of rosin resources [162].

*Introduction of Rosin*

The main components in rosin are monoterpene, sesquiterpene and two-terpenoids chemicals. By chemical method the resin can be divided into two parts: the acidic materials (resin acid) and neutral materials. Resin acid accounts for about 85% to 90% of the whole weight, while the neutral materials accounts for 6-10%. In the neutral materials, low boiling point part and high boiling point part each accounts for 3-5%, respectively. Pinaceae plants resources are enriched in China and distributed all over the country. There are more than 20 species of rosin-producing Pinaceae plants. According to statistics, pine forest area of China is more than 13,000,000 hectares, and the recoverable pine oil is about 1,500,000 tons/year. At present, pine resin production is 500,000 tons/year, ranking first in the world. Pine resin can be processed into rosin and turpentine rosin, which are useful chemical raw materials. Presently, the demand of rosin in the world is increasing, while Asia accounts for 60% of world export trade, as a result, China plays a vital role in rosin market of Asian, even the world.

According to different origin, rosin can be divided into 3 groups: gum rosin produced from turpentine, extracted rosin gained by extraction from sliced pine stump, floating oil rosin recovered from wastewater of Kraft Pulping. The gum rosin has the best quality and the yield of which in 1997~1999 stabilized in 400,000 tons per year [163, 164].

At present, gum rosin (mainly used in adhesives, inks, paints and food industries, the annual output of China is about 80,000 tons), polymerized rosin (used for the ink, adhesives, paint and so on, the annual output of China is only about 3,000 tons), light-colored rosin, paper sizing agent, edible rosin, terpene resin (mainly used for adhesives, coatings and so on), synthetic fragrance (to synthesize terpineol, linalool, rose enol, sandalwood, neroli alcohol, etc.), synthetic drugs (such as camphor, synthetic borneol, synthetic terpineol, synthetic fiber, etc.) all have shown good market prospects. In particular, rosin resin and terpene resin in the future may become the leading products of rosin, turpentine deep processing [165]. With the increasing market demand, enhancement of business strength and the improvement of technological level, there have great developing space for the deep processing products of rosin and turpentine in China.

*Preparation Process of Rosin*

Main Process of the Resin Processing

Main processes of resin processing include: resin melting, refining and distillation. Firstly, to melt the resin and turn it into liquid by adding turpentine and water, then remove mostly impurities by filtration, wash away the dark soluble matters and remove iron compounds by adding depigmenting agents such as oxalic acid. Then, wash the liquid resin again and then do the purification process to further remove impurities and mostly water, the yield liquid turpentine is distillated in the intermittent distillation tower or continuous distillation tower with over-heated steam, and the residues are rosin [166].

Modified Rosin and Deep-Processed Products

Rosin has been widely used in various industrial sectors with its perfect features such as anti-corrosion, insulation, bonding, etc. However, rosin itself has some shortcomings, such as the incidental crystallization of raw rosin which forbids its utilization in paint and insulated materials; the susceptible oxidation in the air which causes darker color and decrease the solubility of solvent; the high acid value which make it easy to react with the heavy metal salts in the paint and the coated metal surface. In order to overcome all these abovementioned shortcomings, researchers have carried out a number of modifications and deep processing for rosin [167, 168].

There are two chemical reactive centers in the resin acid molecules: double bond and carboxyl.

1) Double bond involved modified reactions: abietic-type acids in the case of heat or affected by inorganic acid, organic acids are susceptible to isomeric reaction to form L-shaped pimaric acid structure with conjugated double bonds and achieve a dynamic balance. These conjugated double bonds under certain conditions can generate a variety of rosin adducts, which can develop a series of modified rosin resin. ① Rosin be hydrogenated, dehydrogenated, or disproportionated: after hydrogenation, dehydrogenation or disproportionation, the aromatic ring can become more stable, so as to achieve the result of

decreasing the sensitivity of oxidation in the air with the elimination of conjugated double bonds. With the improvement of physical properties, the use of rosin can be widely expanded. ② Polymerized rosin: there are conjugated double bonds in the abietic-type acids molecules, and in suitable reactive conditions, they can aggregate with each other to generate a light yellow thermoplastic resin – polymerized rosin, which has a strong practical value. ③ Peroxides and epoxides of rosin acid: in order to improve the stability to oxygen, increase the softening point, and maintain the stability of color, it is an effective way to turn the rosin acid into their peroxides and epoxides. Meanwhile, it is an effective way to prepare chemical active rosin derivatives.

2) Carboxyl reactions in rosin: ① Resinate: Resinate can be formed once the hydrogen atom on the carboxyl of resin acid is replaced by the metal atom. The current application of resinate includes the alkali metal resinate, heavy metal resinate. ② Resin acid ester: The high acid value and low thermal stability of rosin can be improved after esterification. ③ Monohydric alcohol esters: Monohydric alcohol esters include ethyl and methyl ester, non-drying liquid, which can not only be used for synthetic rubber resin solvent, but also be used as the plasticizer for a variety of resins. ④ Trihydric alcohol esters: Glycerol rosin ester, prepared by the rosin and glycerine, is the most widely used rosin ester, which have been long used in the coatings industry. ⑤Tetrahydroxy alcohol esters: Take pentaerythritol as the representative example, its appearance is similar to rosin with more surface gloss, larger molecular weight, low acid value, high hardness degree and high melting point, which entitle it wide application in adhesives and inks industries.

Industrial Development Prospects of Rosin

Chinese rosin industry is an important part of the world rosin industry, but there are still some shortcomings such as the low value-addition of products, single product line and so on. As a result, in the future development of the rosin industry, researches should concentrated on developing a variety of high value added products according to different components of rosin and finally to achieve the total utilization of rosin resources [168].

#### *5.3.4.2. Chinese Lacquer*

Chinese lacquer (CL), also known as natural lacquer, crude lacquer, raw lacquer, and national lacquer. CL is prepared from the sticky white latex by cutting the sumach bark, it is an ancient natural resin coating. As early as the Shang Dynasty (about 17 BC to the beginning of the 11 BC), China had begun to prepare elaborate lacquer paint using lacquer. Sumac mainly grow in the eastern and central Asia, beside China, Vietnam, Korea, Japan, Myanmar and other countries also have Sumac. Sumac is widely distributed in China from Gansu Province in the west to Shandong Province in the south, where have the humid climate and fitting temperature for the growth of sumac. The major producing areas of sumac are Hunan, Hubei, Guangdong, Guangxi, Sichuan, Yunnan, Guizhou, Shaanxi, and Henan provinces. The lacquer appears as sticky white latex when cutting from the sumac bark, the color of which can soon turn into gold, red, purple, and finally dark brown. After a long time, the sticky latex can turn into solid state. Fresh lacquer has tiny acid fragrance, however, storing for too long time it would become corrupted smelly. Processed CL refers to the cooked lacquer through sunshine, stirring, oxidation reaction after mixed with tung oil.

*Characteristics and Composition of the CL*

CL can directly be used to paint objects by forming hard film which is solid, water-resistant, oil-resistant, anti-organic solvents, insulative, corrosion-resistant and lustrous. However, in the coatings industry, lacquer is seldom directly used because the slow solidified speed, causing skin allergies before the solidification, deep color, crisp, poor sunlight-resistant of lacquer film, as well as the poor antioxidants and strong caustic alkali resistance.

CL is a natural, stable oil-water colloidal dispersion system with complex components which is varied from the origin. But in general CL is constituted by the urushiol, laccase, gumminess and water. Its interfacial film is similar to the structure of biological membrane, which is the enzymatic reaction place of lacquer. The membrane is mainly constituted of amphoteric glycoprotein, in which the continuous phase is urushiol and the discrete phase is constituted of amphoteric blue glycoprotein, laccase, hydrophilic lacquer polysaccharides, water and ash [169].

1) Urushiol: Urushiol is yellow viscous liquid, soluble in a variety of organic solvents and does not dissolve in water, the specific gravity is 0.9689. Urushiol is the main component of lacquer, accounting for about 50-80% of the total weight. According to the structure analysis and urushiol homologues, it is proved that urushiol is a catechol substitute with a long side-chain, and the side-chain position, number of carbon atom, unsaturation degree are different from tree species. Urushiol refers to general designation of these mixtures of the different branch chains. Urushiol is very active and prone to chemical reaction, it can polymerize under heating and oxidation, which result in the difficulty of using ordinary physical methods to separate them, neither crystallization nor distillation.

2) Laccase: laccase is a kind of Cu-contained polyphenol oxidase exists in lacquer, and is also the important part of the lacquer because it can catalyze the solidification of lacquer at room temperature. Laccase presents in the nitrogen-containing parts of lacquer, it can not be dissolved in neither water nor organic solvent, but can dissolve in urushiol. However, laccase can not dissolve any more after being separated from lacquer. The content and activity of laccase directly affects on the desiccation capacity of lacquer and quality of lacquer films.

3) Water: Water is major component of the lacquer accounting for 15-35% of the total weight. Usually higher content of water destroys the lacquer quality. Water is a necessary component in the desiccation process of lacquer, however, the water content should be around a suitable range of 4-6%. If the water content is too high, the urushiol content might be relatively low which results in the poor quality of lacquer film, adhesive performance and easy to corrosion. Drying process requires the environment to maintain relative humidity at 80%, too low humidity will cause a long period of drying or even can not dry, while too high humidity will make the film easily wrinkled and without luster.

4) Gumminess: gumminess is a kind of automatic excreted colloidal liquid when the bark is hurt, and it is the main component in the lacquer which is insoluble in organic solvents but soluble in water. Gumminess is also known as polysaccharides with similar capacities to Arab glue, such as in appearance of yellowish- white solid, fragrant. It accounts for 3.5% of the total weight of lacquer. The components of gumminess are mainly polysaccharide, including d-galactose, 1-arabinose, d-xylose, l-rhamnose, and the hydrolyzed products, such as d-galacturonic acid and d-guluronic acid. Gumminess is not the necessary ingredient for desiccation of lacquer, however, it is the fine natural emulsifier and stabilizer which plays a vital role in the formation of stable and uniform emulsion of large components of lacquer.

Meanwhile, Gumminess has important effect on the desiccating velocity and the performance of lacquer film.

*Preparation Process of CL*

Raw lacquer materials generally are not directly used but painted after been refined. Since that there are always 20-40% of water and dust in the collected raw lacquer, they can meet the various painting requirements after necessary treatments. Modern methods of lacquer refining are more or less the same all over the country. The three main products are raw lacquer, black lacquer and transparent lacquer. The latter two are more difficult in refining technologies. Before the film formation, the raw lacquer is transformed into polymers or net-like structure high-molecular compounds in the pre-oxidized polymerization step through the catalysis function of laccase. And then gradually removing water and adding the colorant, drier and other resins to enhance the transparency and black degree. A section of the refining process includes filtration, cool, sun curing, fine filter, and barreling.

1) Filtration: The purpose of filtration is to remove all the mechanical impurities in the raw lacquer. The classical method of filtration is firstly filtrates with flax roughly, and then filtrates with slender white cloth exactly, finally refined and filtrated with fine white cotton. Centrifuges can also be used to filter raw lacquer. The quality of the fine lacquer has much to do with the fineness and roughness of filtrated lacquer, therefore, the final refined lacquer must be filtrated with cotton.

2) Cooling and stirring: Cooling and stirring refer that put the filtrated raw lacquer on a plate or a basin or in the lacquer machine, then stirring constantly after adding a certain amount of water so that the urushiol can be oxidized into urushiol quinone under the catalytic role of laccase and further polymerized into urushiol dimers and even urushiol polymers. This step has great effect on the physical and chemical properties of the film's surface, and the detailed parameters should be optimized by considering multiple conditions, such as the duration of cooling and stirring, the nature of the raw materials, and requirements of the refining lacquer and so on.

3) Adding colorant etc: To meet the needs of some certain products, it is necessary to add some colorant or transparent agent in the refining process. For example, to achieve the requirement of such "right black", it is often using 3~5% iron hydroxide; in order to increase the transparent of lacquer, it is usually adding yellow gardenia juice, gamboges 5~10%, turpentine 15%, etc.

4) Sun-curing and dehydration: Sun-curing is also known as transparence or dehydration, which is based on the cooling process of lacquer. Sun curing requires continual stirring and the duration depends on the weather and temperature. However, the duration of sun curing can not be too long, otherwise the liquid lacquer polymerize into polymer which is hardened into a useless jelly. The best sun curing end point is that the exterior and interior lacquer color is uniform and the bubbles are clear and light.

*Prospect of CL Resource*

CL is a kind of natural composite polymer with complex structure [43]. Through modern biotechnology a large number of biologically active substances can be extracted from lacquer which can either be used in the manufacture of medicinal products or fine chemicals. For example, the polysaccharide is the main component of the lacquer, which has special immune function of promoting the growth of leukocyte. With the special molecular structure of urushiol, it can be used as the fine raw material to prepare functional polymer materials. Laccase plays a role in the protection of plants. When the tissues of plants are hurt, laccase can act as the catalyst to catalyze the oxidation of a variety of substances, occlude the wound and suppress or kill the pathogenic bacteria. In addition, laccase is responsible for the electron transfer of continuous reactions, the establishment of a continuously reactive chain, and the oxidized polymerization of the substrates.

The in-depth research and development of high-value added products, aiming at different components of the lacquer, will provide us brand-new ideas of total utilization of lacquer, widen the existing species and uses of lacquer based products, and arouse an great opportunity of new development for the old industry.

## REFERENCES


[1] R, B. H., Biomass Refining. *Science* 1982, 218, (4573), 643.

[2] Huang, y.; Gao, z.; Huang, h.; Wei, p., Biorefining——new industrial mode of realizing sustainable development. *Chinese Journal of Bioprocess Engineering* 2006, 4, (3), 1-8.

[3] Zeng, A. P.; H, B., Bulk chemicals from biotechnology: the case of 1, 3-propanediol production and the new trends. *Advance in Biochemical Engineering Biotechnology* 2002, 74, 239-259.

[4] Song, X.; Wen, Q.; Guo, H., Cellulose based functional materials. *Chinese Polymer Bulletin* 2002, (4), 47-52.

[5] Chen, H., *Cellulose biotechnology*. chemical Industry Press: Beijing, 2005.

[6] Bao, H.; Wang, J.; Huang, X.; Shen, Q., Cellulose and hemi-cellulose based advance materials. *Journal of Cellulose Science and Technology* 2007, (4), 66-76.

[7] Zhang, J.; Cao, J.; Wang, S., Modification research on acetate cellulose. *Journal of Dong Hua University* 2004, 30, (6), 120-124.

[8] Mei, J.; Chen, J., The present and future trend of acetate cellulose. *Journal of Cellulose Science and Technology* 1999, 7, (4), 56-62.

[9] A, I.; L, Y., The effect of different molecular weight PEG additives on cellulose acetate asymmetric dialysis membrane performance. *Journal of Membrane Science* 2006, 280, (1-2), 920-927.

[10] Liu, L. The fractionation of straw and its high value conversion [doctoral dissertation] Chinese Academy of Sciences Beijing, 2006.

[11] B, I.; S, S., Development and Performance of Some Porous Cellulose Acetate Membranes for Reverse Osmosis Desalination. *Journal of Applied Polymer Science* 1970, 14, 2559-2568.

[12] T, H.; Y, M., Reuse of carwash water with a cellulose acetate ultrafiltration membrane aided by flocculation and activated carbon treatments. *Desalination* 2004, 169, (3), 257-267.

[13] Qin, J. J.; Li, Y.; Lee, L. S.; Lee, H., Cellulose acetate hollow fiber ultrafiltration membranes made from CA/PVP 360 K/NMP/water. *Journal of Membrane Science* 2003, 218, (1), 173-183.

[14] G, A.; K, S.; R, M.; D, M.; M, R.; V, M., Studies on cellulose acetate and sulfonated poly (ether ether ketone) blend ultrafiltration membranes. *European Polymer Journal* 2004, 40, (4), 751-762.
[15] R, H.; E, F.; S, R. M.; A, D., Properties of cellulose acetate nanofiltration membranes application to brackish water desalination. *Desalination* 2004, 167, (1-3), 403-409.
[16] K, A., Dielectric properties of cellulose acetate reverse osmosis membranes in aqueous salt solutions. *Journal of Membrane Science* 1990, 50, (1), 71-84.
[17] P, D. A.; T, C. M.; C, B. J., Cellulose acetate reverse osmosis membranes: optimization of the composition. *Journal of Applied Polymer Science* 2006, 100, 4052-4058.
[18] A, B.; L, G. J.; E, G.; G, L.; M, T., Ammonium removal from aqueous solutions by reverse osmosis using cellulose acetate membranes. *Desalination* 2005, 184, (1-3), 149-155.
[19] Ma, X.; Shen, L.; Yang, Z.; Cao, J., Study on synthesis of cellulose acetate used in cigarette with domestic wood pulp. *Synthetic Fiber in China* 2004, 33, (5), 10-12.
[20] Yin, H.; Liu, Y., Applications of activated carbon cellulose in cigarette filter. *Chemical Engineering* 2000, (6), 66-67.
[21] Wang, C.; Gao, W.; Wang, H., The filtration mechanism and improvement of filtration of cigarette filters with cellulose acetate. *Artificial Fibre* 2003, 33, (4), 23-25.
[22] Yang, Z., Exploration of curled technology of smoke used cellulose acetate. *Tobacco Science & Technology* 2001, (9), 9-12.
[23] Xu, D., *Cellulose derivatives.* . Chemical Industry Press: Beijing, 2001; p 47-57.
[24] Wu, Y.; Li, W.; Shen, X., *Organic chemistry*. University of Science and Technology of China Press: Hefei, 2002.
[25] Chen, J.; Zhou, Z., low-pollution technology research on cotton linter pulp. *China Pulp & Paper* 1999, 18, (3), 41-46.
[26] Wei, G., The status quo and countermeasures of China's paper industry pollution prevention and control. *World Pulp and Paper* 2000, 19, (1), 44-46.
[27] Wu, J., Biotechnology of pulp and paper wastewater. *Paper Science & Technology* 2002, 21, (1), 34-38.
[28] L, N. H.; I, R. D.; R, S. Acetylation of Wood-cellulose materials. 94113778.3, 1995.
[29] Mei, J.; Ou, Y., Studies on polymerization and the properties of cellulose acetate at high temperature. *Journal of Cellulose Science and Technology* 2000, 8, (1), 9-16.
[30] Gao, J.; Tang, L., *Cellulose scienc*. Science Press: Beijing, 1996.
[31] R, W.; M, R.; D, G.; J, S., Process design and costing of bioethanol technology: A tool for determining the status and direction of research and development. *Biotechnology Progress* 1999, 15, (5), 794-803.
[32] L, M. C.; A, C. P., Derivatization of cellulose in lithium chloride and N-N-dimethylacetamide solutions. *Polymer(Guildford)* 1987, 28, (13), 2317-2323.
[33] Wu, J.; Zhang, H.; Zhang, J.; He, J., Homogeneous acetylation and regioselectivity of cellulose in a new Ionic liquid *Chemical Journal of Chinese Universities* 2006, 27, (3), 592-594.
[34] Yan, L.; Li, W.; Liu, S.; Qi, Z.; Pang, H., Preparation of acetate cellulose using solid superacid SO4^2-/Fe2O3 as catalyst. *Journal of University of Science and Technology of China* 1994, 36, (4), 422-425.
[35] Yu, q.; Li, G., Preparation of acetylated wood and its thermonplastics research. *Chemistry and Industry of Forest Products* 1994, 14, (2), 33-38.

[36] Zhu, J., Studies and development of acetylated wood. *Forest Science* 1994, 30, (3), 272-279.
[37] Lang, S. C.; Matsumura, H. A method of preparing cellulose acetate. 93106682.4, 1994.
[38] Sun, X. F.; Sun, R.; Sun, J. X., Acetylation of Rice Straw with or without Catalysts and Its Characterization as a Natural Sorbent in Oil Spill Cleanup. *Journal of Agricultural and Food Chemistry* 2002, 50, (22), 6428-6433.
[39] M, S.; T, T. Process for manufacturing cellulose acetate. 6984731, 2006.
[40] A, B.; B, S.; J, L.; R, S.; J, W., Process for obtaining cellulose acetate from agricultural by-products. *Carbohydrate Polymer* 2006, 64, 134-137.
[41] Zhang, J.; Chen, H., Preparation of cellulose acetate from crop straw. *Chinese Journal of Chemical Engineering* 2007, 58, (10), 2548-2553.
[42] E, J. 332203, 1918.
[43] Yang, Z.; Su, M.; Gao, X., *Foundation and application of cellulose aether*. South China University of Technology Press: Guangzhou, 1990; p 6.
[44] Shao, z., *cellulose ester*. Chemical Industry PresS: Beijing, 2007; p 9.
[45] Yin, X. Production methods of diatomite slow-release fertilizer and soil amendments. 1534006, 2004.
[46] Wang, Z.; Bao, Y.; Jin, Z.; Zhao, Y.; Xin, Y., Analysis of the effect of. biological active water retention agent on the improvement of simulate deserted land. *Journal of Northeast Forestry University* 2003, 31, (3), 27-28.
[47] Wang, z.; Bao, Y.; Feng, X. Preparation method of biological active water retention agent used in deserted land. 1498941, 2004.
[48] Mou, Q.; Mou, L. A preparation method of fertilizer used in deserted land. 1257855, 2000.
[49] Zhou, D.; Mei, C., Crop straw based materials in the 21st century. *Journal of Nanjing Forestry University(Natural Science Edition)* 2000, 24, (5), 1-4.
[50] P, S.; F, A.; H, C.; A, N.; G, S., Cement-bonded straw board subjected to accelerated processing. *Cement and Concrete Composites* 2004, 26, (7), 797-802.
[51] Lin, L. Z.; Yao, Y. G.; N, S., Liquefaction mechanism of beta-O-4 lignin model compound in the presence of phenol under acid catalysis Part 1. Identification of the reaction products. *Holzforschung* 2001, 55, (6), 617-624.
[52] company, D. d. g. m. x., Modern wood composite manufacturing techniques. *World Plastics* 2007, 25, (12), 67-68.
[53] Zhang, G., Researches on the biomass fiber / plastic composite porous foam Technology. *China Wood-Based Panels* 2006, 12, 8-11.
[54] M, K.; E, J.; V, T., Liquefaction of lignocellulosic biomass and its potential applications. *Journal of Biotechnology* 2007, 131, (2s), 23.
[55] Tong, C.; Pan, X., Liquefaction of lignocellulose raw materials. *China Forest Products Industry* 2000, 27, (6), 3-5.
[56] Ma, T.; Zheng, Z.; Zhang, H.; Lin, M., Mechanism of phenol in acid lignocellulosic liquefication. *Journal of Southwest Forestry College* 2006, 26, (2), 93-96.
[57] Xie, S. Study on selective liquefaction with phenol for fraction separation of straw. institute of Process Engineering , Chinese Academy of Sciences, Beijing, 2008.
[58] Lin, L. Z.; Yao, Y. G.; M, Y.; N, S., Liquefaction mechanism of cellulose in the presence of phenol under acid catalysis. *Carbohydrate Polymers* 2004, 57, (2), 123-129.

[59] Chen, H., *Theories and applications on the ecological value-increment of straw resource*. Chemical Industry Press: Beijing, 2006.

[60] Zhang, Q.; Zhao, G., Wood liquefaction in phenol and polyhydric alcohol. *Journal of Beijing Forestry University* 2003, 25, (6), 71-76.

[61] Z, L. L.; S, N.; G, Y. Y.; M, Y.; N, S., Liquefaction mechanism of beta-O-4 lignin model compound in the presence of phenol under acid catalysis Part 2. Reaction behavior and pathways. *Holzforschung* 2001, 55, (6), 625-630.

[62] Lin, L. Z.; Yao, Y. G.; N, S., Liquefaction mechanism of lignin in the presence of phenol at elevated temperature without catalysts - Studies on beta-O-4 lignin model compound .3. Multi-condensation. *Holzforschung* 1997, 51, (4), 333-337.

[63] Lin, L. Z.; Yao, Y. G.; M, Y.; N, S., Liquefaction mechanism of lignin in the presence of phenol at elevated temperature without catalysts - Studies on beta-O-4 model compound .1. Structural characterization of the reaction products. *Holzforschung* 1997, 51, (4), 316-324.

[64] Lin, L. Z.; S, N.; Yao, Y. G.; M, Y.; N, S., Liquefaction mechanism of beta-O-4 lignin model compound in the presence of phenol under acid catalysis Part 2. Reaction behavior and pathways. *Holzforschung* 2001, 55, (6), 625-630.

[65] Zheng, Z.; Zhang, H.; Gu, J., Research development of biomass liquefaction. *Yunan Chemical Technology* 2004, 31, (5), 27-30.

[66] F, O.; T, Y.; N, H.; A, T.; A, I., Various utilization of liquefied wood products. in：H B Liu, H Y Zhan,Y I Xie. *Emerging Technologies of Pulping & Papermaking* 2002, 74-81.

[67] He, F.; Wang, M., *Carbon fibre and its composite materials*. Science Press: Beijing, 1995.

[68] Ma, X.; Zhao, G., A review on wood-derived rayon carbon fibers. *China Wood Industry* 2006, 20, (6), 1-4.

[69] Wu, Q.; Pan, D., A New Cellulose Based Carbon Fiber from a Lyocell Precursor. *Textile Research Journal* 2002, 72, (5), 405.

[70] K, S.; K, S.; H, O.; M, I.; T, F.; S, H., Integrated process for total utilization of wood components by steam-explosion pretreatmen. *Biomass and Bioenergy* 1998, 14, (3), 195-203.

[71] F, K. J.; S, K.; A, V. R.; D, G. R.; L, C. A.; W, G., Lignin-based carbon fibers for composite fiber applications. *Carbon* 2002, 40, (15), 2913-2920.

[72] K, O.; T, Y.; N, S.; A, O., Preparation of thin carbon fibers from waste wood-derived phenolic resin. *Journal of Materials Science* 2005, 40, (14), 3847-3848.

[73] F, H. L. Process for hydrating olefinic aid&lyde. 243110, 1948.

[74] L, M. M. A. S. B. Process for making 1, 3-diols from epoxides. 4935554, 1990.

[75] N, A. D. W. Method of preparing 1,3-propanediol. 5015789, 1991.

[76] J, T. H. B.; Sauer, J. Two-stage process for the production of 1,3-propanediol by catalytic hydrogenation of 3-hydroxypropanal. 6297408, 2001.

[77] H, S. L.; R, W. P.; J, B. P. Process for preparing alkanediols. 5841003, 1998.

[78] F, K. J.; G, J. T. One-step production of 1,3-propanediol from ethylene oxide and syngas with a catalyst with a N heterocyclic ligand. 2003204118, 2003.

[79] A, M. M. Process for making 1,3-propanediol from epoxides. 4873379, 1989.

[80] A, M. M. Process for making 1, 3-diols from epoxides. 4873378, 1989.

[81] R, B. J.; M, M. J. Catalysts for producing I, 3-diols and/or 3-hydmxyaldehydes, and processes for making and using same. 5225387, 1993.

[82] M, B. J. R. M. J. Catalysts for producing l, 3-diols and／or 3-hydroxyaldehydes, and processes for making and using same. 5030766, 1991.

[83] Feng, K.; Lv, Z.; Li, Q., A review of 1,3-propanediol preparation process technology. *Synthetic Fiber in Dustry* 2007, 30, (4), 46-49.

[84] D, A.; T, H. Process for the preparation of 1,3-propanediol by the hydrogenation of hydroxypropionaldehyde. 5364984, 1994.

[85] D, B. C. A. Process for preparing 3-hydroxyalkanals. 5962745, 1999.

[86] N, A. D. W. Method for the production of 1,3-propanediol. 5171898, 1992.

[87] A, E. H. S. K. T. Hydration of acrolein to hydracrylaldehyde. 3536763, 1970.

[88] D, U. J.; L, N.; D, A. R. Method for the manufacture of l, 3-pmpanediol. 524713, 1993.

[89] Bai, x.; Xue, L.; Han, D., Catalytic hydration of acrolein on acidic ion-exchange resin. *Petrochemical Technology* 2003, 32, (6), 458-461.

[90] D, B. C. A. Process for the production of 1,3-propanediol. 6140543, 2000.

[91] H, Y. H. S.; T, Y. Process of producing hydroxyalkanal. 639920, 1996.

[92] H, Y. H. S.; T, Y. Process of producing hydroxyalkanal. 5811591, 1997.

[93] D, H. T. A. Process for the production of 3- hydroxyalkanal. 5276201, 1994.

[94] N, K. T. K. K. Process for producing l, 3-propanediol. 2004097764, 2004.

[95] D, U.; A, R. D. Method for the manufacture of 1, 3-propanediol. 5093537, 1993.

[96] D, A.; T, H. Process for the preparation of 1,3-propanediol by the hydrogenation of hydroxypropionaldehyde. 5364984, 1993.

[97] N, H. T. W.; D, A. Process for the Production of l,3-propanediol. 5334778, 1994.

[98] R, V.; T, H. Two-stage process for the production of 1,3-propanediol by catalytic hydrogenation of 3-hydroxypropionaldehyde. 0014041, 2000.

[99] Yang, J.; Wang, X. A method of 1,3-propanediol preparation. 1475472A, 2004.

[100] H, B.; K, M.; P, Z. A.; D, D. W., Microbial production of 1, 3-propanediol *Applied Microbiology and Biotechnology* 1999, 52, (3), 289-297.

[101] Liu, D.; Chen, K., A review of microbial fermentation of 1,3-propanediol preparation. *Synthetic Fiber in China* 2005, 34, (9), 11-15.

[102] K, M.; P, Z. A.; D, D. W., High concentration and productivity of 1, 3-propanediol from continuous fermentation of glycerol by Klebsiella pneumoniae. *Enzyme and Microbial Technology* 1997, 20, (2), 82-86.

[103] Wang, J.; Liu, H., Recent research advances in bioconversion to 1,3-pmpanediol. *Chemistry Bulletin* 2001, 64, (10), 621-625.

[104] L, H. S.; Wagner Process for making 1,3-propanediol from carbohydrates using mixed microbial cultures 5599689, 1997.

[105] Chen, G.; Zhao, Y.; Yao, S., Research on the 1, 3 - propanediol fermentation with mixed substrate. *Journal of Sichuan University(Engineering Science Edition)* 2007, 5.

[106] Li, F. f.; Liu, D., Flocculation of Fermented Broth with 1,3 –Propanediol. *Microbiology* 2004, 31, (3), 30-35.

[107] Xiu, Z.; Liu, H.; Zhang, G. A method of producing bio-1,3- propanediol in two steps. 1165623C, 2004.

[108] Liu, D.; Chen, K.; Liu, H.; Lin, R. A method of producing 1,3-propanediol and 2,3-butanediolwith raw starch substrates. 1710086A, 2005.

[109] Wang, L.; Gong, Y.; Liu, D. Desalination process of 1,3- propanediol broth based on electrodialysis. 1522997A, 2004.

[110] Liu, D.; Hao, J.; Li, F. A method of separating and extracting 1,3-propanediol and other by-products from fermentation broth. 1304345C, 2007.

[111] N, L. L.; Vasantha Bioconversion of a fermentable carbon source to l, 3-propanediol by a single microorganism. 5686276, 1997.

[112] C, C. D.; E, A. N.; M, H.; J, S. A., Metabolic engineering of propanediol pathways. *Biotechnology Progress* 1998, 14, (1), 116-125.

[113] Rao, Z.; Zhu, G. A construction method of Saccharomyces cerevisiae producing 1,3-propanediol with glucose substrate. 101130782A, 2008.

[114] Ma, Z.; Rao, Z.; Shen, W., Construction of recombinant Saccharomyces cerevisiae producing 1,3-propanediol by one step method. *Acta Microbiologica Sinica* 2007, 5.

[115] Norbeck, J.; Pahlman, A. K.; Akhtar, N.; Blomberg, A.; Adler, L., Purification and characterization of two isoenzymes of DL-glycerol-3-phosphatase from Saccharomyces cerevisiae. *Journal of Biology Chemistry* 1996, 271, (23), 13875-13881.

[116] E, N. C.; Gatenby Method for the production of 1 ,3-propanediol by recombinant microorganisms. 6013494, 2000.

[117] E, N. C.; M, W. G., Metabolic engineering for the microbial production of 1, 3-propanediol. *Current Opinion in Biotechnology* 2003, 14, (5), 454-459.

[118] U, G. T.; C, S. S., How green are green plastics? *Scientific American* 2000, 283, (2), 37-41.

[119] Ye, l., Research advances in producing PHB, a biodegradable plastic by biotechnology. *Biotechnology Information* 1998, (3), 9-16.

[120] A, K. A.; M, S. K.; N, M. S.; T, C., Biotechnological conversion of agro-industrial wastewaters into biodegradable plastic, poly β-hydroxybutyrate. *Bioresource Technology* 2007, 98, (18), 3579-3584.

[121] Preffer, Recycling, solid waste manage. *Engineering* 1992, 72-84.

[122] H, S.; M, V. L. M. C., Production of polyhydroxyalkanoates by mixed culture: recent trends and biotechnological importance. *Biotechnology Advances* 2004, 22, (3), 261-279.

[123] A, S.; G, S. H., Physiology and molecular genetics of poly(beta-hydroxy-alkanoic acid) synthesis in Alcaligenes eutrophus. *Molecular Microbiology* 1991, 5, 535-542.

[124] Yu, D., Overview of microbial Production of poly-β-hydroxybutyrate. *Da Hua Technology* 2002, (3), 1-2.

[125] Yang, M.; li, H., Study of β-hydroxybutyrate (PHB) fermentation with Alcaligenes eutrophus. *Food and Fermentation Industries* 1992, 25, (3), 1-4.

[126] Yu, P.; Li, J., Fermentation optimization ofβ-hydroxybutyrate (PHB) fermentation with Alcaligenes eutrophus. *Journal of Chinese Institute of Food Science and Technology* 2007, 7, (1), 59-63.

[127] Huang, j.; Qin, H.; Yang, J., Fermentation of β-hydroxybutyrate in circulation air-lift reactor. *Journal of Northwest University (Natural Science Edition)* 2004, 34, (2), 199-202.

[128] Huang, J.; Qin, H.; Wei, Y., Dynamic models for poly-β-hydroxybutyrate fermentation proces. *Chemical Engineering* 2005, 33, (1), 44-47.

[129] Y, P.; C, N.; C, S., Production of polyhydroxyalkanoates, a family of biodegradable plastics and elastomers, in bacteria and plants. *Bio/Technology* 1995, 13, (2), 142-150.
[130] C, N.; Y, P.; C, S., Targeting of the polyhydroxybutyrate biosynthetic pathway to the plastids of Arabidopsis thaliana results in high levels of polymer accumulation. *Proceedings of the National Academy of Sciences of the United States of America* 1994, 91, (26), 12760-12764.
[131] K, B.; I, B.; J, K.; V, M.; C, N.; Y, P.; G, T.; N, T. R.; L, W., Transgenic Arabidopsis plants can accumulate polyhydroxybutyrate to up to 4% of their fresh weight. *Planta* 2000, 211, (6), 841-845.
[132] Y, P.; J, G. K., Production of Polyhydroxyalkanoates in Transgenic Plants. *Biotechnology of Biopolymers: From Synthesis to Patents* 2005.
[133] E, J. M., Cotton crop improvement through genetic engineering. *Critical Reviews in Biotechnology* 1997, 17, (3), 185-208.
[134] P, S.; F, S.; D, S.; D, S., Production of a biodegradable plastic polymer, poly-_-hydroxybutyrate, in transgenic alfalfa. *Crop Science* 2002, 42, 919-927.
[135] M, W.; J, Z.; J, S., Polyhydroxybutyrate synthesis in transgenic flax. *Journal of Biotechnology* 2004, 107, (1), 41-45.
[136] A, L.; C, E.; J, H. H.; C, J.; U, K. H., Polyester synthesis in transplastomic tobacco (Nicotiana tabacum L.): significant contents of polyhydroxybutyrate are associated with growth reduction. *Plant Cell Reports* 2003, 21, (9), 891-899.
[137] D, S. K.; P, P. O., Polyhydroxyalkanoate Polymers and Their Production in Transgenic Plants. *Polyhydroxyalkanoate Polymers and Their Production in Transgenic Plants* 2002, 4, (1), 29-40.
[138] P, P. M.; A, P. L.; K, N. L.; M, B. S., Spatio-temporal characterization of polyhydroxybutyrate accumulation in sugarcane. *Plant Biotechnology Journal* 2007, 5, (1), 173-184.
[139] J, C.; Y, L. S.; K, H., Cloning of the Alcaligenes latus Polyhydroxyalkanoate Biosynthesis Genes and Use of These Genes for Enhanced Production of Poly (3-hydroxybutyrate) in Escherichia coli. *Applied and Environmental Microbiology* 1998, 64, (12), 4897.
[140] Y, L. S., Plastic bacteria? Progress and prospects for polyhydroxyalkanoate production in bacteria. *Trends in Biotechnology* 1996, 14, (11), 431-438.
[141] Wang, F.; Y, L. S., Production of poly (3-hydroxybutyrate) by fed-batch culture of filamentation-suppressed recombinant Escherichia coli. *Applied and Environmental Microbiology* 1997, 63, (12), 4765-4769.
[142] G, B.; B, S.; RM, L., A rapid gas chromatographic method for the determination of poly-b-hydroxybutyric acid in microbial biomass. *Eur J Appl Microbiol Biotechnol* 1978, 6, 29-37.
[143] M, S. K.; N, M. S.; K, R. K. M.; T, C., Production of biodegradable plastics from activated sludge generated from a food processing industrial wastewater treatment plant. *Bioresource Technology* 2004, 95, (3), 327-330.
[144] Garg S K J, A., Fermentative production of 2, 3-butanediol: a review. *Bioresource Technology* 1995, 51, (2-3), 103-109.
[145] F, J. N. B.; Tsao, M. C.; T, G., Production of 2, 3-butanediol from D-xylose by Klebsiella oxytoca ATCC 8724. *Biotechnology and Bioengineering* 1984, 26, (4), 362-369.

[146] N, Z. J.; J. Olsson, L., Fuel ethanol production from lignocellulose: a challenge for metabolic engineering and process integration. *Applied Microbiology and Biotechnology* 2001, 56, (1), 17-34.

[147] L, Y. E. K. C.; Saddler, N.; N, J., Production of 2, 3-butanediol by Klebsiella pneumoniae grown on acid hydrolyzed wood hemicellulose. *Biotechnology Letters* 1982, 4, (11), 741-746.

[148] Y, S. J. N.; Mes-Hartree, E. K. C.; M.Levitin; Brownell, N.; H., H., Utilization of Enzymatically Hydrolyzed Wood Hemicelluloses by Microorganisms for Production of Liquid Fuels. *Applied and Environmental Microbiology* 1983, 45, (1), 153-160.

[149] D, Y. E. K.; L.Louis-Seize; G.Saddler; N., J., Butanediol production from cellulose and hemicellulose by Klebsiella pneumoniae grown in sequential coculture with Trichoderma harzianum. *Applied and Environmental Microbiology* 1985, 50, (4), 924-929.

[150] X, C. N.; Gong, Y. K.; S.Tsao, C.; T, G., Production of 2, 3-butanediol from pretreated corn cob by Klebsiella oxytoca in the presence of fungal cellulase. *Applied. Biochemistry. Biotechnology* 1997, 63, 129-139.

[151] Chen, H.; Sun, F. A method of producing 2,3-butanediol based on directly enzymatic straws fermentation. 101153291, 2006.

[152] Sun, F.; Chen, H., Evaluation of enzymatic hydrolysis of wheat straw pretreated by atmospheric glycerol autocatalysis. *Journal of Chemical Technology & Biotechnology* 2007, 82, (11), 1039-1044.

[153] Sun, F.; Chen, H., Enhanced enzymatic hydrolysis of wheat straw by aqueous glycerol pretreatment. *Bioresource Technology* 2008, 99, (14), 6156-6161.

[154] Wang, Y.; Wang, N.; Zhao, J.; Guo, Z., Starch-based Environmentally Biodegradable Polymer Materials. *Materials Review* 2005, 19, (4), 12-15.

[155] Qiu, W.; Qiu, X.; Wang, F., *Starch plastic——degradable starch research and application*. Chemical Industry Press: Beijng, 2002.

[156] Zhang, K.; Fu, X., Research development on starch blend material. *Chemical Engineering* 2006, (5), 22-24.

[157] Wang, Z.; Fu, X., Research progress of the application of starch in polymer material. *Chemical Engineering* 2006, (3), 29-32.

[158] J, G. G. 1489050.

[159] Gao, J.; Wang, W., Starch based biodegradable plastic materials. *Polymer Materials Science and Engineering* 1998, 14, (4), 16-20.

[160] Zhang, L., *Modified starch*. South China University of Technology Press: Guangzhou, 1999.

[161] Sun, L.; Zhang, H., Scientific research and development progress of starch-based new type environmental protection material. *Chemical Technology Market* 2002, 25, (12), 28-31.

[162] Song, Z., Fine chemicals use of rosin(I)— constitution and properties of resin. *Journal of Chemical Industry of Forest Products(Bimonthly)* 2002, 36, (4), 29-33.

[163] Song, Z., Fine chemicals use of rosin (VI)—— development and trend home and abroad. *Journal of Chemical Industry of Forest Products(Bimonthly)* 2003, 37, (3), 43-45.

[164] Ye, D.; Zhang, Y., The Presence Application and Developing of Resin Industry. *Jiangxi Chemical Industry* 2003, (3), 65-66.

[165] Sun, D., Steam method of melting technology of rosin lipid. *Guangdong Forestry Science and Technology* 1994, (3), 48-50.

[166] Liu, F.; Zhang, Z., Current situation and trends of rosin modification. *Jiangxi Chemical Industry* 2001, (4), 3-6.

[167] Zhang, S., Refining and extraction of rosin. *Guangxi Chemical Industry* 2001, 30, (4), 24-27.

[168] Wang, J.; Nie, Y., Selections of international competitive strategies on chinese gum rosin industry. *Problems of Forestry Economics* 2007, 27, (4), 345-348.

[169] Lian, P., Chemical cmposition and filming mechanism of the oriental lacquer. *Journal of Shanxi Normal University (Natural Science Edition)* 2004, 32, 99-101.

*Chapter 6*

# The Process Engineering in Plant-Based Energy

## 6.1. Introduction

Biomass is the name given to all living matter on the earth. It is a rather simple term for all organic materials synthesized by plants (including algae). Plants capture solar energy to fix carbon by converting $CO_2$ and water to sugars which are stored in three types of polymers: starches, cellulose and hemicellulose. Biomass is typically composed of 75-90% (wt) sugar polymers, and the other 10-25 % (wt) being composed of mainly lignin. Noting that biomass is a term used to describe any material of biological origin, including plant materials such as trees, grasses, agricultural crops, and so on. Other biomass components, which are generally presented in minor amount, include triglycerides, sterols, alkaloids, resins and waxes. The energy in biomass from plant matter originally generates from solar energy through the process of photosynthesis. The energy, which is stored in plants and animals, or in the wastes that they produce, is called biomass energy [1].

Biomass energy is one of the earliest sources of energy utilized by human. It is used to meet a variety of energy needs, including heating home, fueling vehicles, generating electricity, even providing chemical products by thermochemical technology and biotechnology. Biomass energy is considered as petroleum substitute for several reasons. The main reason is that biomass energy can contribute to sustainable development. Biomass energy is increasingly arousing great interests from an energy security perspective. Resources are often locally available and feasibly convert into secondary energy carriers without high capital investments. Moreover, biomass energy can play an important role in reducing greenhouse gas (GHG) emission, since that the biomass can not only fix the carbon but also can be utilized to produce products in a sustainable way to offsets the GHG emissions produced by fossil fuel.

The increasing energy demand will speed up the exhaustion of the finite fossil fuel. In China, it is claimed that imported oil amounted to 31% to meet energy demand in 2000 and the demand would reach 45-55% in 2010. In recent years, China has paid more and more attention on developing its energy industry to meet the increasing energy demand for the growth of national economy and the improvement of living standard. The Law of Renewable Energy published in China has put in practice in January 1, 2006, and the Chinese government issued that biofuels from renewable sources would account for 5% of primary energy by 2010 and 10% by 2020 in legal form. It also provides for the appropriate fiscal and taxation policies to support bio-energy subsidies, such as elastic loss subsidy, raw material subsidies, demonstration project allowance, tax exemption.

The exploitation and utilization of biomass energy have attracted much interest from the governments around the world. The "Aggressive Biofuel Growth" scenario aims to replace 10% of gasoline production with biofuels by 2010, 15% by 2015, and 20% by 2020 throughout most of the world, except for adjustments in line with other projections for Brazil, the European Union, and the United States. Brazil and the United States are the largest producers of ethanol for transportation, accounting for about 90% of the world's production. Ethanol has displaced 40% of gasoline usage in Brazil but only 3% in the United States. The primary feedstock for ethanol is sugarcane in Brazil and maize in the United States. The European Union, especially France and Germany, is the largest producer of bio-diesel, accounting for 88% of the world's production, followed by the United States, which accounts for 8%. Globally, biodiesel production is only about one-tenth of total ethanol production. Rapeseed is the primary feedstock for bio-diesel in the European Union. Other than Brazil, some developing countries have sizable biofuels programs at present such as Colombia, India, and Thailand, but many others are interested in initiating (or have initiated) small pilot programs [2].

As different domestic conditions compared with these countries above, the use of food or oil crops for bio-energy production is inapplicable to the special national situation of China, which is an agricultural country with large population and less land. To avoid the occurrence of "the struggle with food and land" during bio-energy industrial development, the Chinese government issued the discussion on promoting the healthy development guiding opinion of maize deep-processing industry in 2007, clearly stated that lignocellulose (particularly agricultural residues) resources substitute the food or starch as the raw materials for bio-energy feedstock.

This chapter mainly introduces the process engineering for lignocellulose to bio-fuel.

## 6.2. Plant-Based Fuel Pattern and Conversion Process

In all the renewable sources of energy, plant is unique in which it is effectively stored solar energy. Essentially, the use of plant is the reversal of photosynthesis. In nature, all biomass ultimately decomposes to its molecules with the release of heat. The release of energy from the combustion of biomass imitates natural processes. Therefore, the energy obtained from biomass, in contrast to fossil fuels, is a form of renewable energy and without adding carbon dioxide to the environment in principle.

### 6.2.1. Product Categories

Biomass can be used directly (e.g. burning wood for heating and cooking) or indirectly by converting it into solid, liquid or gaseous fuel (e.g. bio-briquette from stalk, alcohol from sugar crops or biogas from animal waste).

#### *6.2.1.1. Solid Fuel*

The typical of solid fuel from biomass is straw briquette. Other materials that can be processed for bio-briquette include trees and agricultural residues such as the residue of sawdust, wood shavings, sawdust, wood flour, etc [3]. Straw briquette technology refers to press the straw, in which moisture content is about 10%, into all kinds of shaped fuel such as bar-formed, pellet-formed and block-formed under high pressure (50 ~ 200 MPa) and high temperature (150 ~ 300 °C). It is in the process of commercialization in China presently [4]. Briquettes can be produced with a density of about 1000–1300 kg/m$^3$ from loose biomass of bulk density 10–20 kg/m$^3$, and the energy density is about 3×104 kJ/kg, which equivalent to medium coal. However, the process of high temperature and high pressure waste a lot of energy. In addition, to achieve the moisture content demand of raw materials (about 12%), the drying pretreatment before compressed process is necessary which also consumes much energy.

#### *6.2.1.2. Gaseous Fuel*

Gaseous bio-fuel includes biogas from anaerobic digestion (AD), and low and medium heating value gases from thermal processes [5]. While thermal gasification is in the early stages of commercial deployment, anaerobic digestion processes are already commercial and widely deployed, either in designed processes for specific environmental problems or in landfills, which are managed to capture the methane that is naturally produced. Anaerobic digestion has been used for many years in the treatment of sewage and animal manures to mineralize the carbon in order to reduce the volume of waste sludge for disposal. A wide range of agricultural, industrial, and urban activities generate residue streams with high organic loadings that are suitable for anaerobic treatment, including intensive animal husbandry (excreta, and bedding materials), food processing (sugar production and vegetable preparation), breweries and distilleries, and materials production (such as pulp and paper, pharmaceutical manufacture, and sewage treatment). The carbon is converted into methane and carbon dioxide in a ratio of about 60:40 (v/v) with a heat value of about 55%–60% that of natural gas. Thermal processes produce various compositions of dry gas containing hydrogen, carbon monoxide, and methane, as the fuel gases, in combination with nitrogen and carbon dioxide, as the major constituents. Thermal gases and biogas can be used directly as fuel in gas burners, or in prime movers such as internal combustion engines and gas turbines. Thermally produced gases, after purification and possibly water gas shift to adjust the $H_2$:CO ratio, can be described as a syngas (a mixture of $H_2$ and CO), which can be used to manufacture methanol, ammonia, Fischer Tropsch liquids, or hydrogen for use in fuel cells.

#### *6.2.1.3. Liquid Fuel*

There are two plant-based liquid fuels in the market today, ethanol and biodiesel. Ethanol is produced by means of fermentation of sugar cane in Brazil (11 Mm$^3$), and from starches in the United States (6 Mm$^3$) and Europe. A much smaller amount of biodiesel (usually an ester of plant lipids such as soybean or rape seed oils) is used in the USA and Europe. The major research and development area is in the production of ethanol from lignocelluloses (such as wood, straw, and grasses), which contain cellulose (40% to 50%) and hemicellulose (25% to 30%), with considerable potential for producing ethanol (about the same yield per tonne as corn) and more stable price structure than food prices [5].

The production of biodiesel from vegetable oil represents another means of producing liquid fuels from plant, and one which is growing rapidly in commercial importance. Commercially, vegetable oils used to produce biodiesel including rapeseed, sunflower and soybean oil, and animal fats. These oils and fats are typically composed of C14-C20 fatty acid triglycerides (constituting about 90-95% of the oil by weight). In order to produce a fuel that suitable for use in diesel engines, these triglycerides are converted to the respective alkyl esters and glycerol by base-catalyzed transesterification with short chain alcohols.

#### *6.2.1.4. Electrical Energy*

The main technologies of biomass power generation have three types : direct combustion, biomass gasification and biogas [6]. Boiler is the key equipment for biomass power generation. Biomass gasification for power generation refers to the process that biomass firstly converted into combustible gas, and then promoted the gas-fired power generation equipment for power generation. The raw materials for biogas preparation come from industrial wastes or agricultural residues. The biogas obtained from anaerobic digestion drives power generating set. The waste heat from generating set can be used for biogas production. Generally, the integrated thermal efficiency for biogas power generation is up to 80% or so, much higher than the normal 30% - 40% of the power generation efficiency.

### 6.2.2. Conversion Process

Various processes can be used to convert biomass to energy. The biomass can be burned, transformed into a fuel gas through partial combustion, into a biogas through fermentation, into bio-ethanol through biochemical processes, into biodiesel, into a bio-oil or into a syngas from which chemicals and fuels can be synthesized. Wood combustion, bio-ethanol production from either sugarcane or corn, and biodiesel production from oilseeds are currently the most economically significant processes but still need significant improvements. A detailed review of many processes that can convert biomass into fuels and chemicals shows that no individual process is without drawbacks. As a result, it is recommended that a bio-refinery should be the best solution to combine and integrate various processes to maximize the economic and environmental benefits, and minimize the waste and pollution [7].

The conversion technologies for utilizing biomass include four basic categories: bioconversion, pyrolysis, gasification and liquefaction, which can be classified into two broad pathways that thermo-chemical decomposition processes and biological digestion processes. Biological processes are essentially microbe digestion and fermentation. High moisture herbaceous plants, marine crops and manure are most suitable for biological digestion. Thermochemical conversion processes can be subdivided into gasification, pyrolysis, and direct liquefaction. The last two processes are usually confused with each other, both of which are thermo-chemical processes to convert feedstock organic compounds into liquid products. In the case of liquefaction, macro-molecule feedstocks are decomposed into low molecule fragments in the presence of a suitable catalyst. As for pyrolysis, on the other hand, a catalyst is usually unnecessary, and the decomposed low molecular fragments are further converted into oily compounds through homogeneous reactions in the gas phase.

#### *6.2.2.1. Bioconversion*

Although the cost of lignocellulosic biomass is far lower than that of sugar and starch crops, the bioconversion of cellulose and hemicellulose to monosaccharides with 5 and 6 carbons, such as glucose, is harder to accomplish comparing with the conversion of starch which is presently used as the raw material of bio-ethanol [8]. The bioconversion process comprises four key technologies as following: (1) pretreatment of raw lignocellulosic biomass; (2) hydrolysis of he micellulose and cellulose to monomer sugars; (3) fermentation of sugars into biofuels or chemicals; (4) products recovery and concentration. One of the advantages of bioconversion with lignocellulosic biomass is the opportunity to create a bio-refinery, producing value-added co-products. For instance, sugars may be subjected to bacterial fermentation under aerobic and anaerobic conditions, producing a variety of other products including lactic acid, which in turn may be processed into plastics and other products. The noncarbohydrate components of lignin also have potential for use in value-added applications.

#### *6.2.2.2. Pyrolysis*

Pyrolysis is the thermal degradation of biomass in the absence of oxidizing agent whereby the volatile components of a solid carbonaceous feedstock are vaporized in primary reactions by heating, leaving a residue consisting of char and ash. Pyrolysis always produces gas, vapour that can be collected as a liquid and solid char. Fast pyrolysis for liquids production is of particular interest currently. Fast pyrolysis occurs in few seconds or less. Therefore, not only chemical reaction kinetics but also heat and mass transfer processes, as well as phase transition phenomena, play important roles [9]. Fast pyrolysis processes are designed and operated to maximize the liquid fraction at up to 75% wt on a dry biomass feed basis. The char may be sold or used internally to provide heat for the process. The gas has a medium heating value and can be used internally to provide process heat, re-circulated as an inert carrier gas or exported for example for feed drying. The liquid is a homogenous mixture of organic compounds and water in a single phase with the fuel properties. Vacuum pyrolysis also gives high yields of liquids of up to 60% wt on dry feed, however, the vapour residence time is short as in conventional fast pyrolysis system, the solid heating rate is low and the solid residence time is also very high [10].

#### *6.2.2.3. Gasification*

Gasification is a thermochemical process that generates gaseous and fuel-rich products. Regardless of how the gasifier is designed, two processes must take place in order to produce a useable fuel gas. In the first stage, pyrolysis releases the volatile components of the fuel at temperatures below 600 °C. The unvaporized by-product of pyrolysis is called char and mainly consists of fixed carbon and ash. In the second gasification stage, the carbon remaining after pyrolysis is either reacted with steam or hydrogen or combusted with air or pure oxygen. Gasification with air results in a nitrogen-rich, low BTU fuel gas. Gasification with pure oxygen results in a higher quality mixture of carbon monoxide and hydrogen which is virtually without nitrogen. Gasification with steam is more commonly called "reforming" and results in a hydrogen and carbon dioxide rich "synthetic" gas (syngas). Typically, the exothermic reaction between carbon and oxygen provides the heat energy to drive the pyrolysis and char gasification reaction [11].

There are two main types of gasifiers: fixed beds and fluidized beds. The poor radial heat transfer in fixed beds limits their use in small gasifiers. since h larger gasifiers, maintaining a high enough temperature for gasification everywhere in the reactor would lead to hot spots where slagging could occur, especially with biomass feedstocks that are more prone to slagging than coal. Fluidized bed gasifiers provide a uniform temperature and greatly reduce the risk of slagging [7].

Nowadays, there are more than 40 factories and enterprises that provide biomass gasification equipment and facilities in China. Some of excellent demonstration projects have been built and operated successfully in towns and rural areas of China. Recently, some commercialization of straw gasification system for central gas supply with low calorific values (CV) have presented in China. By the end of 2000, 388 sets of straw gasification system for central gas supply had been built, which provides biogas about $1.50\times10^8$ $m^3$ and consumes $8.7\times10^7$ tons of straws. It is the effective way that treats large quantities of straw, improves environmental sanitation and increases living standard in rural areas. It is assumed that 1 kg straw can produce 2 $m^3$ of biogas and a household with 4 people only requires 5–6 $m^3$ biogas/day to meet its need of living. Generally speaking, the biogas cost is only about 0.11 RMB/$m^3$ in China presently [4].

#### *6.2.2.4. Liquefaction*

Biomass liquefaction includes biochemical conversion to produce ethanol and thermochemical conversion to produce bio-oil. In fact, hydrolytic liquefaction is the process of producing ethanol, which is fermented by microorganism from materials of high starch content such as steamed and saccharified corn or potato. Thermochemical liquefaction process breaks down biomass such as straw and wood residue into charcoal and hydrocarbons under the condition of medium temperature (500–600 °C), high heating rate (104–105 °C/s) and very short stay time (~2 s). A mixture of gas, liquid (bio-oil) and solid products can be produced while the proportion of each component varied depending on the reaction conditions. Presently, the straw liquefaction technology in China is still on the stage of experiment [4].

## 6.3. Bio-Fuel Efficiency Estimated System

### 6.3.1. Energy Content of Bio-Fuel

Biomass is not the ideal form for fuel use. The heat content calculated on a dry biomass must be corrected for the natural water content that can reduce the net heat available by as much as 20 % in direct combustion applications [1]. Gasification to low calorific gas carries an additional net energy loss and conversion to synthetic gas and liquid fuels results in still greater reduction of net energy, to perhaps 30 % of the original heat content. The energy content of different plant materials determines their calorific value (heat content). This calorific value depends on the percentage of carbon and hydrogen, which are the main contributors to the heat energy value of a biomass material.

The characteristics of important woods fuel indicate that the density of woods varies between 400 and 900 kg/$m^3$, and the energy content (kcal/kg) varies between 4200 and 5400.

To get the maximum energy, the plant materials should be air dried because the amount of energy contained in the plant varies with the amount of moisture content. If the combustible materials are required for the energy recovery process, the amount of water in the plant material will affect the recoverable energy. Energy can be obtained from biomass in three ways: (1) solid waste which can be burnt; (2) syngas and biogas which can be used to generate heat/electricity; (3) bio-fuel which includes ethanol, methanol, biodiesel and their derivatives.

### 6.3.2. Energy Conversion Efficiency

Net energy yield is estimated by reducing bio-fuel output but inputting energy into feedstock production, transportation, and bio-fuel conversion. In reality, energy inputs may be in the form of heat, electricity, gaseous, or liquid fuels. We have, for the sake of simplicity, assumed that all energy inputs are in the same form as the output, i.e., if we produce bio-ethanol, all the inputs are bio-ethanol. This assumption underestimates the net energy yield to the extent that the required energy input is heat process since that could have been generated directly from biomass at higher conversion efficiency. Energy inputs into plant at agriculture and feedstock transportation process is to a large extent determined by fertilizers and motor fuels. Usually, lignocellulosic bioenergy crop production requires lower energy inputs than that of traditional agricultural alcohol crops because of lower machinery use, reduced nutrients requirement, and limited pesticide requirement after the establishment period [12].

The CV of a material is an expression of energy content or heat value, released when burnt in air. The CV is usually measured in terms of the energy content per unit mass, or volume which is MJ/kg for solids, MJ/L for liquids, or MJ/Nm3 for gases. The CV of a fuel can be expressed in two forms, the gross CV (GCV), or higher heating value (HHV) and the net CV (NCV), or lower heating value (LHV). The HHV is the total energy content released when the fuel is burnt in air, including the latent heat contained in the water vapour and therefore represents the maximum amount of energy potentially recoverable from a given biomass source. The actual amount of energy recovered will vary with the conversion technology, as well with the form of that energy i.e. combustible gas, oil, steam, etc. In practical terms, the latent heat contained in the water vapour cannot be used effectively, and therefore, the LHV is the appropriate value to use for the energy available for subsequent use [131].

Using biomass for the generation of heat and power is of increasing importance. Biomass fuels can be experimentally characterized by elemental analyses, determination of higher heating values and determination of softening behavior of the produced ash. All these procedures are regulated by standards to ensure quality and comparability of the measurement results. The heating value of biomass is an important parameter for the planning and the control of power plants using this type of fuel. The HHV, also called gross calorific value, is the enthalpy of complete combustion of a fuel with, for instance, all carbon converted to $CO_2$, and all hydrogen converted to $H_2O$. The HHV is given for standard conditions (101.3 kPa, 25°C) of all products and includes the condensation enthalpy of water. The LHV, also called net heating value, is obtained when the condensation enthalpy of water is not included. While in the U.S.A. the HHV is generally used, the LHV is more common in European countries [14].

In China, biomass is widely used for cooking and heating by burning with a thermal efficiency only between 10% and 30%. Alternatively, converting biomass into gaseous and aqueous fuels, or electricity is possible a more efficient way for biomass utilization.

#### *6.3.2.1. Direct Combustion*

Direct combustion of biomass is the main and traditional way for utilization of biomass energy. Biomass contains less sulfur and ash but more hydrogen than coal. It is not only clean but also convenient in gaseous or liquid form compared with other fossil-based fuel. As biomass is a low-carbon fuel and can absorb produced $CO_2$, therefore it becomes likely to be an attractive clean development mechanism (CDM) option for reducing greenhouse gas (GHG) emission. Biomass combustion is a complex process consisting of consecutive heterogeneous and homogeneous reactions. The main steps of the process are drying, liquefaction, gasification, char combustion, and gas-phase oxidation. The time used for each reaction depends on the fuel size and properties, on temperature, and on combustion conditions [15]. The net energy available from biomass when it is combusted ranges from about 8 MJ/kg for green wood to 20 MJ/kg for dry plant matter, as compared with about 27 MJ/kg for coal. In the industrialized countries, the efficiency of direct combustion is rarely less than 70%, while is less than 50% in many developing countries, and much less than 20% in household cooking use [5].

With the increasing living standard in China recently, some changes have taken place in the structure of rural energy, but straw is still one of the main fuels in rural, accounting for 33–45% of energy consumption for livelihood. However, the wide-ranging utilization of straw as energy for livelihood in rural areas of China was inefficient stoves, which caused serious ecological and environmental problems. In 1980, the Chinese government has launched the National Improved Stoves Pilot County Program (NISPCP) throughout the country with the aim of improving energy efficiency. The efficiency of improved stoves today has reached to over 20–25%, while the old ones only 10–12%. By the end of 2000, more than 189 million rural households substituted improved stoves for their traditional ones, which accounted for 78.4% in the rural households. An improved stove can save about 1–1.5 tons of firewood annually, amounts to 100–200 RMB. To build a new and improved stove costs 60–100 RMB, and the cost of the firewood saved every year amounts to over 100–200 RMB, which indicates that the investment for new stoves are recovered in one year [4].

#### *6.3.2.2. Gasification*

During the past decade, there has been great progress in the development of gasification technology in China. Many kinds of biomass gasification processes have been developed for treating different materials on various purposes. The first generation is the up-draught fixed bed gasifier, with gasification efficiency of over 75% and maximum energy output of about 10 million kJ/h. This system normally converts wood residues into gaseous fuel suitable for boiler fuel. The second generation is the down-draft fixed bed gasifier, with gasification efficiency around 75% and maximum energy output of 4 million kJ/h. This system is mainly used for treating crop straw and wood wastes for supplying domestic cooking gas, and for wood drying process in factories. The most recent generation is the circulating fluidized bed (CFB) gasifier with efficient maintains at about 75% but the maximum energy output can increase to 40 million kJ/h. This system has been operated successfully for treating wood and crop wastes for fuel boiler or to generate electricity. According to available biomass resources and energy demands in China, biomass gasification technology has been mainly applied in three fields. The first is biomass wastes treated by gasification combining with heating equipment (such as drying equipment and boilers) directly. In this case, coal or oil can be substituted by the gas produced and thus benefited from conserving energy. The second is straw gasification for domestic cooking application. Such systems can make use of most of the agricultural waste, convert it to domestic cooking gas, and supply rural areas by setting up small gas stations with pipe network. This technology takes an important role in upgrading the living conditions of rural areas in developing countries. The third is biomass gasification and power generation system (BGPG), which can provide power and electricity by making use of different kinds of biomass in any desired scale. It is suitable for domestic and small industrial applications particularly in rural area, such as rice mill, timber mill and food processing. BGPG is the most promising biomass technology that will be widely applied in developing country because of its flexibility and economy [16].

#### *6.3.2.3. Pyrolysis*

Biomass power generation (direct combustion) and pyrolysis process are different [17] Pyrolysis process reaction is as follows:

Main pyrolysis reaction:

$$C_nH_mO_k \rightarrow (m/2)H_2 + kCO + (n - k)C \quad (6.1)$$

Partial reaction:

$$C + 2H_2 = CH_4 \quad (6.2)$$

$$C + H_2O = CO + H_2 \quad (6.3)$$

$$C + \left(\frac{1}{2}\right)O_2 = CO \quad (6.4)$$

$$C + O_2 = CO_2 \quad (6.5)$$

$C + CO_2 = 2CO$(6.6)

$$CO + H_2O = CO_2 + H_2 \quad (6.7)$$

$$CO + H_2O = CO_2 + H_2 \quad (6.8)$$

An intriguing prospective benefit of biomass pyrolysis is the possible reduction of emissions of $NO_x$ and sulfur at coal-fired plants. Wood fuel contains less nitrogen than coal does, and the moisture in the wood serves to cool the coal combustion process and reduce the formation of thermal $NO_x$.

Although the pyrolysis process/woody biomass system generates $CO_2$ as shown in formula (6.5), (6.7) and (6.8), a planned forestation effort can balance this effect and recapture GHGes to prevent global warming.

In case of direct combustion of biomass systems, $CO_2$ emission coefficient is estimated as follows:

CO2 emission coefficient=

$$\left(\frac{\text{the amount of } CO_2 \text{ discharge at the time of (fuel manufacturing process + fuel combustion by the fuel)}}{\text{(heat value of fuel)}}\right)$$

$$=\left(\frac{\text{(g C/h)}}{\text{Mcal/h}}\right) \tag{6.9}$$

When the amount of $CO_2$ emitted at the time of fuel combustion does not act as environmental load in formula (6.9), the $CO_2$ discharge coefficient is zero in most cases. According to Nedos estimate, the total amount of $CO_2$ discharged from the fuel manufacturing process is extremely low at 518 kg C/h. If we consider a 6000 kW scale plant with 40% overall efficiency, the heating value of the fuel will be 12,900 M cal/h.

In case of pyrolysis process/woody biomass system also the amount of $CO_2$ emitted at the time of fuel combustion does not act as environmental load in formula (6.9), therefore contributes to reduce $CO_2$ emission.

#### *6.3.2.4 Bioconversion*

Various kinds of plant-based products, not only including fuel but also chemical, fodder, food, medicine and so on, can be produced through bioconversion process. Obviously the difference of raw materials, pretreatment and fermentation process varies from different final products. Studies of life cycle assessment (LCA) demonstrate that a plant-based product has great potential to reduce both the consumption of non-renewable energy resources and GHG emissions. However, the production of biomass requires agricultural land and is often associated with adverse environmental effects such as eutrophication of surface and ground water. Decision making in favor of or against plant-based and conventional fossil product alternatives therefore often requires weighing of ecological impacts. Kim et al [18] demonstrated that utilizing biomass for bio-fuels would save nonrenewable energy and reduce GHG. Manish et al [19] compared the bio-hydrogen production processes and method steam methane reforming (SMR) on the basis of net energy ratio, energy efficiency and GHG emissions. It was found that biohydrogen production processes reduce GHG emissions and non-renewable energy use by 57–73% and 65–79%, respectively, as compared to the SMR process. Efficiency of biohydrogen processes increase significantly when by-products are considered. Dai et al [20] introduced the net energy value (NEV) and net renewable energy value (NREV) to assess the energy and renewable energy efficiency of the corn and cassava fuel ethanol system during their life cycle. The results showed that NEVs of corn and cassava ethanol are 1.472 MJ/L and 2.417 MJ/L respectively, and NREVs of corn and cassava ethanol are 1.474 MJ/L and 2.459 MJ/L respectively. Corn and cassava ethanol utilizations as energy and renewable energy are feasible, and cassava ethanol has better feasibility than corn ethanol. Yield and chemicals use are key factors to improve NEV and NREV. Better byproducts utilization and substitution of organic fertilizer for chemical fertilizer help to improve energy efficiency and renewable capability. As the complexity of structures and components for lignocellulosic resources, and various plant-based products from different conversion methods, it is difficult to give comprehensive evaluation of lignocellulosic biomass conversion energy efficiency. Thus, researches on evaluation system not only promote to the development of bioconversion technologies and the integration process, but also provide reference for economic analysis in conversion process.

### 6.3.3. Mass Balance Analysis

#### *6.3.3.1. Direct Combustion*

Burning of biomass needs certain amount of air. Combustion products are smoke and unburned solid particles and ash. The unburned solid particle is calculated to be negligible as its volume fraction share is less. And the ash content is obtained usually directly from biomass fuel industry analysis data. Therefore, the combustion mass balance evaluation is actually based on the calculation of the amount of air and smoke [21].

#### *6.3.3.2. Gasification*

Gasification is a process that devolatilizes solid or liquid hydrocarbons, and converts them into a low or medium BTU gas [22]. Gasification has several advantages over traditional combustion. It takes place in a low oxygen environment that limits the formation of dioxins and of large quantities of SOx and NOx. Furthermore, it requires just a fraction of the stoichiometric amount of oxygen necessary for combustion.

The basic gasification reactions that must be considered are:

1) $C + O_2 \rightarrow CO_2$ -393 kJ/mol (exothermic)
2) $C + H_2O \rightarrow CO + H_2$ +131 kJ /mol (endothermic)
3) $C + CO_2 \rightarrow 2CO$ +172 kJ/mol (endothermic)
4) $C + 2H_2 \rightarrow CH_4$ -74 kJ/mol (exothermic)
5) $CO + H_2O \rightarrow CO_2 + H_2$ -41 kJ/mol (exothermic)
6) $CO + 3H_2 \rightarrow CH_4 + H_2O$ -205 kJ/mol (exothermic)

All of these reactions are reversible and their rates depend on the temperature, pressure and concentration of oxygen in the reactor. Gasification is a thermochemical process that generates a gaseous, fuel-rich product. The product of gas resulting from biomass gasification contains various tars, particulates, halogens, heavy metals and alkaline compounds depending on the fuel composition and the particular gasification process.

#### *6.3.3.3. Pyrolysis*

Pyrolysis is a process that converts biomass at temperatures around 500 °C in absence of oxygen to liquid (bio-oil), gaseous and solid (char) fractions. With flash pyrolysis techniques (fast pyrolysis) the liquid fraction can be maximized (up to 70% of the thermal biomass input). Bio-oil contains about 40% wt of oxygen and is corrosive and acidic. Crude bio-oil can in principle (after some modifications and only for better quality oils) be used for firing engines and turbines. The oil can also be upgraded (e.g. via hydrogenation) in order to reduce the oxygen content. Zhang et al [23]carried out pyrolysis experiments of raw materials including corn stalk, rice husk and pine sawdust for liquid fuel production in a self-made small fluidized bed reactor. The relationship between input and output is shown in the Table 6.1.

**Table 6.1. Mass balance analysis [23]**

| Input kg/h | | Output kg/h | | |
|---|---|---|---|---|
| Material | input | bio-oil | gas | solid carbon |
| Corn stalk | 1.872 | 1.000 | 0.580 | 0.292 |
| Rice husk | 3.940 | 1.544 | 1.072 | 1.324 |
| Pine sawdust | 2.340 | 1.460 | 0.460 | 0.420 |

### *6.3.3.4. Bioconversion*

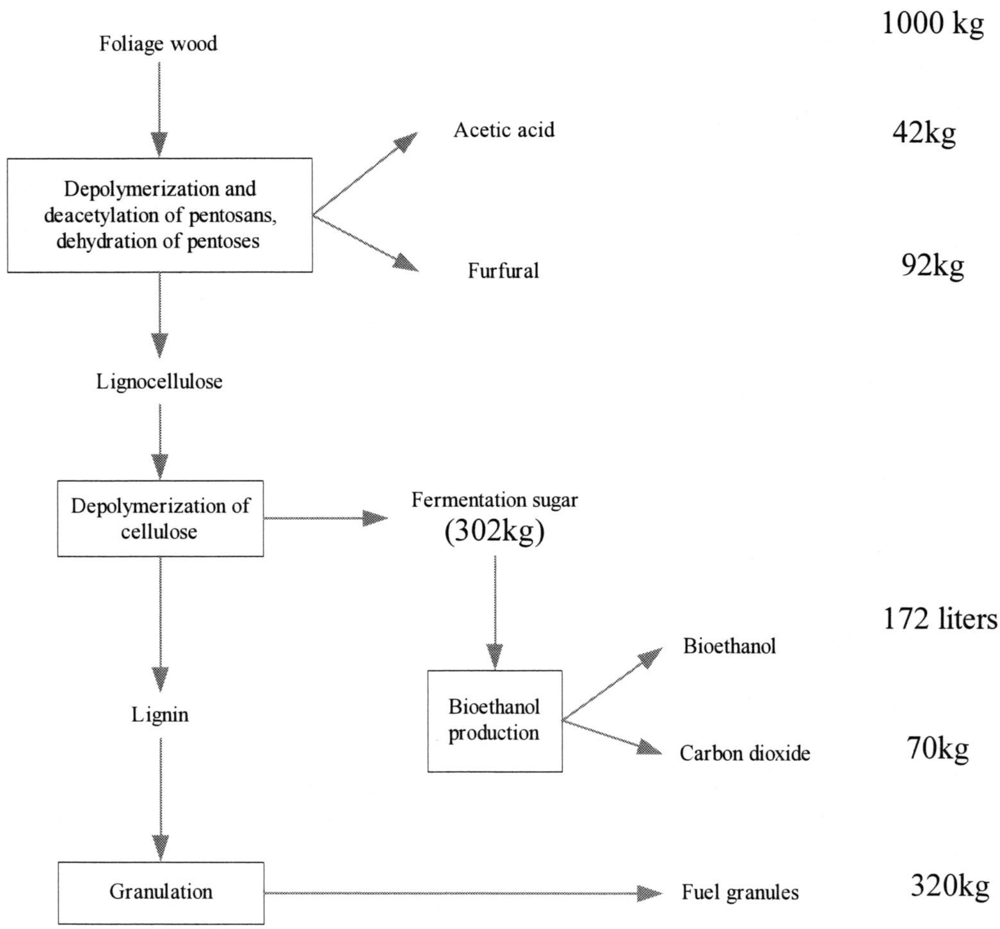


Figure 6.1. The manufacturing scheme and output of furfural, bioethanol and other products from foliage wood [24]

Lignocellulose consists of three main components: cellulose, hemicellulose and lignin. Traditionally, cellulose was hydrolyzed into glucose for microorganism fermentation, while most of hemicellulose and lignin are wasted without efficient utilization. If the plant-based products from hemicellulose and lignin were taken into account, the mass utilization efficiency of bioconversion is the highest among biomass conversion technologies. Gravitis et al [24] reported the new process of simultaneous obtaining of furfural and bioethanol from foliage wood and agricultural wastes. The production process (Fig 6.1) does not create any solid waste products. Since 1997 for the first time in the world's industrial practice the new technology yielding furfural and fermentable sugars further processed into bioethanol has been realized in Russia with capacity 4300 tons/a of furfural and 8800 tons/a of bioethanol.

## 6.4. Directional Multi-Level Conversion to Bio-Fuel

Presently, the existing methods of conversion (both thermochemical conversion and bioconversion) can not achieve the goal of complete exploitation and utilization for lignocellulosic biomass resources. For instance, the lignocellulosic material is only regarded as fossil fuels substitutes for energy production and never been taken into account its conversion to higher value-added products. For bioconversion we often only pay attention to the main fermented products and neglect by-product in the pretreatment and fermentation processes. If these by-products can be transformed into high value-added products, it will significantly enhance the economic efficiency of the total bioconversion process. Therefore, it is necessary to base on directional multi-level conversion thought and combine of bio-chemical and biological technology to achieve biomass high value utilization. Figure 6.2 shows straw directional multi-level conversion to bio-based products with bio-fuel for leading product.

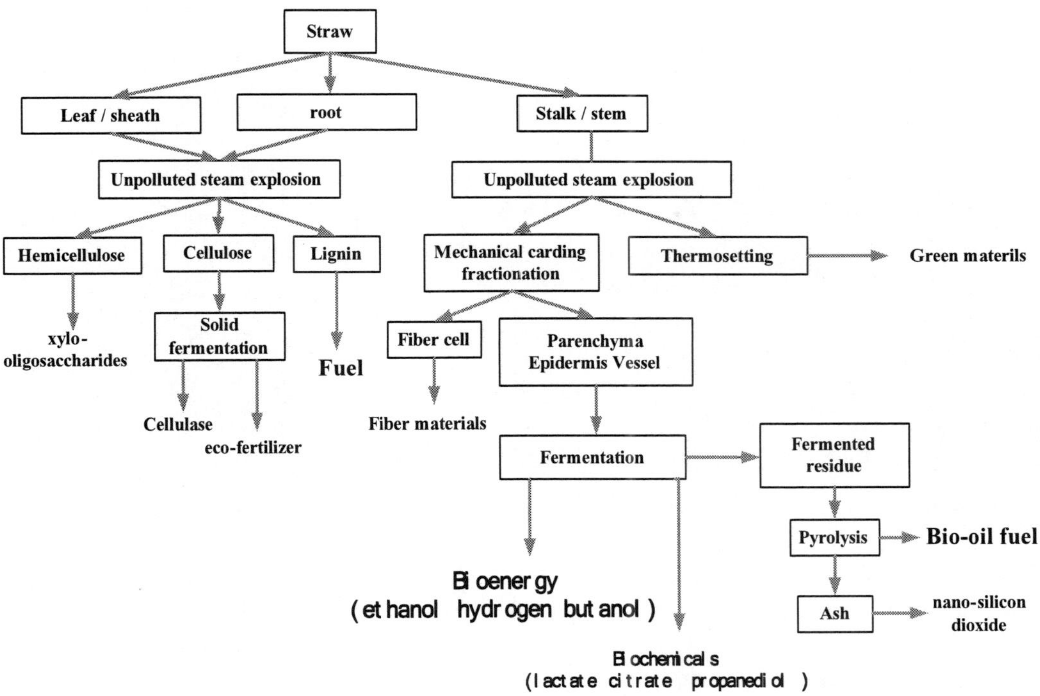


Figure 6.2. Straw directional multi-level conversion to bio-based products with biofuel for leading product

# 6.5. BIO-FUEL

Conversion routes for producing transportation fuel from biomass are plentiful. The potential of biotechnology is for the most part well known. Biotechnological process generally occurs under mild conditions. Biocatalysts, substrates, intermediates and by-products, as well as the product itself, are biodegradable. By using modern biotechnological approaches to convert lignocellulosic biomass into commodity products is fundamental significance. Four typical bio-fuels including ethanol, butanol, hydrogen and biodiesel are briefly introduced in this section.

## 6.5.1. Ethanol

### *6.5.1.1. Introduction*

Bio-ethanol is an attractive alternative fuel since that it is a renewable plant-based resource and it has the potential to reduce particulate emissions in compression-ignition engines and is completely oxygenated thereby. It can be blended with petrol or used as neat alcohol in dedicated engines, taking advantage of the higher octane number and higher heat of vaporization. Bio-ethanol currently accounts for more than 94% of global bio-fuel production, with 60% from sugar cane and 40% from other crops. Brazil and the United States are the two countries that produce large quantities of fuel ethanol from sugar cane and maize respectively, and they together account for about 70% of the world bio-ethanol production. In 2006, global production of bio-ethanol has reached 13.5 billion gallons, up from 12.1 billion gallons in 2005. China also pays much more attention to bio-ethanol production and utilization. Data showed that United State and Brazil were top 2 with 4.85 billion and 4.49 billion gallons of bio-ethanol production respectively. China with 1.02 billion gallons production ranked third among of the world bio-ethanol producers in 2006.

Presently, the raw materials for bio-ethanol production mainly are sugar or grain (starch) around the world. With the increasing demand for bio-fuel and the special situation of large population with less land in China, the conflict between food production and energy consumption is inevitable. However, the amount of agriculture residues biomass is approximately 15 tons with 3.7 billion tons bio-ethanol productivity every year in China. It should be called for the exploitation of lignocellulose feed-stocks, such as agricultural and forest residues for the ethanol production, and the future large-scale use of ethanol will most certainly have to be based on production from lignocellulosic materials.

### *6.5.1.2. Key Technology and Process*

Lignocellulosic biomass can be utilized to produce ethanol by mainly two steps: hydrolysis of cellulose to produce reducing sugars and fermentation of sugars to ethanol. Although considerable researches have been made to improve the hydrolysis of lignocellulosic materials, the cost of ethanol production from lignocellulosic materials is still relatively high. The main challenges are the low yield and high cost of hydrolysis process [25]. To achieve the industrialization of cellulosic ethanol, a number of technological breakthroughs and cost reductions in individual process steps are required.

*Pretreatment*

In the context of biological processing of cellulosic biomass to fermentable sugars and then to ethanol or other products, pretreatment generally refers to the disruption of the naturally resistant carbohydrate-lignin shield that limits the accessibility of enzymes to cellulose and hemicellulose [26]. The purpose of pretreatment is to remove lignin and hemicellulose, reduce cellulose crystallinity, and increase the porosity of the materials. Presently, physical, chemical, physico-chemical, and biological process have been used for pretreatment of lignocellulosic materials (Table 6.2).

The most investigated pretreatment method is steam explosion with and without addition of acid catalyst. Acid-catalyzed steam pretreatment is one of the effective methods to fractionate wood components before enzymatic saccharification [27]. Agricultural residues, such as corn stover, have low lignin content and are more easily to be pretreated than hardwood and softwood resides containing higher amount of lignin and having a much more recalcitrant cell wall structure. Steam explosion pretreatment without acid addition has been proved that can enough meet the stalk pretreatment requirements and also avoid acid pollution. Our work [28] approved that about 94% cellulose would be recovered from wheat straw by steam explosion pretreatment without any chemicals addition. Sugars recovery rate of hemicellulose was 80% by water countercurrent extraction and the lignin yield was 75% by acid precipitation.

**Table 6.2. Common methods for lignocellulose pretreatment [25]**

| Methods | Examples |
|---|---|
| Physical Pretreatment | Chipping, milling or grinding |
| Chemical pretreatment | Acid hydrolysis, Alkaline hydrolysis, Organosolv process, Oxidation delignification, |
| Physico-chemical pretreatment | Steam exploded, Ammonia fiber explosion(AFEX), $CO_2$ explosion |
| Biological pretreatment | Brown-,white-and soft-rot fungi *Pleurotus ostreatus, Phanerochaete sordida, Pycnoporus cinnabarinus, Sporotrichum pulverulentum* |

*Cellulolytic Enzymes and Hydrolysis*

The first step for cellulosic ethanol production is to decompose cellulosic biomass into sugars for microorganism fermentation. Cellulose can be hydrolysed into glucose either enzymatically by cellulases or chemically by acids. Enzymatic hydrolysis has better technical advantage than acid hydrolysis in milder reaction condition (pH 4.5, 45-50 °C) and less equipment corrosion. The high temperature and dilute acid hydrolysis also increase the sugar degradation rate but causes inhibition for the subsequent fermentation stage.

(1) The Enzyme System

Cellulose is a linear polymer consisting of D-anhydroglucopyranose joined together by β-1,4-glucosidic bonds with a degree of polymerization (DP) from 100 to 20,000. A hydro cellobiose is the repeating unit of cellulose. It is now clear that the enzyme known as "cellulase" is really a complex of enzyme that work synergistically to attack native cellulose. The widely accepted mechanism for enzymatic cellulose hydrolysis involves synergistic actions by *endoglucanase* (EC 3.2.1.4), *exoglucanase or cellobiohydrolase* (EC 3.2.1.91), and β-glucosidase (EC 3.21.21). Endoglucanase hydrolyze accessible intramolecular β-1,4-glucosidic bonds of cellulose chains randomly to produce new chain ends; exoglucanases processively cleave cellulose chains at the ends to release soluble cellobiose or glucose; and β-glucosidase hydrolyzes cellobiose to glucose in order to eliminate cellobiose inhibition. These three hydrolysis processes occur simultaneously [29].

(2) Substrate-Related Factors Affecting Hydrolysis

Enzymatic hydrolysis is generally considered as consisting of three steps: the adsorption of cellulase enzymes onto the surface of the cellulose, the subsequent breakdown of cellulose into fermentable sugars through the synergistic action of the cellulase enzymes, and the desorption of the cellulase enzymes from the lignocellulosic residue into the supernatant [30]. The susceptibility of cellulosic substrates to enzymatic hydrolysis is thought to depend on a number of substrate structural features including cellulose crystallinity, the degree of cellulose polymerization, the lignin and hemicellulose content. It has been frequently suggested that the surface area of the substrate available to cellulases is the most influential factor determining the hydrolysis rate, whereas cellulose fibers are embedded in a lignin-polysaccharide matrix, which lead cellulase is hard to access the surface area of the substrate and degrade the cellulose efficiently. Steam pretreatment has been proved to be an effective method to fractionate the hemicellulose and lignin components and enhance the hydrolysis of the cellulosic residue [31].

The cellulase hydrolysis rate declines rapidly with the increasing conversion of the substrate. One proposal is that with hydrolysis proceeding the substrate becomes enriched in more recalcitrant cellulose since that the less recalcitrant amorphous cellulose is hydrolyzed. Other possible contributing factors to the declining hydrolysis rate include enzyme adsorptive loss to lignin; deactivation of enzyme through thermal, mechanical, and chemical actions. Addition of surfactants during hydrolysis, such as nonionic Tween 20 [32], Tween 80 [33], polyoxyethylene glycol [34], is capable of modifying the cellulose surface property and minimizing the irreversible binding of cellulose on cellulose.

(3) Microorganisms Producing Cellulose-Degrading Enzymes

A large number of microorganisms have been reported to produce cellulose decomposing enzymes, but fungi get the most attention because of their fair production rate and complex cellulolytic enzymes systems. Most of fungi are able to produce the necessary enzymes for the degradation of lignocellulosic materials. They belong to the Ascomycetes, Deuteromycetes, or Basidiomycetes groups [35]. Filamentous fungi, typically Trichoderma sp., are the preferred source of industrial cellulase preparation because their capacity for extracellular production is greater than that of cellulolytic bacteria.

Cellulose can also be degraded by both aerobic and anaerobic bacteria. The most studied bacteria with respect to their cellulase systems are species of *Clostridium, Cellulomonas, Bacillus and Pseudomonas*. However, the cellulolytic enzyme systems of bacteria are not directly comparable to those of fungi. Bacteria often produce cellulases in small amounts (<0.1 g/L) and degradation of cellulose seems to take place by a cluster of multi-enzyme complexes which are difficult to disrupt without loss of total activity as well as of the individual components.

Filamentous prokaryotes such as Actinomycetes are also important members of the microbial community responsible for the degradation of cellulose. Mesophilic species of *Streptomyces* and thermophilic species of *Thermomonospora* and *Thermoactinomyces* are the most studied of these organisms.

Moreover, cellulose-to-ethanol biotransformation can be conducted by various anaerobic thermopilic bacteria, such as *Clostridium thermocellm*, as well as by some filamentous fungi, including *Monilia* sp, *Neurospora crassa* , *Neurospora* sp., *Zygosaccharomyces rouxii*, *Aspergillus viride* and *paecilomyces* sp. Studies on the fermentation process utilizing these microorganisms have shown longer duration (3-12 days) with a poor yield (0.8-60 g/L of ethanol) as result of low resistance of microorganisms to higher concentration of ethyl alcohol. Another disadvantage of this process is the production of various by-products, primarily acetic and lactic acids [36].

*Hydrolysis and Fermentation Strategies*

End-product inhibition of cellulases is well known and cellobiose is thought to be a stronger inhibitor of cellulases than glucose in most cellulases system. Several methods have been proposed to overcome the end-product inhibition resulting from the rapid accumulation of sugar during hydrolysis, which include the increase of enzymes concentrations, the supplementation of cellulases with β-glucosidase activity, separation of sugars from the hydrolysate by ultrafiltration, or simultaneous saccharification and fermentation of the substrate [30].

(1) Separate Hydrolysis and Fermentation (SHF)

Enzymatic hydrolysis performed separately from fermentation step is known as SHF. The advantage of SHF is the ability to carry out each step under optimal conditions, i.e. enzymatic hydrolysis at 45-50 °C and fermentation at about 30 °C. The major drawback of SHF is that the released sugars inhibit the activity of cellulase and β-glucosidase during hydrolysis. Logen Corporation, a major manufacturer of industrial enzymes in Canada, developed an SHF process comprising a dilute-acid-catalyzed steam explosion and the removal of the major part of the acetic acid released during the pre-treatment, the use of *Saccharomyces cerevisae* as a fermenting organism, distillation of broth, bioethanol dehydration and disposal of stillage in landfill [37].

(2) Simultaneous Saccharification and Fermentation

Simultaneous saccharification and fermentation (SSF) is the process in which the enzymatic degradation of lignocellulose is combined with the fermentation of glucose obtained from the hydrolysis of these polysaccharides [37]. The key of the SSF from biomass is its ability for rapidly converting the sugars into ethanol as soon as they are formed diminishing their accumulation in the medium. Sugars are much inhibitory for conversion process than ethanol. SSF offers an easier operation and a lower equipment requirement than the sequential process since no hydrolysis reactors are needed. Nevertheless, the major drawbacks of the SSF from biomass include the different optimal conditions, mainly pH and temperature for the hydrolysis of cellulose and fermentation. Cellulases work in an optimal way at 40-50 °C and pH of 4-5 whereas the fermentation of hexoses with *S. cerevisiae* is carried out at 30 °C and pH of 4-5, and fermentation of pentoses is optimally performed at 30-70 °C and pH of 5-7. Moreover, the ethanol concentration in final culture is toxic to cellulase activity. Wu and Lee [38] reported that ethanol inhibits the cellulase from *T. reesi* progressively and linearly up to 65 g/L, for a typical SSF temperature of 38 °C, the cellulase lost 9%, 36%, and 64% of its original activity at ethanol concentrations of 9 g/L, 35 g/L and 60 g/L respectively.

(3) Nonisothermal Simultaneous Saccharification and Fermentation

The enzymatic hydrolysis reaction in the SSF process is operated at a temperature lower than the optimum level of enzymatic hydrolysis. This forces the enzyme activity to be far below its potential, which results in high requirement of enzyme. In order to overcome this problem, nonisothermal simultaneous saccharification and fermentation process (NSSF) was suggested [39]. In this process, saccharification and fermentation occur simultaneously but in two separate reactors at different temperatures. The lignocellulose is retained inside in a hydrolysis reactor and hydrolyzed at the optimum temperature for the enzymatic reactions (e.g. 50°C). The effluent from the reactor is recirculated through a fermentor, which runs at its optimum temperature (e.g. 30°C). The cellulase activity is increased 2-3 times when the hydrolysis temperature is raised from 30°C to 50 °C.

(4) Simultaneous Saccharification and Co-Fermentation

The simultaneous saccharification and co-fermentation (SSCF) process is considered as an improvement to SSF. In SSF bioreactor, only hexoses are converted to ethanol, and pentoses can be fermented in another bioreactor with different microorganism. Therefore, two bioreactors and two biomass production setup is required in SSF. In SSCF process, it is suggested to ferment both hexoses and pentoses in a single bioreactor with a single microorganism. The hydrolyzed hemicellulose during pretreatment and the solid cellulose are not separated after pretreatment, allowing the hemicellulose sugars to be converted to ethanol together with SSF of the cellulose. Teixeira et al. [40] studied the SSCF with a recombinant strain of *Z.mobilis* for ethanol production from hybrid polar wood and sugarcane bagasse, which resulted in ethanol yields of 92.8% and 91.9% of theoretical from pretreated hybrid poplar wood and sugarcane bagasse, respectively.

#### *6.5.1.3. Energy Efficiency Analysis*

Glucose of 100 g produces 51.4 g ethanol and 48.8 g carbon dioxide by *S. cerevisiae* anaerobic fermentation. As a part of sugar to be used for the growth of yeast, the ethanol yield is slightly lower than the theoretical value. During the fermentation process, about 91% energy content from glucose can be preserved in ethanol. The energy conversion efficiency of lignocellulosic biomass to bioethanol can be estimated according to the glucose content in lignocellulosic raw material.

Actually, a major controversy has arisen about the energy balance of the ethanol production from corn with first generation processes. The consensus seems that the energy balance is slightly positive, i.e. about 10-15% of the energy contained in the bioethanol comes from the solar energy captured by corn plants, while the rest comes from fossil fuels used in the production and processing of the corn kernels. Nearly 2/3 of the required energy is consumed by the conversion of the kernels into ethanol, and the subsequent ethanol purification [7].

#### *6.5.1.4. Prospect*

The industrial-scale bio-ethanol production needs significant change at three levels. The first change is the substrate from traditional corn or other starch substrate to non-grain biomass. The second change is the transformation process from traditional polluted technologies to the clear technology, from a single product to the comprehensive products. The third change is the integration of micro industrial biotechnology and macro industrial ecology technology into the bioethanol industry.

Although great progresses have been made in the past several years in all aspects of lignocellulosic conversion to ethanol, the primary obstacle for commercialization of ethanol produced by fermentation is still its high cost of production relative to the local cost of gasoline. Effects will continue toward improving the enzymatic hydrolysis with cost-efficient enzymes and producing more robust pentose fermenting microorganisms with higher productivity. Research on genetic engineering for modified microorganisms that can convert both hexose and pentose to ethanol will be focused. Meanwhile, it is necessary to screen and select robust fermenting organisms, which are more tolerant to inhibitors and ferment all sugar in the raw material in concentrated hydrolysate at high productivity and with high ethanol concentration. At present, as to domestic and foreign enterprises started massive capital influx bioethanol field, the production of cellulose ethanol is no longer a dream, it is getting faster and faster to the pace of industrialization.

### 6.5.2. Acetone and Butanol

#### *6.5.2.1. Introduction*

Butanol is an important chemical and organic solvent with many applications in the plastic industry, a food-grade extractant chemical in the food and flavor industry, and a fuel extender. In practice, the relatively high CV in comparison with ethanol, the low vapour pressure and low miscibility with water provide clear advantages. Butanol has nearly the same energy value as gasoline and can be used in existing fuel delivery infrastructure and in existing automobile engines without modification and at 100% concentration. Butanol also contributes to clean air by reducing emissions and unburned hydrocarbons in tail pipe exhaust.

The historic acetone-butanol-ethanol (ABE) fermentation is one of the oldest known industrial fermentations. It was ranked second only to ethanol fermentation by yeast in its scale of production, and is the largest anaerobic fermentation industry. However, since the 1950's industrial ABE fermentation has declined continuously due to increasing substrate cost, and almost all butanol is now produced via much cheaper petrochemical routes. The oil crisis in 1973 led to renewed interest in novel fermentation and product recovery technologies as well as in the metabolism and genetics of the bacterial species involved.

Though butanol currently is produced by the petrochemical route in worldwide, research has progressed in the microbiology, biochemistry, and genetics of AB-producing bacteria as well as in substrate preparation, sterilization, fermentation and downstream processing technology. Nowadays using renewable resources to ferment butanol is becoming very attractive for the production of chemicals and liquid fuels. Because of the potential for carcinogen carryover, the use of the petroleum-derived butanol is not desirable for food applications. Butanol from biomass fermentation will be more popular in the view of health and environmental protection.

#### *6.5.2.2. Key Technology and Process*

The high cost of lignocellulose converting to fermentable sugar has been identified as a major factor affecting the economic viability of the AB fermentation. Butanol is highly toxic to biological systems at quite low concentration, which means that the level of solvents obtainable in the final fermentation broth was only about 2%. The recovery of these solvents with low concentration by distillation was expensive because of the large amount of energy required. Efforts have been made to understand the selection of hyper-butanol-producing strains, utilization of lignocellulosic biomass hydrolysates, and cost reduction of solvent recovery with the aim of improving butanol productivity, concentration and yield.

*Clostridium Recombination*

Clostridium are gram-positive spore-forming bacillary anaerobes and diverse in both physiology and genetics. Among solventogenic clostridia, *Clostridium acetobutylicm*, *Clostirudium berijierinckii*, *Clostridium saccharobutylicum*, and *Clostridium saccharoperbutylacetonicum* are primary solvent producers. Fermentation of these microorganisms is biphasic-involving acidogenesis, the production of acids (acetic and butyric acids), followed by solventogensis, which is the production of solvent (ABE in a ratio 3:6:1) from the previously produced acids during fermentation [41]. Major side products of the acid and solvent formation are large amounts of the gases $CO_2$ and $H_2$, obtained from the fermenters in molar (and volume) ratio of approximately 1.5:1. The genome of the solvent producing C.acetobutylicun ATCC 824 has been sequenced, which can facilitate the optimization of the butanol fermentation process and competitive butanol production by fermentation come within reach [42].

Progress has been made over the past decade with respect to genetic manipulation of the solventogenic Clostridia Mutagens such as hydrogen peroxide, nalidixic acid, ethyl methanesulfonate, N-methyl-N-nitro-N-nitrosoguanidine, and UV irradiation have been used to induce mutations in soventogenic Clostridia [43]. Annous et al [44] reported that *C.acetobutylicum* mutants BA101 was isolated by using N-methyl-N-nitro-N-nitrosoguanidine together with selective enrichment on the glucose analog 2-deoxyglucose. Amylolytic enzyme production by C.acetobutylicum BA 101 was 1.8- and 2.5-folds higher than that of the ATCC 824 strain grown in starch and glucose, respectively. Chen et al [45] found that *C.beijerinckii* BA 101 produced 32.6 g/L total solvents with butanol at 20.9 g/L when grown in MP2 medium containing 60 mM sodium acetate and 8% glucose during batch fermentation. These values represent the highest solvent and butanol concentrations produced by a solventogenic Clostridium strain when grown in batch culture.

Recombinant DNA technology, in addition to traditional mutagenesis and selection, has been employed to modify targeted metabolic pathways in the solventogenic clostriadia. Manipulation of the genes encoding the acid- and solvent- producing enzymes in C.acetobutylicum has demonstrated higher than any previously reported values. Nair et al [46] reported that inactivation of solR gene (coded for a putative repressor for butanol and acetone formation) via homologous recombination yielded a mutant with 3.2-fold increase in the final butanol concentration. Inactivation of pta gene reduced phosphotransacetylse (PTA) and acetate kinase (AK) activities and significantly reduced acetate production, while inactivation of buk gene reduced butyrate kinase (BK) activity and significantly reduced butyrate production and increased butanol production [47].

Although progress has been made, the use of recombinant DNA technology has so far not yield a hyper-butanol-producing industrial strain, probably because of lacking understanding of the global regulation of butanol production and unique physiology of the solventogenic Clostridia. As butanol is highly toxic to the solventogenic Clostridia, metabolic engineering of various microorganisms not typically associated with butanol production but resistant to butanol toxicity has been suggested. In addition, simultaneous fermentation and recovery to relieve butanol toxicity might be the best short-term solution given the currently available solventogenic Clostridia strains [48].

*Feedstock Utilization*

So far, *C.acetobutylicum* is able to use polymeric substrates such as starch and xylan, but not cellulose, for growth. The degradation of starch is mediated by α-amylase and genetic data, which indicates the presence of at least one additional α-amylase gene. Larch wood xylan is only partially hydrolyzed by endoxylanases and a β-D-xylosidase. C.beijerinckii also grows on starch and employs the catalytic activities of glucoamylase and α-amylase [49]. However, direct fermentation of cellulose-containing biomass to butanol in a consolidated bioprocess with a cellulolytic and solventogenic bacterium remains a future task. Only fairly active mini-cellulosomes were found in *C.acetobutylicum* ATCC 824 and attempts to correct reading frame errors and improvement of promoters did not lead to substantial cellulase production [50].

At present, one of the major bottlenecks hampering the economic viability of *C.acetobutylicum* is the cost of substrates, accounting for up to 60% of the total production costs. The high cost of substrates, including molasses, whey permeate, corn and starchy roots, is major factor affecting the economic viability of butanol production by fermentation. Lignocellulose is the most abundant renewable resource on the planet, and has great potential as a substrate for fermentation. For the use of lignocellulosic substrates, several strategies have been studied, such as the use of hydrolyastes, con-cultures with true cellulolytic organisms, or the addition of cellulases to the fermentation medium.

Many researchers have been carrying out degradation of lignocellulose to sugars for butanol fermentation since 1980's. Yu et al reported that *C.acetobutylicum* was grown on acid hydrolysates of steam-exploded aspen wood chip with final butanol yields of 9.0 g/L (0.26 g of butanol per g of sugar consumed). However, it should be noted that pretreatment of agricultural residues, such as corn fiber, with acid/alkali results in fermentation inhibitors. Marchal et al [51] described a one-step hydrolysis and fermentation process involving the use of cellulase from *T. reesei* and conversion of alkali-pretreatment wheat straw into butanol and acetone by *C. acetobutylicum*. The results obtained for solvent concentration (17.3 g/L) and overall conversion time (36h) demonstrated an improved performance over the separate hydrolysis and fermentation operation. Querishi et al studied the butanol production from wheat straw hydrolysate in batch cultures using *C. beijerinckii* P260. When wheat straw hydrolysate supplement with 60 g/L glucose, the resultant medium containing 128.3 g/L sugars was successfully fermented (due to product removal) to produced 47.6 g/L ABE, and the culture utilized all the sugars.

In China, the research on the substrate substitution for butanol fermentation is also attracting much attention. Chen et al [52] reported that 12.8 g/L total solvent concentration and 29.9% yield were obtained in batch broth by *C. acetobutylicum* grown on the rice straw enzymatic hydrolysate. Li et al [53] investigated enzymatic hydrolysis of steam-exploded corn stalk in a membrane reactor coupled with fermentation of acetone and butanol by *C. acetobutylicum* AS1.132. A butanol yield of 0.14 g/g (cellulose plus hemicellulose) and the maximum butanol productivity of 0.31 g/(L•h) were obtained with the dilution rate of 0.075/h. The conversion rates of cellulose and hemicellulose were 72% and 80%, respectively.

*Continuous Fermentation*

Continuous culture technique can be used to improve reactor productivity and to study the physiology of the culture in a steady state. Reports have indicted that continuous cultures can be utilized with the same efficiency as batch cultures as regards solvent concentration and yield but with improved efficiency as regards productivity. However, as a precaution of single-stage continuous processes for the industrial-scales, solvent production is not stable and decline with time, with a concomitant increase in acid formation, although high solvent yields can be achieved for various time periods [41]. Fick et al [54] maintained the stable continuous culture of *C. acetobutylicum* on a complex medium containing 40 g/L glucose for two months. The solvent yield was 13 g/L but less efficient use of the substrate. Two-stage or multistage continuous fermentation systems have been investigated in an attempt to separate the propagation phase from the production stage. A laboratory-scale two-stage system was reported by Bahl et al [55] Solvent concentration of 18.2 g/L (12.78 g/L butanol) with a yield of 0.34 g/g and a productivity of 0.55 g/L per h were obtained from *C. acetobutylicum* DSM1731 in a two-stage phosphate-limited chemostate. The Soviet Union in the 1960s had been a multi-level continuous fermentation (7 to 11 fermentation tank) solvent production [56] China in the same period has also carried out relevant research. Shanghai Institute of Plant Physiology and Ecology, Chinese Academy of Science started to the acetone butanol multi-level continuous fermentation with a solvent yield of 2.3 times higher than the batch fermentation.

Generally, the materials for immobilization of acetone alcohol Clostridium cell were absorbents such as clay brick, porcelain and etc. Qureshi et al [57] immobilized cells of *C. beijerinckii* onto clay brick particles by adsorption and achieved reactor productivity on the order of 15.8 g/L•h. The studies of using immobilized cells combined with advanced solvent recovery technologies for the fed-batch or continuous fermentation has also carried out in recent years.

*Solvent Recovery Process*

Recently, some new on-line product recovery technologies are available, such as membrane-based systems, liquid/liquid extraction, adsorption and gas stripping. There is no answer to the question of which technology is most suitable (Table 6.3). Membrane-based system shows a high selectivity for solvents, but might suffer from clogging and fouling and seem to be more suitable for immobilized cells. Liquid/liquid extraction also has a high selectivity, but emulsions might form that render the process less suitable. On the other hand, gas stripping does not lead to complete remove of solvents from the fermentation broth, but nor does it suffer from particulate substrates or from clogging or fouling by biomass. All these procedures can be designed to allow on-line product recovery, so that butanol toxicity, the major problem of the traditional process, is reduced.

High concentration of substrates and cells results in high productivity. However, a sugar concentration of more than 60g/L in a nonintegrated process without product removal is not used, which lead to a low concentration of the cells of solvent-clostridium due to the butanol inhibition. Usually, the maximum concentration of total solvents (ABE) in the fermentation broth is 20 g/L, of which butanol is only about 13 g/L. Ezeji [58] et al devised the integration of fed-batch and gas stripping system to recovery solvent in ABE fermentation, a total of 500 g of glucose was used to produced 232.8 g of solvents in 1 L culture broth. Qureshiet al [59] discussed the separation of butanol from aqueous solutions and fermentation broth by adsorption of silicalite, resins, bone charcoal, activated charcoal, bonopore, and polyvinylpyridine. Silicalite appeared to be the more attractive to concentrate butanol from dilute solutions (5 to790-810g/L) and less energy requirement for butanol recovery (1,948 kcal/kg butanol).

**Table 6.3. Novel product recovery techniques for clostridial solvent fermentation [49]**

| Method | Principle | Advantage | Disadvantage |
|---|---|---|---|
| Adsorption | Adherence of solvents to e.g. silicalite or ion-exchange resins, heat regeneration | – | High price of material, low capacity, low selectivity, possible fouling |
| Gas stripping | Heating of effluent, purging with gas, condensation of solvent/water vapours | Simple to perform, low chance of clogging or fouling | Low selectivity, no complete removal of solvents, more energy required compared to membrane-based processes |
| Liquid/liquid extraction | Contact of water-immiscible solvent with fermentation broth, recovery of dissolved acetone/butanol/isopropanol by distillation | High capacity, high selectivity, low change of clogging or fouling | Expensive to perform, possible forming of emulsions |
| Membrane evaporation | Selective diffusion of solvents across a porous membrane, recovery of evaporated vapours by applying vacuum and condensation | Smaller membrane area required compared to pervaporation, simple to perform | Possible clogging or fouling |
| Perstraction | Similar to liquid/liquid extraction, with a membrane separating fermentation broth and extractant | High selectivity, simple to perform | Large membrane area required, possible clogging or fouling |
| Pervaporation | Selective diffusion of solvents across a non-porous membrane, recovery of evaporated vapours by applying vacuum or sweep gas | High selectivity compared to membrane evaporation, simple to perform | Lower membrane flux compared to membrane evaporation, possible clogging or fouling |
| Reverse osmosis | High-pressure separation of dilute aqueous solution into a concentrated one and pure water by use of a semipermeable membrane, distillation of concentrated solution | Lower costs than conventional distillation | Possible clogging or fouling |

In order to improve efficiency and reduce the absorption adsorbent costs, the biomass absorption method to produce anhydrous ethanol appeared at domestically and abroad [60]. Hassaballh et al [61] reported the use of cornmeal as adsorber to condense 85% (w/w) alcohol vapor and the concentration of product up to 99.8% (w/w) . Westgate et al [62] indicated that starch was a well-established adsorption agent for drying ethanol, and the purity of ethanol and energy consumption were both promising.

#### *6.5.2.3. Energy Efficiency Analysis*

During AB production, about 50 % (w/w) of the fermented sugars were converted to $CO_2$ and $H_2$, and 33-39 % to solvents [56]. In a Russian AB plant, the 4.7% (w/v) of sugar equivalents in the incoming fermentation substrate (a mixture of starch, maltodextrins, sucrose, and pentoses) were reduced to about 0.5%. At the end of an average fermentation, the broth contained 6.4 g of acetone, 10.0 g of butanol, and 1.5 g of ethanol per liter. In laboratory experiments, the *clostridia* fermented pentose hydrolyzates containing the relative amount of sugars in 2.7 tons starch to 1 ton of solvents. Per mole of hexose (glucose), about three moles of $CO_2$ and H2 are formed; during the fermentation, 1.7 tons of gases are formed per 1 ton of solvents (1,649 kg $CO_2$, 97%w/ w, and 51 kg $H_2$, 3 % w/w). A simple calculation using an average yield is shown in Table 6.4, although the sugar content (in "starch equivalents") for a mixture of hexoses and pentoses is difficult to quantify.

**Table 6.4. End products of the AB process and the combustion energy of the products**

| Product | Weight (kg)[a] | Combustion energy (kJ/g) | Net calorific value per ton of substrate (MJ/t)[b] |
|---|---|---|---|
| Butanol | 207±10 | 33.1 | 6,852 |
| Acetone | 100±8 | 28.5 | 2,850 |
| Ethanol | 56±15 | 26.7 | 1,495 |
| $CO_2$ | ≈600 | – | – |
| $H_2$ | ≈19 | 119.8 | 2,276 |
| Total AB | | | 13,273 |

[a]Kilogram of product per ton of starch equivalent

[b]Kilojoules of combustion energy (net calorific value) in end product per ton of starch equivalent

#### *6.5.2.4. Prospect*

Improvements and developments in both the process technology and the overall performance of the microorganisms utilized in AB fermentation could result in a marked improvement in the economic competitiveness of the fermentation route for solvent production. As direct fermentation of cellulosic biomass to butanol in a consolidated bioprocess with a celluolytic solventogenic bacterium remains a future task, butanol fermentation process still need to be improved in a number of areas including the following : 1) the optimization of solvent production with alternative fermentation substrates derived from waste- and lignocellulose-based feedstock; 2) Improvement of butanol tolerance by manipulation of growth and production conditions; 3) development of cheap and efficient alternative methods of solvent recovery; 4) full integration in a biorefinery concept making as much use as possible of by-products.

### 6.5.3. Hydrogen

#### *6.5.3.1. Introduction*

The main advantages of hydrogen as the future fuel mainly due to its high conversion efficiency, renewable, high heating value (122kJ/g) and not directly (during the process of energy generation) contributing to the green house effect. Hydrogen can be safely transported in pipelines. In comparison, the energy yield of methane is 50.1 kJ/g and of ethanol is only 26.5 kJ/g. Hydrogen can also be used for many industrial applications, including synthesis of ammonia, alcohols and aldehydes as well as hydrogenation of petroleum. Methane, in contrast, is of little commercial value, except as a fuel. In addition, as hydrogen can be produced from many natural sources, it is expected to have stable price in the future, independent of the fluctuation in price and availability of single sources. Hydrogen also allows flexibility in balancing centralized and decentralized power supply [63].

Many countries in the world have started to implement a diversified energy strategy, making efforts on research to develop hydrogen energy. Industrialized nations have increased their investment in development of hydrogen energy by at least 20.5% each year in the past five years, according to a survey by the Department of Energy of USA. Japan is one of the earliest countries in the world starting studies on hydrogen energy. Now, in Japan, the development of fuel cells and technologies on the manufacturing, transportation and storage of hydrogen has been basically mature. The United States has paid much attention to hydrogen energy. In 2003, the Bush administration invested 1.7 billion US dollars in a hydrogen fuel initiative which focused on the industrialized production, storage and application of hydrogen energy. The European Union has also stepped up the development and application of hydrogen energy. The five Nordic nations, Norway, Denmark, Finland, Sweden and Iceland have established the Nordic Energy Research recently to improve their capability in producing bio-hydrogen. China has developed a sample vehicle driven by hydrogen fuel cells, marking that China has the capability to develop hydrogen-powered fuel cell engines.

The discovery and development of less expensive methods of bulk production of hydrogen will accelerate the establishment of a hydrogen economy. At present hydrogen is commonly produced by extraction from hydrocarbon fossil fuels via a chemical path. Hydrogen may also be extracted from water by electrolysis or biosynthesis in an algae bioreactor, from biomass by pyrolysis or microorganism fermentation. These methods are less developed for bulk generation in comparison to chemical paths derived from hydrocarbons.

*Hydrogen Production from Hydrocarbon*

Most of today's hydrogen is produced at oil refineries or by the chemical industry, largely using steam to reform natural gas. Steam methane reforming is the most common and cheapest way to produce hydrogen at present. It involves the heating of methane ($CH_4$), of which natural gas is mostly composed, in a catalytic reactor. This strips away the hydrogen atoms, and steam is then added to the process to free up more hydrogen, with carbon dioxide as a byproduct. Roughly 48% of worldwide hydrogen production comes from this fully commercial process. It must be stressed that operation of the hydrogen plant itself was the source of the majority of the GHG emissions (65%) with the remaining emissions coming from the plant's construction and from natural gas production and transport. Coal can also be reformed to produce hydrogen, through gasification. This is a commercial procedure as well, but less efficient and emits more $CO_2$ than steam methane reforming. The size of the world's remaining coal reserves has prompted some scientists to suggest that coal would be the main feedstock for hydrogen, which allow countries like China to move to the fuel sooner [64].

*Electrolytic Hydrogen Production*

At present, roughly 4% of the world's hydrogen is derived from the electrolysis of water. This process involves the use of electricity to split water into hydrogen and oxygen atoms, and is already cost-effective for producing extremely pure hydrogen in small amounts. But electrolysis remains expensive at larger scales, primarily because of the electricity, which currently costs on average three to five times as much as a fossil fuel feedstock [64].

*Biohydrogen*

One of the promising hydrogen production approaches is conversion from biomass. Alternative thermochemical (pyrolysis and gasification) and biological (biophotolysis and fermentation) processes can be practically applied to produce hydrogen.

However, hydrogen produced from biomass in thermolysis process mostly contains different gases like $O_2$, CO, $CH_4$ and some amount of moisture. Sometimes, presence of these gases drops the heating value of hydrogen and causes problems in efficient burning of fuels [65]. Biological hydrogen production process is found to be more environment friendly and less energy intensive as compared with thermochemical and electrochemical processes. Hydrogen may be produced in biosystem, which includes two ways of light-drive process and anaerobic fermentation, the former is theoretically perfect process with transforming solar energy into hydrogen by photosynthetic bacteria directly. However, due to the low utilization efficiency of light and difficult design of light reactor, this method is hard to be applied in practice. The latter carries out anaerobic fermentation by the hydrogenogens, which has many advantages, such as rapid, simple, easy operation, and hydrogen production by renewable resources and organic waste. However, the yield and rate of hydrogen production are still low at present [66]. To achieve the goals of practical biohydrogen processes, advanced low cost bioreactors, systems with oxygen-tolerant hydrogenate and high efficiencies (greater than 10% total solar conversion) need to be developed and engineered. On the basis of these facts, this section focuses on the development of biohydrogen technologies.

#### *6.5.3.2. Key Technology and Process*

The phenomenon of biological hydrogen production was observed one century ago. When the oil crisis broke out in 1970s, the technology started receiving attention, especially in hydrogen production by photosynthetic process. Researchers have started to investigate hydrogen production with anaerobic bacteria since 1980s. There are three types of microorganisms for hydrogen generation: cyano-bacteria, anaerobic bacteria, and fermentation bacteria. Biological hydrogen can be generated from plant by biophotolysis of water using microalgae (green algae and cyanobacteria), fermentation of organic compounds, and photodecomposition of organic compounds by photo-synthetic bacteria [67]. The process of biological hydrogen production can be broadly classified into two distinct types. One is light dependent and the other is light independent process. Light mediated processes include direct or indirect biophotolysis and photo fermentation, whereas dark fermentation is the major light independent process.

*Light-Driven Processes*

Light-driven processes are still at the conceptual stage because of their questionable economics and remain to be demonstrated at full-scale. Three types of light-driven processes are described below:

(1) Direct biophotolysis process can take place in anaerobically adapted green microalgae. H2 production is a biological process which utilizes solar energy and photosynthetic systems of algae to convert water into chemical energy.

$$2H_2O + \text{solar energy} \rightarrow 2H_2 + O_2$$

The two photosynthetic systems responsible for photosynthesis process are: ( i ) Photo System I (PS I ) which produces reductant for $CO_2$ and ( ii ) Photo System II (PS II ) which splits water to evolve $O_2$. It was also clear that this process was rather feeble, fragile and fraught with difficulties mainly due to $O_2$ inhibition of the hydrogenase reaction. However, recent work on the mechanism and three-dimensional structure of the reversible hydrogenase used in H2 production provided scant hope for any eventual development of a reversible hydrogenase reaction sufficiently resistant to $O_2$ to be able to operate under practical conditions of direct biophotolysis [68].

(2) In indirect biophotolysis, $H_2$ is produced as following:

$$6H_2O + 6CO_2 \rightarrow C_6H_{12}O_6 + 6O_2$$

$$C_6H_{12}O_6 + 12H_2O \rightarrow 12H_2 + 6CO_2$$

Many types of green algae and cayanobacteria, besides having the ability to fix $CO_2$ via photosynthesis, also have the ability to fix nitrogen from the atmosphere and produce enzyme that can catalyze the second $H_2$ generating step. The result showed that the specific growth rate and growth yield based on $CO_2$ decreased with an increase in light intensity or $CO_2$ concentration. $CO_2$ uptake rate per unit cell decreased linearly with the initial $CO_2$ concentration in the gas phase. $CO_2$ injection during the growth phase of the cell resulted in an increase in cell concentration as well as $H_2$ production per gram of cell [69].

The overall biochemical pathways for the photo fermentation process can be expressed as follows:

$$(CH_2O)_2\ \text{Ferredoxin} \rightarrow \text{Nirogenase} \rightarrow H_2$$

Photosynthetic bacteria have long been studied for their capacity to produce significant amounts of $H_2$. The advantage of their use is in the versatile metabolic capabilities of their organisms and the lack of Photo system II (PS II ), which automatically eliminates the difficulties associated with $O_2$ inhibition of $H_2$ production. Phototrophic bacteria require an organic or inorganic electron source to drive their photosynthesis. They can utilize a wide range of cheap compounds. These photoheterotrophic bacteria have been found suitable to convert light energy into $H_2$ using organic waste as substrate in batch processes, continuous cultures, or immobilized whole cell system using different solid matrices like carrageenan, agar gel, porous glass, and polyurethane foam [70].

(3) Dark-fermentation process

Hydrogen production via dark fermentation is a special type of anaerobic digestion comprising only hydrolysis and acidogensis. It leads to the production of hydrogen, carbon dioxide and some simple organic compounds (such as volatile fatty acids (VFA) and alcohols). These readily degradable organic compounds can be used for further methane production. *Clostridium* sp. and *Enterobacter* sp. are frequently reported to produce hydrogen from carbohydrates in pure culture. *Rhodopseudomonas* sp., *Citrobacter* sp. and *Bacillus* sp. have also been used. *Caldicellulostiruptor saccharolyticus* has been used under hyperthermophilic conditions. Although it is possible to achieve a higher hydrogen yield with use of a pure culture, it is less useful for industrial applications, because the pure culture can easily be contaminated by various hydrogen utilizers, such as methanogens or sulfate-reducing bacteria. In addition, a mixed culture is more stable during the process of hydrogen production from bio-waste. Regarding mixed cultures, *Clostridium* sp. and *Thermoanaerobacterium* sp. are the species most often indicated under mesophilic and thermophilic condtions, respectively.

So far, especially phototrophic microorganisms and mesophilic and moderate thermophilic heterotrophic microorganisms have been considered, but not thermophiles growing above 60 °C. Van Neil et al [71] investigated two extreme thermophiles growth and hydrogen production, about 83% of the theoretically hydrogen production were obtained. The maximum hydrogen production rates were 11.7 and 5.1 mmol/g dry weigh/h by *Caldicellulosiruptor. saccharolyticus* and *Thermotoga. elfii* cultures kept thermostatically at 70 °C and 65°C, respectively.

(4) Two-step process of dark- and photo-fermentation

A combination of dark and photo fermentation in a two-stage hybrid system can improve the overall yield of hydrogen [72]. The synergy of the process lies in the maximum utilization of the substrate which otherwise fails to achieve complete conversion due to thermodynamic limitation. In the first stage, the biomass is fermented to acetate, carbon dioxide and hydrogen in a thermophilic dark fermentation. In a separate photo bioreactor acetate is converted to hydrogen and carbon dioxide. The combination could be expected to reach as close as possible to the theoretical maximum production of 12 mol of $H_2$ (mol glucose)-1 equivalent ccording to the following reactions:

( i ) Stage I —— Dark fermentation (facultative anaerobes)

$$C_6H_{12}O_6 + 2H_2O \rightarrow 2CH_3COOH + 2CO_2 + 4H_2$$

( ii ) Stage II —— Photofermentation (photosynthetic bacteria)

$$2CH_3COOH + 4H_2O \rightarrow 8H_2 + 4CO_2$$

Several studies have been undertaken to increase the hydrogen yield by the combination process of dark- and photo-fermentation. Claassen et al [73] used potato steam peels for hydrogen production by the combination of heterotrophic and photo-heterotrophic fermentation step. Eroğlu et al [74] introduced two-stage hydrogen production process from live mill wastewater. The first two-stage process involved dark-fermentation by activated sludge cultures and the second two-stage process involved photo fermentation by *Rhodobacter sphaeroides*. Yokoi et al [75] reported that a high yield of hydrogen of 7.0 mol $H_2$/mol glucose from the starch remaining in the starch residues was attained in two-step repeated batch cultures containing *C. butyricum* and *E. aerogenes* , and by *Rhodobacter* sp. M-19. In two stage processes, the major bottleneck lies in the scaling up the photo-fermentation process. The improvement of these processes will definitely improve overall $H_2$ yields, as well as the economy of the process.

*Fermentative Hydrogen Production from Lignocellulose*

Lignocellulosic biomass is widely available as a cheap raw material for biological energy production. The utilization of renewable lignocellulose for biological hydrogen production is a global trend especially for the major agricultural country with large population like China. However, as we all known that this goal is difficult to achieve. Successful biological conversion of lignocellulose to hydrogen strongly depends on its biodegradability by fermentative bacteria. The main focus in ongoing research is on delignification and hydrolysis of lignocellulosic biomass [63]. Various studies using lignocellulose as substrate have been carried out for microbial $H_2$ production. Lay found that the heat-shock digester in which clostridia predominated the hydrogen-producing bacteria could transform microcrystalline cellulose into hydrogen. However, the hydrogen significantly inhibited the hydrogen-producing activity of sludge when initial microcrystalline cellulose concentrations exceeded 25.0 g/L [76]. Himeshima et al [77] developed a photoinduced hydrogen production system coupling cellulose hydrolysis by cellulase and glucose dehydrogenase and hydrogen production with platinum nanoparticles as a catalyst using the visible light photosensitization of Mg chlorophyll-a from *Spilurina*. Taguchi et al [78] investigated continuous hydrogen production by fermentation of Avicel hydrolysate using *Clostridium* sp. strain No.2.

Although these researches on the use of cellulosic chemicals have made progress, it still can not be proved the utilization of natural raw material. Fan et al [79] reported the conversion of acid-pretreated wheat straw into biohydrogen gas by cow dung compost. The maximum cumulative hydrogen yield of 68.1 ml $H_2$/g TVS after 126.5 h. De Vrije et al [80] tested the hydrolysate of pretreated *Miscanthus*, a common plant in Netherland, as substrate to promoted biohydrogen production by the extreme thermophilic bacterium Thermotoga elfii. Our lab [81] studied the biohydrogen production by simultaneous saccharification and fermentation from steam-exploded corn straw (SECS) using *Clostridium butylicum* AS1.209. The maximal hydrogen yield was 68 ml/g SECS and the conversion rate of substrate was 32% with the values of 72% hemi-cellulose and 60% cellulose.

*Reactor Design*

The critical challenges of the hydrogen fermentation are low hydrogen conversion efficiency and unstable production. One of the important factors governing the hydrogen conversion efficiency is process engineering, such as bioreactor design and operating parameters.

Technical systems for the production of phototrophic microorganisms are termed photobioreactors (PBR) [82]. The design of the technical and technological basis for PBR is the most important issue for economic success in the field of phototrophic biotechnology. Until recently, open ponds systems were the most important design principle for microalgae production. However, for the future applications, open pond systems for large-scale production seem to have a lower innovative potential than closed systems. Despite a certain variability in sharp, the most common technical design for open pond systems are raceway cultivators driven by paddle wheels and are usually operated at water depths of 15-20 cm. At these water depths, biomass concentration is up to 1,000 mg/L and productivities are 60-100 mg / (L day-1). Closed PBRs are characterized by the regulation and control of nearly all the biotechnologically important parameters as well as by the following fundamental benefits: a reduced contamination risk, no $CO_2$ losses, reproducible cultivation conditions, controllable hydrodynamics and temperature, and flexible technical design. Closed PBRs are currently tested for microalgae cultures in the following configurations: (1) the vertical column reactor, (2) flattened, plate-type systems and (3) the tubular reactors [83]. Flat panel reactors and tubular rectors show better efficiencies. It is worthwhile looking further into these two types of reactors, and scaling up of them for practical purposes.

Any of the current anaerobic reactor designs can be used for liquid substrates, e.g. UASB (up flow anaerobic sludge blanket), fluidized bed reactor, fixed-bed bioreactor or CSTR (completely stirred tank reactor). CSTR is the most popular continuously working reactors at laboratory scale because of its simplicity of building and operation. However, this system does not allow for high organic loading rates (OLR) and high biomass concentrations. The volumetric hydrogen production range is, therefore, relatively low [63]. Camilli et al [84] compared the performance of the three reactors for biohydrogen production via dark anaerobic fermentations. The CSTR showed both optimal $H_2$ yield (0.30 L H2/g carbohydrate ) and production rate (4.50 mmol $H_2$/L reactor, h) when operated at hydraulic retention time 32.9 h. The best performance of CSTR was probably due to the mechanical stirring that promoted both $H_2$ and $CO_2$ removal from the fermentation broth, therefore reducing feed-back inhibition phenomena.

#### *6.5.3.3. Energy Efficiency Analysis*

*Efficiency in Light-Driven Processes*

The efficiencies are based on photosynthetic active radiation (PAR). Sunlight is the ultimate energy source for microalgae. Although the wavelength range of solar radiation is very broad, the PAR between 400 and 700 nm can be used by microalgae. On an energy basis, 43% of the solar radiation is in the PAR region. For purple bacteria PAR region is 400–950 nm. On an energy basis 65.8% of the solar radiation is in this PAR region. This means that based on the total solar spectrum they would be 0.43 times the value. Photochemical efficiencies for the photoautotrophic hydrogen production are only 3–10%, when the oxygen is totally and immediately removed [83].

The performance of photosynthetic bacteria with respect to biological hydrogen production must be evaluated on the basis of several parameters. One of them is the yield coefficient of $H_2$ produced relatively to the carbon source consumed. Another important parameter is the efficiency of light, taking into account the $H_2$ production rate and the $H_2$ energy content, as well as the absorbed light. If the assumption can be made that the number of photons needed to produce 1 mole of hydrogen, not depending very much on the wavelength (in the utilizable spectrum of wavelengths), and it is assumed that all available light is absorbed in the bioreactor, then the efficiencies (with different kinds of artificial light) can be compared to the theoretical efficiency as calculated above (about 10% or more).

However, the main bottleneck of biological hydrogen production seems not to be the $H_2$ yield itself but light efficiency. This will determine the costs of PBR and their operation.

*Efficiency in Dark Fermentation*

In anaerobic conditions, the reaction that the organic matters (e.g glucose) break down to hydrogen can be described by equation as follow:

$$C_6H_{12}O_6+H_2O \rightarrow 2CH_3COOH+2CO_2+4H_2$$

The maximal theoretical yield for fermentative biohydrogen production processes operating in the absence of light is calculated to be 4 mol of $H_2$ per mole of glucose (33% of stoichiometric), but in practice only 10%-20% of the glucose energy content is recoverable as $H_2$ because of the limited metabolic energy derived from $H_2$ fermentations [84].

### *6.5.3.4. Prospect*

Research progress for identifying the major techno-economic bottlenecks of various bioprocesses for commercial production of hydrogen appears promising. However, high hydrogen yield remains to be the ultimate goal and challenge for the biohydrogen research and development. Enhancement of hydrogen yield may be possible by using suitable microbial strains, process modification, efficient bioreactor design and also redirection metabolic pathway by genetic and metabolic engineering technique. Biological hydrogen production, employing renewable biomass may be a potential answer to overcome some economic constrains to fulfill many of our energy needs [72].

## 6.5.4. Bio-Diesel

### *6.5.4.1. Introduction*

Bio-diesel is defined as the mono alkyl esters (methyl and ethyl esters) of long chain fatty acids derived from vegetable oils or animal fats, for use in compression-ignition (diesel) engines. It is formed from transesterification of vegetable oils with methanol (or ethanol). Bio-diesel has many merits as a renewable energy resource that includes being derived from a renewable, domestic resource, thereby relieving reliance on petroleum fuel, and it is biodegradable and non-toxic [85].

Like petroleum diesel, bio-diesel operates in compression-ignition engines. Bio-diesel is most often blended with petroleum diesel in rations of 2% (B2), 5% （B5）, or 20% （B20）. It can also be used as pure bio-diesel (B100). Bio-diesel fuels can be used in regular diesel vehicles without making any changes to the engines, although older vehicles may require replacement of fuel lines and other rubber components. It can also be stored and transported using diesel tank and equipment. Since bio-diesel is oxygenated, it is a better lubricant than diesel fuel, increasing the life of engines, and combusts more completely. Indeed, many countries are introducing bio-diesel blends to enhance the lubricity of low-sulfur diesel fuels. The higher flash point of bio-diesel makes it a safer fuel to use, handle, and store. With its relatively low emission profile, it is an ideal fuel for use in sensitive environments, such as heavily polluted cities [86].

### *6.5.4.2. Key Technology and Process*

Bio-diesel, an alternative biodegradable diesel fuel, is derived from triglycerides by transesterification with alcohols. The direct use or blending, microemulsions, thermal cracking and transesterification of vegetable oils have been explored as potential alternatives to diesel. The interest in the use of renewable fuel started with the direct use of vegetable oils as a substitute for diesel. However, their direct use in compression ignition engines was restricted due to high viscosity which resulted in poor fuel atomization, incomplete combustion and carbon deposition on the injector and the valve seats causing serious engine fouling. Other constraints of the direct application of vegetable oil were its low volatility and polyunsaturated character. To overcome these constraints, the processes like pyrolysis, micro-emulsification, transesterfication, etc. were especially developed. Pyrolysis of the vegetable oil resulted in products with low viscosity, high cetane number, acceptable amount of sulphur, water and sediments and copper corrosion values but unacceptable characteristics in the terms of their ash contents, carbon residues, and pour points. Similarly, micro-emulsion of vegetable oil decreased the viscosity of the oil but resulted in irregular injector needle sticking, heavy carbon deposits and incomplete combustion during 200 h laboratory screening endurance test. Out of these three methods, transesterification is the most feasible process adopted so far for the decreasing of viscosity. It consists of a sequence of three consecutive reversible reactions where triglycerides are converted to diglycerides which are further converted to monoglycerides followed by the conversion of monoglycerides to glycerol. In each step an ester is produced and thus three ester molecules are produced from one molecule of triglyceride. It gives glycerol as a by-product which has a commercial value[87].The transesterification can be conducted either in the presence or absence of a catalyst. The usual catalysts used are alkalis (NaOH, KOH), acids (sulfuric acid, HCl) or enzymes (lipases). The kinetics of acid-catalyzed and alkali-catalyzed reactions has been well studied and these processes have been commercialized. The transesterification of triglycerides by non-catalytic supercritical methanol (SCM), ethanol, propanol and butanol has also proved to be the most promising process. This section lays emphasis on the technical aspects of bio-diesel production by transesterification.

*Chemical Catalytic Transesterification*

Acids used for transesterification include sulfuric, phosphoric, hydrochloric, and organic sulfonic acid. Although transesterification by acid catalysis is much slower than that by alkali catalysis, acid-catalyzed transesterification is more suitable for glycerides because of relatively high free fatty acid contents and more water. Alkalis used for transesterification include NaOH, KOH, carbonates, and alkoxides such as sodium methoxide, sodium ethoxide, sodium propoxide, and sodium butoxide. Alkali-catalyzed transesterification speed approximately 4000 times faster than that catalyzed by the same amount of an acidic catalyst, and is thus most often used commercially [88]. The limits of alkali-catalytic transesterification are due to the sensitivity that this process has to purity of reactants, to the fatty acid, as well as to the water concentration of the sample.

Although chemical transesterification using an alkali-catalysis process gives high conversion levels of triglycerides to their corresponding methyl ester in short reaction times, the reaction has several drawbacks: it is energy intensive, it is difficult to recovery glycerol, it has to be removed the acidic or alkaline catalyst from the product, alkaline wastewater requires treatment, and free fatty acids and water interfere with the reaction, etc.

*Enzyme Catalyzed Transesterification*

Enzymes have been proposed to overcome the drawbacks facing the conventional chemically catalyzed bio-diesel production methods, and have shown promising results. Most lipases used as catalysts in organic synthesis are of microbial origin, such as *Candida rugosa, Psuedomonas fluorescens, Rhizopus oryzae, Burkholderia cepacia, Aspergillus niger, Thermomyces lanuginosa* and *Rhizomucor miehei*. Microbial lipases are easy to obtain, by fermentation processes and a few basic purification steps [85]. Köse et al [89] investigated the transesterification of the refined cotton seed oil by using immobilized enzyme from *Candida antarctica*, commercially called Novozym 435 in a solvent-free medium. Maximum methyl ester yield was 91.5% at the optimal methanolysis conditions as follows: 30% enzyme based on oil weight; oil/alcohol molar ratio 1:4; temperature 50 °C and reaction time 7 h. The degree of methanolysis was low in reaction systems due to the irreversible inactivation of the lipase by contacting with insoluble methanol (MeOH). To avoid serious deactivity of lipase in the presence of a high concentration of methanol, Shimada et al [90] developed a stepwise methanolysis system with immobilized *Candida antarctica* lipase. Two step batch methanolysis was most effective for the production of biodiesel fuel from waste oil: the first-step reaction was conducted in the presence of 1/3 molar equivalent of MeOH for the stoichiometric amount, and the second-step reaction was performed by adding 2/3 molar equivalent of MeOH. The conversion of waste oil to biodiesel fuel reached >90% in the two reaction systems, and the lipase catalyst could be used for >100 days without activity decrease.

Since the cost of lipase production is the main hurdle to commercialization of the lipase-catalyzed process, several attempts have been made to develop cost-effective systems. Combining a whole cell biocatalyst with the use of a recombinant microorganism can be expected to considerably decrease the cost of lipase production. Another useful approach to reduce the production cost is to use solvent-tolerant lipase. Further enhancement of lipase production may be achieved by genetic engineering.

*Transesterification in Supercritical Fluid*

The reaction mechanism of transesterification in supercritical fluid is same to the acid- or alkali transesterification. The difference lies in the transesterification of vegetable oils in supercritical methanol without using any catalyst. As for reactions in supercritical methanol and ethanol, nearly complete conversions can be achieved in a very short time (2-4 min). Saka et al [91] have developed a catalyst-free method for biodiesel fuel production by employing supercritical methanol. The supercritical treatment at 350 °C,43 MPa and 240 s with a molar ratio of 42 in methanol is the optimum condition for transesterification of rapeseed oil to biodiesel. Demirbas [92] investigated the change in yield percentage of methyl esters with supercritical methanol method with a molar ration of 41 with the critical temperature of 512.4 K and critical pressure of 8.0 MPa, which concluded that increasing reaction temperature, especially supercritical temperature had a favorable influence on ester conversion. Kusdiana et al [93] investigated the effect of water on the yield of methyl ester in transesterification of triglycerides and methyl esterification of fatty acids as treated by catalyst-free supercritical methanol. The present of water at a certain amount could enhance the methyl esters formation, as complete conversions were always achieved regardless of the content of water. In addition, Madras et al [94] investigated that biodiesel was enzymatically synthesized in supercritical carbon dioxide, but only 30% conversions were obtained in the enzyme-catalyzed reactions in supercritical carbon dioxide.

Compared with the catalytic processes under barometric pressure, the supercritical methanol process is non-catalytic, simplifying purification process of products, shortening reaction time, environmental friendly and lower energy consumption. However, the reaction requires temperature of 527-675 K and pressure of 35-60 MPa, which increased equipment investment costs. It is hard to achieve the large-scale industrial production in the short term.

*Microalgae for Biodiesel Production*

Microalgae have been suggested as very good candidates for fuel production because of their advantages of higher photosynthetic efficiency, higher biomass production and faster growth compared to other energy crops. Additionally the land crops may require substantial energy inputs for irrigation, planting, fertilization, and harvesting, these can be greatly minimized with an aquatic crop, with a well-designed system. Developing high lipid content microalgae or "bioengineering microalgae" for bio-fuels production would become new and promising way in bio-fuels production in the future. The National Renewable Energy Laboratory (NREL) of the USA has identified a gene that plays a important role in controlling Acetyl-CoA carboxylase (ACCase) activity which is a rate-limiting enzyme in fatty acid synthesis in plants and animals, and has studied naturally occurring genetic mutations in algae strains that affect oil synthesis [86]. Researches at Tsinghua University of China [95] developed an integrated method for bio-diesel production from heterotrophic microalgal oil by acidic transesterification. High lipid content (55%) accumulated in *Chlorella protothecoides* cells. The microalgal was extracted from these heterotrophic cells by using n-hexane. The best process combination was 100% catalyst quantity with 56:1 molar ration of methanol to oil at 30 °C.

Although there is significant interest in algal bio-diesel, it is important to keep in mind that this is still years away from being ready for actual commercial implementation. Two approaches appear possible to achieve high oil algae, one is using of extremophile algae which can tolerate extreme conditions, the other is using of PBR for keeping invasive strains out, and optimizing the growth environment. Unfortunately, the focus does not appear to be in developing lower cost PBR to bring down the capital cost for building a "PBR farm", which ultimately will present a barrier to commercialization.

*Pyrolysis of Single Cell Oils for Bio-Diesel Production*

The main barrier to the production of bio-diesel is the shortage of oil or fat. Finding new oil or fat resource becomes a hot research task. Using lignocellulosic biomass as the substrate to produce single cell oil may be a potential oil resource for the production of biodiesel because plenty of lignocellulosic biomass is produced every year and it is the main resource for producing renewable energy. Our lab [96] studied the pyrolysis behaviors of fermented mass, SSF from steam-exploded wheat straw by *Microsphaeropsis* sp., in tubular furnace pyrolysis equipment. n-Hexadecanoic acid, hexadecanoic acid, methyl ester, 9-octadecenoic acid and methyl ester were found in the hexane-extracted fraction of liquid products. The harvest rates of n-hexadecanoic acid, methyl ester and 9-octadecenoic acid, methyl ester increased from 6.27% to 30.0% and from 4.18% to 10.26%, respectively, when pyrolysis was performed with permutit as catalyst by loading rate of 20% (W/W) of the fermented mass. GC/MS of n-hexane fraction of pyrolysis bio-oil from fermented mass is shown in

Figure 6.3. This research found fatty acid methyl eaters, which were used as bio-diesel, could be produced by pyrolysis from the mass containing lipid for the first time, hence providing a novel way for the production of bio-diesel.

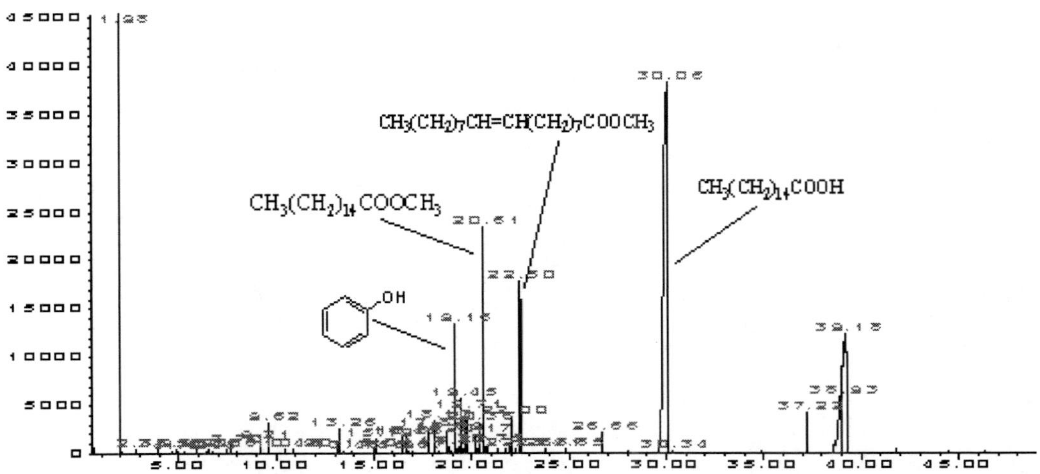


Figure 6.3. GC/MS of n-hexane fraction of pyrolysis bio-oil from fermented mass

### *6.5.4.3. Energy Efficiency Analysis*

An entire fuel-cycle energy balance assessment should include not only the energy content of bio-diesel and the energy consumed during its manufacturing process, but also the energy absorbed/released by all the necessary processes to obtain the final product. Studies, carried out for bio-diesel, demonstrated that the energy overall balance (including extraction, refining and esterification) is positive. It has to be pointed out that the overall energy balance depends on the use given to the rape-seed straw. It may be cut and utilized as an energy source or left and ploughed in soil. Several fuel-cycle energy balances for rape-seed, depending on the rape-seed yield per hectare, are shown in Table 6.5 [97].

To optimise the overall process, seed pressing should be carried out close to the crop growing area. This reduces the transportation energy requirements. Bio-oil is then transported to centralised refining plants, where economies of scale are earned. In addition, the use of dry residuals and crushed seed, still containing 5-10 % oil, may also be burnt to produce electricity. However, this possibility has not been considered in the present techno-economic analysis. As regard to the sunflower, an energy balance for bio-diesel from this crop gives similar figures. One hectare of sunflower produces less bio-diesel, but on the other hand more straw is obtained and less fertilisers and agro-chemicals are needed.

**Table 6.5. Energy balances for bio-diesel from rape-seed in MJ/ha, depending on the rape-seed yield per ha.**

| Parameters | ETSU [1] | ARC [2] | Levington [3] | Levington |
|---|---|---|---|---|
| Seed yield(t/ha) | 3.2 | 3.6 | 4.08 | 4.08 |
| **Energy inputs** | | | | |
| Agriculture | -4300 | -4300 | -4687 | -4945 |
| Fertilisers | -12800 | -12800 | -7190 | -7190 |
| Agro-chemicals | -600 | -600 | -337 | -337 |
| Seeds | -200 | -200 | -35 | -35 |
| Storage/packaging | -300 | -300 | -282 | -282 |
| Transport | -774 | -871 | -723 | -1122 |
| Processing | -16071 | -18080 | -17251 | -17251 |
| **Total input** | **-35045** | **-37151** | **-30506** | **-31162** |
| **Final outputs** | | | | |
| Bio-fuel | 45800 | 51525 | 54346 | 54346 |
| Cake | 3700 | 4136 | 1316 | 1316 |
| **Total output without straw** | **49500** | **55688** | **55662** | - |
| ***Energy output/Energy input(straw out)*** | ***1.41*** | ***1.50*** | ***1.82*** | - |
| **Net energy balance without straw** | **14455** | **18537** | **25157** | - |
| Straw | 38400 | 43200 | 0 | 60000 |
| **Total output with straw** | **87900** | **98888** | - | **115662** |
| ***Energy output/Energy input(straw in)*** | ***2.51*** | ***2.66*** | - | ***3.71*** |
| **Net energy balance with straw** | **52855** | **61737** | - | **84500** |

[1] Energy Technology Support Unit, 1996 study.

[2]Adapted by ECOTEC Research and Consulting the ETSU study with higher yield of rape-seed per ha, according to the more recent standard agricultural pocketbooks (ARC 1999) data.

[3]Levington Agriculture, 2000 study.

Sources: Adapted from British Association for Bio Fuels and Oils (BABFO) (www.biodiesel.co.uk/emissions_from_liquid_biofuels.htm, www.biodiesel.co.uk/levington.htm)

### *6.5.4.4. Prospect*

Bio-diesel has become more attractive recently because of its environmental benefits and the fact that it is made from renewable resources. The remaining challenges are its high cost and limited availability of fat and oil resources. There two aspects of the cost of bio-diesel one is the raw material (fats and oils) cost and the other is processing cost. The cost of raw materials accounts for 60 to 75% of the total cost of bio-diesel. The use of used cooking oil can decrease the cost significantly, while, the quality of used cooking oils is destroyed obviously. Studies are needed to find a cheaper way to utilize used cooking oils to make bio-diesel. There are several choices such as removing free fatty acid from used cooking oil before transesterification, using acid catalyzed transesterification, or using high pressure and high temperature. In term of production cost, there are also two aspects, the transesterification process and by-product (glycerol) recovery. A continuous transesterification process is one choice to decrease the production cost. The foundation of this process is shorter reaction time and greater production capacity. The recovery of high quality glycerol is another way to drop production cost. Because little water is present in the system, the bio-diesel glycerol is more concentrated. Unlike the traditional soap glycerol recovery process, the energy required to recover bio-diesel glycerol is low due to the elimination of the evaporation process. In addition, the process also is simpler than soap glycerol recovery since there is a negligible amount of soap in bio-diesel glycerol. This implies that the cost of recovering high quality glycerol from bio-diesel glycerol is lower than that of soap glycerol and that the cost of bio-diesel fuel can be dropped if a bio-diesel plant has its own glycerol recovery facility [98].

## 6.6. The Future of Bio-Fuels

Over the last few years, bio-fuels have gained worldwide interests for their potential to reduce GHG emissions, improve energy security, and enhance rural development. At the same time, reports on the environment and social costs associated with bio-fuel production have stirred up a storm of controversy. Nevertheless, there remain several silver linings-in terms of on-gonging developments in feedstock selection and production technologies-that may yet allow bio-fuels to fulfill their promise as a viable source of renewable energy [99].

Almost all of the commercially available bio-fuels today are produced from either starch or sugar rich crops, or oilseeds. Producing bio-fuels from these sources is less than ideal competing with food or feed production. Recent research attention has turned to the utilization of dedicated feedstocks for bio-fuel production, including perennial grasses, wood, microalgae and agricultural, forestry or municipal waste.

In order to diversify the bio-fuel feedstock resource base, it is needed to develop process technologies which can convert these next generation feedstocks to liquid fuels. There are two primary pathways for converting biomass to bio-fuel, one is biochemical and the other is thermochemical conversion. Biochemical conversion pathways are used to convert cellulosic biomass to bio-fuel by breaking down the recalcitrant components of plant material-cellulose and hemicellulose-into sugars, which are then fermented to bio-fuels. The limiting factor in term of yield is cellulose breakdown rate, which must be accomplished by pretreatment and enzymatic hydrolysis. A second limiting factor in biochemical conversion pathways is the lack of high-productivity and inhibitor-tolerance microorganism used in industrialization. Metabolic engineering including block of undesirable pathways and induction of gene expression associated with desirable pathways to enhance the production of bio-fuels and organic acids using lignocellulosic biomass is under progress. However, no single cost effective and efficient technology is currently available to meet the challenges of large-scale utilization of lignocellulosic biomass. Therefore, we suggest here an integrated approach including efficient bioreactor design, selection and optimization of physical and chemical conditions for several organisms may be used under co-cultivation conditions. Further, strain improvement for enhancing cellulase biosynthesis using mutagenesis, metabolic engineering and genomics approaches, should be used for the lignocellulosic bioconversion processes into a powerful technology to produce the value added and industrially significant products [100].

In recent years, there has been a policy driven and economically facilitated for bio-fuel production, and only now are the environmental costs of this becoming apparent. A careful evaluation of potential costs and benefits of the range of bio-fuel production processes is needed to shape more informed policy in the future. In particular, policy instruments to enhance the traceability of bio-fuel feedstocks need to be developed to ensure that they are produced in environmentally and socially responsible ways.

## REFERENCES


[1] Demirbas, A., Biomass resource facilities and biomass conversion processing for fuels and chemicals. *Energy Conversion and Management* 2001, 42, (11), 1357-1378.

[2] Hazell, P. B. R.; Pachauri, R. K.; International Food Policy Research, I.; Energy; Resources, I., Bioenergy and Agriculture: Promises and Challenges. *International Food Policy Research Institute*: 2006.

[3] Liu, S. C.; Jiang, J. C., Conversion technology and utilization of biomass energy ( I )——technology and equipment ofbBiomass shaped fuel production. *Biomass Chemical Engineering(Chinese)* 2007, 41, (4), 59-63.

[4] Zeng, X.; Ma, Y.; Ma, L., Utilization of straw in biomass energy in China. *Renewable and Sustainable Energy Reviews* 2007, 11, (5), 976-987.

[5] Overend Ralph, P., *Biomass conversion technologies. National Renewable Energy Laboratory*, Golden, Colorado, USA 2004.

[6] Liu, B. L.; Jiang, J. C., Conversion technology and utilization of biomass energy(VI )——technology and equipment for biomass power generation. *Biomass Chemical Engineering(Chinese)* 2008, 42, (2), 55-60.

[7] Briens, C.; Piskorz, J.; Berruti, F., Biomass valorization for fuel and chemicals production–a review. International *Journal of Chemical Reactor Engineering* 2008, 6, (6), 2.

[8] Balat, M.; Balat, H.; Öz, C., Progress in bioethanol processing. Progress in Energy and Combustion *Science* 2008, 34, (5), 551-573.

[9] Bridgwater, A. V., Renewable fuels and chemicals by thermal processing of biomass. *Chemical Engineering Journal* 2003, 91, (2-3), 87-102.

[10] Bridgwater, A. V.; Toft, A. J.; Brammer, J. G., A techno-economic comparison of power production by biomass fast pyrolysis with gasification and combustion. *Renewable and Sustainable Energy Reviews* 2002, 6, (3), 181-246.

[11] Klein, A. Gasification: An Alternative Process for Energy Recovery and Disposal of Municipal Solid Wastes. *Columbia University*, 2002.

[12] Berndes, G.; Azar, C.; Kåberger, T.; Abrahamson, D., The feasibility of large-scale lignocellulose-based bioenergy production. *Biomass and Bioenergy* 2001, 20, (5), 371-383.

[13] McKendry, P., Energy production from biomass (part 1): overview of biomass. *Bioresource Technology* 2002, 83, (1), 37-46.

[14] Friedl, A.; Padouvas, E.; Rotter, H.; Varmuza, K., Prediction of heating values of biomass fuel from elemental composition. *Analytica Chimica Acta* 2005, 544, (1-2), 191-198.

[15] Nussbaumer, T., Combustion and Co-combustion of Biomass: Fundamentals, Technologies, and Primary Measures for Emission Reduction. *ENERGY AND FUELS* 2003, 17, (6), 1510-1521.

[16] Leung, D. Y. C.; Yin, X. L.; Wu, C. Z., A review on the development and commercialization of biomass gasification technologies in China. Renewable and Sustainable *Energy Reviews* 2004, 8, (6), 565-580.

[17] Iwasaki, W., A consideration of the economic efficiency of hydrogen production from biomass. *International Journal of Hydrogen Energy* 2003, 28, (9), 939-944.

[18] Kim, S.; Dale, B. E., Life cycle assessment of various cropping systems utilized for producing biofuels: Bioethanol and biodiesel. *Biomass and Bioenergy* 2005, 29, (6), 426-439.

[19] Manish, S.; Banerjee, R., Comparison of biohydrogen production processes. *International Journal of Hydrogen Energy* 2008, 33, (1), 279-286.

[20] Dai, D.; Liu, R. H.; Pu, G. Q.; Wang, C. T., Evaluation of energy production efficiency of biomass based fuel ethanol program( chinese). *Transactions of the Chinese Society of Agricultural Engineering* 2005, 21, (11), 121-123.

[21] Liu, R. H.; Nu, W. S.; Zhang, D. L., *Biomass thermo-chemical conversion technology(chinese)[*M]. Beijng:Chemical Industry Press 2005, 55-57,113-118.

[22] Klein, A. Gasification: An Alternative Process for Energy Recovery and Disposal of Municipal Solid Wastes. *Columbia University,* 2002.

[23] Zhang, C. M.; Liu, R. H., Analysis of mass balance and influencing factors for biomass pyrolysis process(chinese). *Journal of Agricultural Mechanization Research* 2006, (10), 144-146.

[24] J, G.; N, V.; J, Z.; A, K.; K, M.; A, S.; M, S., Furfural chemicals and biofuels from agriculture. *Chemicals and Biofuels from Hardwoods*: USA, 2006.

[25] Sun, Y.; Cheng, J., Hydrolysis of lignocellulosic materials for ethanol production: a review. *Bioresource Technology* 2002, 83, (1), 1-11.

[26] W, B. Y. C. E., Pretreatment: the key to unlocking low-cost cellulosic ethanol. *Biofuels, Bioproducts and Biorefining* 2008, 2, (1), 26-40.

[27] Ramos, L.; Breuil, C.; Saddler, J., Comparison of steam pretreatment of eucalyptus, aspen, and spruce wood chips and their enzymatic hydrolysis. *Applied Biochemistry and Biotechnology* 1992, 34-35, (1), 37-48.

[28] Chen, H. Z.; Liu, L. Y., Unpolluted fractionation of wheat straw by steam explosion and ethanol extraction. Bioresource Technology 2007, 98, (3), 666-676.

[29] H, P. Z. Y.; E, H. M.; R, M. J., Outlook for cellulase improvement: Screening and selection strategies. *Biotechnology Advances* 2006, 24, (5), 452-481.

[30] Gregg, D. J.; Saddler, J. N., Factors affecting cellulose hydrolysis and the potential of enzyme recycle to enhance the efficiency of an integrated wood to ethanol process. *Biotechnology and Bioengineering* 1996, 51, (4), 375-383.

[31] B, Y.; E, W. C., Effect of xylan and lignin removal by batch and flowthrough pretreatment on the enzymatic digestibility of corn stover cellulose. *Biotechnology and Bioengineering* 2004, 86, (1), 88-98.

[32] M, A.; T, E.; J, B.; A, W.; M, G.; F, T.; G, Z., The effect of Tween-20 on simultaneous saccharification and fermentation of softwood to ethanol. *Enzyme and Microbial Technology* 2003, 33, 71-78.

[33] S, H. S.; B, D. S. J.; G, C. D., Effect of surfactants on cellulose hydrolysis. *Biotechnology and Bioengineering* 1993, 42, (5), 611-617.

[34] B, K. J.; J, B.; H, B. M.; F, T.; H, J., Use of surface active additives in enzymatic hydrolysis of wheat straw lignocellulose. *Enzyme and Microbial Technology* 2007, 40, (4), 888-895.

[35] Kuhad, R.; Singh, A.; Eriksson, K.-E., Microorganisms and enzymes involved in the degradation of plant fiber cell walls. I*n Biotechnology in the Pulp and Paper Industry,* 1997; pp 45-125.

[36] Lin, Y.; Tanaka, S., Ethanol fermentation from biomass resources: current state and prospects. *Applied Microbiology and Biotechnology* 2006, 69, (6), 627-642.

[37] Cardona, C. A.; Sánchez , Ó. J., Fuel ethanol production: Process design trends and integration opportunities. *Bioresource Technology* 2007, 98, (12), 2415-2457.

[38] Wu, Z.; Lee, Y. Y., Inhibition of the enzymatic hydrolysis of cellulose by ethanol. *Biotechnology Letters* 1997, 19, (10), 977-979.

[39] Wu, Z.; Lee, Y., Nonisothermal simultaneous saccharification and fermentation for direct conversion of lignocellulosic biomass to ethanol. Applied Biochemistry and Biotechnology 1998, 70-72, (1), 479-492.

[40] Teixeira, L.; Linden, J.; Schroeder, H., Simultaneous saccharification and cofermentation of peracetic acid-pretreated biomass. *Applied Biochemistry and Biotechnology* 2000, 84-86, (1), 111-127.

[41] Jones, D. T.; Woods, D. R., Acetone-butanol fermentation revisited. *Microbological reviews* 1986, Vol.50,No.4, 484-524.

[42] J, N.; G, B.; V, O. M.; S, M. K.; Q, Z., Genome sequence and comparative analysis of the solvent-producing bacterium clostridium acetobutylicum. *Journal of Bacteriology* 2001, 183, (16), 4823-4838.

[43] Ezeji, T. C.; Qureshi, N.; Blaschek, H. P., Bioproduction of butanol from biomass: from genes to bioreactors. *Current Opinion in Biotechnology* 2007, 18, (3), 220-227.

[44] A, A. B.; P, B. H., Isolation and characterization of Clostridium acetobutylicum mutants with enhanced amylolytic activity. *Applied and Environmental Microbiology* 1991, 57, (9), 2544-2548.

[45] K, C. C.; P, B. H., Acetate enhances solvent production and prevents degeneration in Clostridium beijerinckii BA101. *Applied Microbiology and Biotechnology* 1999, 52, (2), 170-173.

[46] Nair, R. V.; Green, E. M.; Watson, D. E.; Bennett, G. N.; Papoutsakis, E. T., Regulation of the sol Locus Genes for Butanol and Acetone Formation in Clostridium acetobutylicum ATCC 824 by a Putative Transcriptional Repressor. *Journal of Bacteriology* 1999, 181, (1), 319-330.

[47] Green, M. E.; Bennett, G. N., Genetic manipulation of acid and solvent formation in Clostridium acetobutylicum ATCC 824. *Biotechnology and Bioengineering* 1998, 58, (2-3), 215-221.

[48] Ezeji, T. C.; Qureshi, N.; Blaschek, H. P., Bioproduction of butanol from biomass: from genes to bioreactors. *Curr Opin Biotechnol* 2007, 18, (3), 220-7.

[49] Dürre, P., New insights and novel developments in clostridial acetone/butanol/isopropanol fermentation. *Applied Microbiology and Biotechnology* 1998, 49, (6), 639-648.

[50] Sabathé, F.; Bélaïch, A.; Soucaille, P., Characterization of the cellulolytic complex (cellulosome) of Clostridium acetobutylicum. FEMS *Microbiology Letters* 2002, 217, (1), 15-22.

[51] Marchal, R.; Rebeller, M.; Vandecasteele, J. P., Direct bioconversion of alkali-pretreated straw using simultanesous enzymatic hydrolysis and acetone-butanol fermentation. *Biotechnology Letters* 1984, 6, (8), 523-528.

[52] W, C. S.; X, M.; S, W. L.; H, Z. X., Acetone-butanol fermentation of rice straw enzymatic hydrolysis. *Industrial Microbiology* 1998, 28, (4), 30-34.

[53] Li, D.; Chen, H., Fermentation of acetone and butanol coupled wirh enzymatic hydrolysis of steam exploded cornstalk stover in a membrane reactor. *The Chinese Journal of Process Engineering* 2007, 7, (6), 1212-1216.

[54] M, F.; P, P.; M, E. J., Optimal conditions for long-term stability of acetone-butanol production by continuous cultures ofClostridium acetobutylicum. *Biotechnology Letters* 1985, 7, (7), 503-508.

[55] H, B.; W, A.; G, G., Continuous production of acetone and butanol by Clostridium acetobutylicum in a two-stage phosphate limited chemostat. *Applied Microbiology and Biotechnology* 1982, 15, (4), 201-205.

[56] Zverlov, V.; Berezina, O.; Velikodvorskaya, G.; Schwarz, W., Bacterial acetone and butanol production by industrial fermentation in the Soviet Union: use of hydrolyzed agricultural waste for biorefinery. *Applied Microbiology and Biotechnology* 2006, 71, (5), 587-597.

[57] N, Q.; J, S.; J, L.; P, B. H., Continuous solvent production by Clostridium beijerinckii BA101 immobilized by adsorption onto brick. *World Journal of Microbiology and Biotechnology* 2000, 16, (4), 377-382.

[58] C, E. T.; N, Q.; P, B. H., Acetone butanol ethanol (ABE) production from concentrated substrate: reduction in substrate inhibition by fed-batch technique and product inhibition by gas stripping. *Applied Microbiology and Biotechnology* 2004, 63, (6), 653-658.

[59] N, Q.; S, H.; S, M. I.; A, C. M., Energy-efficient recovery of butanol from model solutions and fermentation broth by adsorption. *Bioprocess and Biosystems Engineering* 2005, 27, (4), 215-222.

[60] Han, X. L.; Lu, F.; Dong, K. L.; Zhao, Y. G., Resarch progress in anhydrous ethanol produciton through biomass absorption. *Liquor-making science and technology* 2007, 151, (1), 84-86.

[61] A, H. A.; H, H. J., Drying of ethanol vapors by adsorption on corn meal. *Biotechnology and Bioengineering* 1990, 35, (6), 598-608.

[62] Westgate, P. J.; Ladisch, M. R., Sorption of organics and water on starch. *Industrial & Engineering Chemistry Research* 1993, 32, (8), 1676-1680.

[63] Jan Bartacek, J. Z. P. N. L. L., Developments and constraints in fermentative hydrogen production. *Biofuels, Bioproducts and Biorefining* 2007, 1, (3), 201-214.

[64] Dunn, S., Hydrogen futures: toward a sustainable energy system. *International Journal of Hydrogen Energy* 2002, 27, (3), 235-264.

[65] Nath, K.; Das, D., Hydrogen from biomass. *Current Science* 2003, 85, (3), 265-271.

[66] Liu, X.; Ren, N.; Song, F.; Yang, C.; Wang, A., Recent advances in fermentative biohydrogen production. *Progress in Natural Science* 2008, 18, (3), 253-258.

[67] Demirbas, A., Biofuels sources, biofuel policy, biofuel economy and global biofuel projections. *Energy Conversion and Management* 2008.

[68] Benemann, J. R., Hydrogen production by microalgae. *Journal of Applied Phycology* 2000, 12, (3), 291-300.

[69] Vijayaraghavan, K.; Soom, M. A. M., Trends in biological hydrogen production—a review. Int. J. Hydrogen Energy, in press, available online at www. sciencedirect. com 2004.

[70] Debabrata, D. A. S.; Khanna, N.; Veziro?Lu, T. N., Recent Developments in Biological Hydrogen Production Processes. *Chemical Industry and Chemical Engineering Quarterly* 14.

[71] van Niel, E. W. J.; Budde, M. A. W.; de Haas, G. G.; van der Wal, F. J.; Claassen, P. A. M.; Stams, A. J. M., Distinctive properties of high hydrogen producing extreme thermophiles, Caldicellulosiruptor saccharolyticus and Thermotoga elfii. *International Journal of Hydrogen Energy* 2002, 27, (11-12), 1391-1398.

[72] Meher Kotay, S.; Das, D., Biohydrogen as a renewable energy resource—Prospects and potentials. *International Journal of Hydrogen Energy* 2008, 33, (1), 258-263.

[73] Claassen, P. A. M.; Budde, M. A. W.; van Noorden, G. E.; Hoekema, S.; Hazewinkel, J. H. O.; van Groenestijn, J. W.; de Vrije, T., Biological hydrogen production from agro-food by-products. *Biological hydrogen product proceedings* 2004.

[74] Eroğlu, E.; Eroğlu, İ.; Gündüz, U.; Türker, L.; Yücel, M., Biological hydrogen production from olive mill wastewater with two-stage processes. *International Journal of Hydrogen Energy* 2006, 31, (11), 1527-1535.

[75] Yokoi, H.; Saitsu, A.; Uchida, H.; Hirose, J.; Hayashi, S.; Takasaki, Y., Microbial hydrogen production from sweet potato starch residue. *Journal of Bioscience and Bioengineering* 2001, 91, (1), 58-63.

[76] Lay, J. J., Biohydrogen generation by mesophilic anaerobic fermentation of microcrystalline cellulose. *Biotechnology and Bioengineering* 2001, 74, (4), 280-287.

[77] Himeshima, N.; Amao, Y., Photoinduced Hydrogen Production from Cellulose Derivative with Chlorophyll-a and Platinum Nanoparticles System. *ENERGY AND FUELS* 2003, 17, (6), 1641-1644.

[78] Taguchi, F.; Yamada, K.; Hasegawa, K.; Taki-Saito, T.; Hara, K., Continuous hydrogen production by Clostridium sp. strain no. 2 from cellulose hydrolysate in an aqueous two-phase system. *Journal of Fermentation and Bioengineering* 1996, 82, (1), 80-83.

[79] Fan, Y.-T.; Zhang, Y.-H.; Zhang, S.-F.; Hou, H.-W.; Ren, B.-Z., Efficient conversion of wheat straw wastes into biohydrogen gas by cow dung compost. *Bioresource Technology* 2006, 97, (3), 500-505.

[80] de Vrije, T.; de Haas, G. G.; Tan, G. B.; Keijsers, E. R. P.; Claassen, P. A. M., Pretreatment of Miscanthus for hydrogen production by Thermotoga elfii. *International Journal of Hydrogen Energy* 2002, 27, (11-12), 1381-1390.

[81] Li, D.; Chen, H., Biological hydrogen production from steam-exploded straw by simultaneous saccharification and fermentation. *International Journal of Hydrogen Energy* 2007, 32, (12), 1742-1748.

[82] Pulz; Pulz, O., Photobioreactors: production systems for phototrophic microorganisms. *Applied Microbiology and Biotechnology* 2001, 57, (3), 287-293.

[83] Akkerman, I.; Janssen, M.; Rocha, J.; Wijffels, R. H., Photobiological hydrogen production: photochemical efficiency and bioreactor design. *International Journal of Hydrogen Energy* 2002, 27, (11-12), 1195-1208.

[84] Camilli, M.; Pedroni, P. M., Comparison of the performance of three different reactors for BioHydrogen production via dark anaerobic fermentations. Proceedings of the international hydrogen energy congress and exhibition. *BHP* 2005, 250.

[85] Sulaiman, A.-Z., Production of biodiesel: possibilities and challenges. *Biofuels, Bioproducts and Biorefining* 2007, 1, (1), 57-66.

[86] Vasudevan, P. T.; Briggs, M., Biodiesel production—current state of the art and challenges. *Journal of Industrial Microbiology and Biotechnology* 2008, 35, (5), 421-430.

[87] Sharma, Y. C.; Singh, B.; Upadhyay, S. N., Advancements in development and characterization of biodiesel: A review. *Fuel* 2008, 87, (12), 2355-2373.

[88] Fukuda, H.; Kondo, A.; Noda, H., Biodiesel fuel production by transesterification of oils. *Journal of Bioscience and Bioengineering* 2001, 92, (5), 405-416.

[89] Köse, Ö.; Tüter, M.; Aksoy, H. A., Immobilized candida antarctica lipase-catalyzed alcoholysis of cotton seed oil in a solvent-free medium. *Bioresource Technology* 2002, 83, (2), 125-129.

[90] Shimada, Y.; Watanabe, Y.; Sugihara, A.; Tominaga, Y., Enzymatic alcoholysis for biodiesel fuel production and application of the reaction to oil processing. *Journal of Molecular Catalysis. B, Enzymatic* 2002, 17, (3-5), 133-142.

[91] Saka, S.; Kusdiana, D., Biodiesel fuel from rapeseed oil as prepared in supercritical methanol. *Fuel* 2001, 80, (2), 225-231.

[92] Demirbas, A., Biodiesel from vegetable oils via transesterification in supercritical methanol. *Energy Conversion and Management* 2002, 43, (17), 2349-2356.

[93] Kusdiana, D.; Saka, S., Effects of water on biodiesel fuel production by supercritical methanol treatment. *Bioresource Technology* 2004, 91, (3), 289-295.

[94] Madras, G.; Kolluru, C.; Kumar, R., Synthesis of biodiesel in supercritical fluids. *Fuel* 2004, 83, (14-15), 2029-2033.

[95] Miao, X.; Wu, Q., Biodiesel production from heterotrophic microalgal oil. *Bioresource Technology* 2006, 97, (6), 841-846.

[96] Peng, X. W., Production of Single Cell Oils from Steam-exploded Straw in Solid-stated Fermentation and Pyrolysis of Fermented Mass for Producing Biodiesel[doctoral dissertation]. Beijing: *Institute of Process Engineering,Chinese Academy of Sciences* 2008.

[97] Enguídanos, M.; Institute for Prospective Technological, S.; Joint Research, C.; European, C., Techno-economic Analysis of Bio-diesel Production in the EU: A Short Summary for Decision-makers. *European Commission Joint Research Centre*: 2002.

[98] Ma, F.; Hanna, M. A., Biodiesel production: a review. *Bioresource Technology* 1999, 70, (1), 1-15.

[99] Koh, L. P.; Ghazoul, J., *Biofuels, biodiversity, and people: Understanding the conflicts and finding opportunities*. Biological Conservation In Press, Corrected Proof.

[100] Kumar, R.; Singh, S.; Singh, O. V., Bioconversion of lignocellulosic biomass: biochemical and molecular perspectives. *Journal of Industrial Microbiology and Biotechnology* 2008, 35, (5), 377-391.

# INDEX

## A

## B

## D

## F

## I

## N

## O

## P

## Q



## R

## S

## T

## U



## V



## W

## X



## Y